A. Sengupta (Ed.)

Chaos, Nonlinearity, Complexity

Studies in Fuzziness and Soft Computing, Volume 206

Editor-in-chief
Prof. Janusz Kacprzyk
Systems Research Institute
Polish Academy of Sciences
ul. Newelska 6
01-447 Warsaw
Poland
E-mail: kacprzyk@ibspan.waw.pl

Further volumes of this series
can be found on our homepage:
springer.com

Vol. 190. Ying-ping Chen
Extending the Scalability of Linkage Learning Genetic Algorithms, 2006
ISBN 3-540-28459-1

Vol. 191. Martin V. Butz
Rule-Based Evolutionary Online Learning Systems, 2006
ISBN 3-540-25379-3

Vol. 192. Jose A. Lozano, Pedro Larrañaga, Iñaki Inza, Endika Bengoetxea (Eds.)
Towards a New Evolutionary Computation, 2006
ISBN 3-540-29006-0

Vol. 193. Ingo Glöckner
Fuzzy Quantifiers: A Computational Theory, 2006
ISBN 3-540-29634-4

Vol. 194. Dawn E. Holmes, Lakhmi C. Jain (Eds.)
Innovations in Machine Learning, 2006
ISBN 3-540-30609-9

Vol. 195. Zongmin Ma
Fuzzy Database Modeling of Imprecise and Uncertain Engineering Information, 2006
ISBN 3-540-30675-7

Vol. 196. James J. Buckley
Fuzzy Probability and Statistics, 2006
ISBN 3-540-30841-5

Vol. 197. Enrique Herrera-Viedma, Gabriella Pasi, Fabio Crestani (Eds.)
Soft Computing in Web Information Retrieval, 2006
ISBN 3-540-31588-8

Vol. 198. Hung T. Nguyen, Berlin Wu
Fundamentals of Statistics with Fuzzy Data, 2006
ISBN 3-540-31695-7

Vol. 199. Zhong Li
Fuzzy Chaotic Systems, 2006
ISBN 3-540-33220-0

Vol. 200. Kai Michels, Frank Klawonn, Rudolf Kruse, Andreas Nürnberger
Fuzzy Control, 2006
ISBN 3-540-31765-1

Vol. 201. Cengiz Kahraman (Ed.)
Fuzzy Applications in Industrial Engineering, 2006
ISBN 3-540-33516-1

Vol. 202. Patrick Doherty, Witold Łukaszewicz, Andrzej Skowron, Andrzej Szałas
Knowledge Representation Techniques: A Rough Set Approach, 2006
ISBN 3-540-33518-8

Vol. 203. Gloria Bordogna, Giuseppe Psaila (Eds.)
Flexible Databases Supporting Imprecision and Uncertainty, 2006
ISBN 3-540-33288-X

Vol. 204. Zongmin Ma (Ed.)
Soft Computing in Ontologies and Semantic Web, 2006
ISBN 3-540-33472-6

Vol. 205. Mika Sato-Ilic, Lakhmi C. Jain
Innovations in Fuzzy Clustering, 2006
ISBN 3-540-34356-3

Vol. 206. A. Sengupta (Ed.)
Chaos, Nonlinearity, Complexity, 2006
ISBN 3-540-31756-2

A. Sengupta
(Ed.)

Chaos, Nonlinearity, Complexity

The Dynamical Paradigm of Nature

Professor A. Sengupta
Department of Mechanical Engineering
Nuclear Engineering and Technology Programme
Indian Institute of Technology Kanpur
Kanpur 208016, India
E-mail: osegu@iitk.ac.in

ISSN print edition: 1434-9922
ISSN electronic edition: 1860-0808
ISBN-10 3-642-42160-1 Springer Berlin Heidelberg New York
ISBN-13 978-3-642-42160-0 Springer Berlin Heidelberg New York

Springer is a part of Springer Science+Business Media
springer.com

Softcover re-print of the Hardcover 1st edition 2006

Typesetting: by the authors and techbooks using a Springer LaTeX macro package
Cover design: Erich Kirchner, Heidelberg

Printed on acid-free paper SPIN: 11402640 89/techbooks 5 4 3 2 1 0

Dedicated to the

♡ synthetic cohabitation of Yang and Yin ♡

induced by

♡ Cha(os-)No(nlinearity-comple)Xity ♡

and to

♡ MPCNS-2004 ♡

that made all this possible

Preface

I think the next century will be the century of complexity. We have already discovered the basic laws that govern matter and understand all the normal situations. We don't know how the laws fit together, and what happens under extreme conditions. But I expect we will find a complete unified theory sometime this century. There is no limit to the complexity that we can build using those basic laws.

Stephen Hawking, January 2000.

We don't know what we are talking about. Many of us believed that string theory was a very dramatic break with our previous notions of quantum theory. But now we learn that string theory, well, is not that much of a break. The state of physics today is like it was when we were mystified by radioactivity. They were missing something absolutely fundamental. We are missing perhaps something as profound as they were back then.

Nobel Laureate David Gross, December 2005.

This volume is essentially a compilation of papers presented at the International Workshop on *Mathematics and Physics of Complex and Nonlinear Systems* that was held at Indian Institute of Technology Kanpur, March 14 – 26, 2004 on the theme *ChaNoXity: The Nonlinear Dynamics of Nature.* ChaNoXity — symbolizing Chaos-Nonlinearity-compleXity — is an attempt to understand and interpret the dynamical laws of Nature on a unified and global perspective. The Workshop's objective was to formalise the concept of chanoxity and to get the diverse body of practitioners of its components to interact intelligently with each other. It was aimed at a focused debate and discussion on the mathematics and physics of chaos, nonlinearity, and complexity in the dynamical evolution of nature. This is expected to induce a process of reeducation and reorientation to supplement the basically linear reductionist approach of present day science that seeks to break down natural systems to their simple constituents whose properties are expected to combine in a relatively simple manner to yield the complex laws of the whole. There were approximately 40 hours of lectures by 12 speakers; in keeping with its aim of providing an open platform for exposition and discourse on the thematic topic, each of the 5-6 lectures a day were of 75 minutes duration so as to provide an adequate and meaningful interaction, formal and informal, between the speaker and his audience.

The goals of the workshop were to

- Create an awareness among the participants, drawn from the research and educational institutions in India and abroad, of the role and significance of nonlinearity in its various manifestations and forms.
- Present an overview of the strong nonlinearity of chaos and complexity in natural systems from the mathematical and physical perspectives. The relevant mathematics were drawn from topology, measure theory, inverse and ill-posed problems, set-valued and nonlinear functional analyses.
- Explore the role of non-extensive thermodynamics and statistical mechanics in open, nonlinear systems.

There were lively and animated discussions on self-organization and emergence in the attainment of steady-states of open, far-from-equilibrium, complex systems, and on the mechanism of how such systems essentially cheat the dictates of the all-pervading Second Law of Thermodynamics[1]: where lies the source of Schrodinger's negative entropy that successfully maintains life despite the Second Law? How does Nature defeat itself in this game of the Second Law, and what might be the possible role of gravity in this enterprise? Although it is widely appreciated that gravity — the only force to have successfully resisted integration in a unified theory — is a major player in the dynamics of life, realization of a satisfactory theory has proved to be difficult, with loop quantum cosmology holding promise in resolving the vexing "big-bang singularity problem". The distinctive feature of this loop quantization is that the *quantum Wheeler-DeWitt differential equation* (that fails to remove the singularity, backward evolution leading back into it), *is replaced by a difference equation, the size of the discrete steps determined by an area gap*, Riemannian geometry now being quantized with the length, area, volume operators possessing discrete eigenvalues. Discrete difference equations, loops of one-dimensional objects (based on spin-connections rather than on the metrics of standard General Relativity) considered in the period-doubling

[1] *The second law of thermodynamics holds, I think, the supreme position among the laws of Nature. If someone points out to you that your pet theory of the universe is in disagreement with Maxwell's equations then so much the worse for Maxwell's equations. If it is found to be contradicted by observation, well, these experimentalists do bungle things sometimes. But if your theory is found to be against the second law of thermodynamics I can give you no hope; there is nothing for it but to collapse in deepest humiliation.* A. Eddington, *The Nature of the Physical World*, Macmillan, New York (1948).

perspective, the extreme nonlinear curvature of big-bang and blackhole chaotic spacetimes: do all these point to a radically different paradigm in the chaos-nonlinearity-complexity setting of discrete dynamical systems?[2] Thus is it "a quantum foam far removed from any classical spacetime, or is there another large, classical universe" on the "other side of the singularity responsible for a quantum bounce from an expanding branch to a contracting branch"?[3] Could this possibly be the outcome of the interaction of our classical real world with a negative partner acting as the provider of the illusory negative entropy, whose attraction manifests on us as the repulsive "quantum bounce" through the agency of gravity? Would the complex structure of "life" and of the universe as we know it exist without the partnership of gravity?

Not all the papers presented at the Workshop appear here; notable exceptions among those who gave three or more lectures are S. Kesavan (Institute of Mathematical Science at Chennai, India) who spoke on *Topological Degree and Bifurcation Theory,* and M. Z. Nashed (University of Central Florida, USA) whose paper *Recovery Problems from Partial or Indirect Information: Perspectives on Inverse and Ill-Posed Problems* could not be included due to unavoidable circumstances. The volume contains three papers by Realpe and Ordonez, Majumdar, and Johal that were not presented at the Workshop. A brief overview of the papers appearing follows.

Francisco Balibrea (Universidad de Murcia, Spain) provides a comprehensive review of the complicated dynamics of discrete dynamical systems in a compact metric space using the notions of Li-Yorke and Devaney chaos, sensitive dependence of initial conditions, transitivity, Lyapunov exponents, and the Kolmogorov-Sinai and topological entropies. **Sumiyoshi Abe** (University of Tsukuba, Japan) surveys the fundamental aspects of nonextensive statistical mechanics based on the Tsallis entropy, and demonstrates how the method of steepest descents, the counting algorithm and the evaluation of the density of states can appropriately be generalized for describing the power-law distributions. **Alberto Robledo** (Universidad Nacional Autonoma de Mexico, Mexico) gives an account of the dynamics at critical attractors of simple one-dimensional nonlinear maps relevant to the applicability of the Tsallis generalization of canonical statistical mechanics. Continuing in this spirit of non-extensivity, **A. G. Bashkirov** (R.A.S. Moscow, Russia) considers the Renyi entropy as a cumulant average of the Boltzmann entropy, and finds that the thermodynamic entropy in Renyi thermostatistics increases with system complexity, with the Renyi distribution becoming a pure power-law under appropriate conditions. He concludes that "because a power-law distribution is characteristic for self-organizing systems, the Renyi entropy can be considered as a potential that drives the system to a self-organized state". **Karmeshu**

---

[2] Gerard 't Hooft, Quantum Gravity as a Dissipative Deterministic System, *Class. Quantum Grav.*, **16**, 3263-3279 (1999)

[3] Abhay Ashtekar, Tomasz Pawlowski, and Parampreet Singh, Quantum Nature of the Big Bang, *ArXiv:* gr-qc/0602086

and Sachi Sharma (J.N.U., India) proposes a theoretical framework based on non-extensive Tsallis entropy to study the implication of long-range dependence in traffic process on network performance. **John Realpe** (Universidad del Valle, Colombia) **and Gonzalo Ordonez** (Butler University, Indianapolis and The University of Texas at Austin, USA) study two points of view on the origin of irreversible processes. While the "chaotic hypothesis" holds that irreversible processes originate in the randomness generated by chaotic dynamics, the approach of the Prigogine school maintains that irreversibility is rooted in Poincare non-integrability associated with resonances. Considering the simple model of Brownian motion of a harmonic oscillator coupled to lattice vibration modes, the authors show that Brownian trajectories require resonance between the particle and the lattice, with chaos playing only a secondary role for random initial conditions. If the initial conditions are not random however, chaos is the dominant player leading to thermalization of the lattice and consequent appearance of Brownian resonance characteristics. **R. S. Johal** (Lyallpur Khalsa College, India) considers the approach to equilibrium of a system in contact with a heat bath and concludes, in the context of non-extensivity, that differing bath properties yield differing equilibrium distributions of the system. **Parthasarathi Majumdar** (S.I.N.P., India) reviews black hole thermodynamics for non-experts, underlining the need for considerations beyond classical general relativity. The origin of the microcanonical entropy of isolated, non-radiant, non-rotating black holes is traced in this perspective in the Loop Quantum Gravity formulation of quantum spacetime. **Russ Marion** (Clemson University, USA) applies complexity theory to organizational sciences and finds that "the implications are so significant that they signal a paradigm shift in the way we understand organization and leadership". Complexity theory, in his view, alters our perceptions about the logic of organizational behavior which rediscovers the significant importance of firms' informal social dynamics that have long been "suppressed or channeled". He feels that a complexity appraoch to organizations is particularly relevant in view of the recent emphasis in industrialized nations toward knowledge-based, rather than production-based, economies. **A. Sengupta** (I.I.T. Kanpur, India) employs the topological-multifunctional mathematical language and techniques of non-injective illposedness to formulate the notion of chanoxity in describing the specifically nonlinear dynamical evolutionary processes of Nature. Non-bijective ill-posedness is the natural mode of expression for chanoxity that aims to focus on the nonlinear interactions generating dynamical evolution of real irreversible processes. The basic dynamics is considered to take place in a matter-negmatter kitchen space of Nature which is inaccessible to both the matter and negmatter components, distinguished by opposing evolutionary directional arrows. Dynamical equilibrium is considered to be represented by such competitively collaborating homeostatic states of the matter-negmatter constituents of Nature, modelled as a self-organizing engine-pump system.

Acknowledgement A project of this magnitude would not have succeeded without the help and assistance of many individuals and organizations. Financial support was provided by Department of Science and Technology and All India Council for Technical Education, New Delhi, National Board for Higher Mathematics, Mumbai, and the Department of Mechanical Engineering IIT Kanpur. It is a great pleasure to acknowledge the very meaningful participation of Professor Brahma Deo in the organization and conduct of the Workshop, the advice and suggestions of Professors N. Sathyamurthy and N. N. Kishore, the infrastructural support provided by IIT Kanpur, and the assistance of Professor Pradip Sinha. I am also grateful to Professors A. R. Thakur, Vice-Chancellor, West Bengal University of Technology Kolkata, and A. B. Roy of Jadavpur University for their continued support to ChaNoXity during and after the Workshop.

April 30, 2006 A. Sengupta
Kanpur

Contents

List of Contributors

Sumiyoshi Abe
Institute of Physics
University of Tsukuba
Ibaraki 305-8571, JAPAN.
suabe@sf6.so-net.ne.jp

Francisco Balibrea
Departamento de Matemáticas
Universidad de Murcia
30100 Espinardo, Murcia, SPAIN.
balibrea@um.es

A. G. Bashkirov
Institute Dynamics of Geospheres,
Russian Academy of Sciences
119334 Moscow, RUSSIA.
abas@idg.chph.ras.ru

R. S. Johal
Department of Physics
Lyallpur Khalsa College
Jalandhar, INDIA.
johal_jld@dataone.in

Karmeshu
School of Computer and System Sciences
Jawaharlal Nehru University
New Delhi 110067, INDIA
karmeshu@mail.jnu.ac.in

Parthasarathi Majumdar
Theory Group
Saha Institute of Nuclear Physics
Kolkata - 700064, INDIA
parthasarathi.majumdar@saha.ac.in

Russ Marion
Department of Leadership
Clemson University
Clemson, SC 29631, USA.
marion2@clemson.edu

M. Zuhair Nashed
Department of Mathematics
University of Central Florida
Orlando, FL 32816, USA.
znashed@mail.ucf.edu

Gonzalo Ordonez
Department of Physics and Astronomy
Butler University
Indianapolis, IN 46208, USA
gordonez@butler.edu

John Realpe
Universidad del Valle, Cali
Colombia.
jrealpe@calima.univalle.edu.co

Alberto Robledo
Instituto de Fisica
Universidad Nacional Autonoma de Mexico
Apartado Postal 20-364,
Mexico 01000 D. F., Mexico
robledo@fisica.unam.mx

A. Sengupta
Department of Mechanical Engineering
Indian Institute of Technology Kanpur
Kanpur 208016, INDIA.
osegu@iitk.ac.in

Sachi Sharma
School of Computer and System Sciences
Jawaharlal Nehru University
New Delhi 110067, INDIA
shachi.sharma@rediffmail.com

1

Chaos, Periodicity and Complexity on Dynamical Systems

Francisco Balibrea

Departamento de Matemáticas. Universidad de Murcia.30100-Murcia (Región de Murcia), Spain
E-Mail: balibrea@um.es

Summary. In the setting of discrete dynamical systems $(\mathbb{X}, f)$ where $\mathbb{X}$ is a compact metric space and f is a continuous self-mapping of $\mathbb{X}$ into itself, we introduce two ways of appreciating how complicated the dynamics of such systems is. First through several notions of chaos like Li-Yorke and Devaney chaos, sensitive dependence of initial conditions, transitivity, Lyapunov exponents, and the second through different notions of entropy, mainly the Kolmogorov-Sinai and topological entropies. In particular Kolmogorov-Sinai is introduced in a very general way. Also we review some known relations among these notions of chaos and entropies.

Complicated dynamics can be also understood via periodic orbits. To this aim we concentrate in the forcing relations among the periods of the orbits in the simplest cases such that $\mathbb{I} = [0, 1]$, $\mathbb{S}^1$ and in other more complicated spaces. Additionally, we resume some results recently obtained for delay difference equations of the form $x_{n+k} = f(x_n)$ for $k \geq 2$.

1.1 Introduction

Roughly speaking we understand by a dynamical system a set of states (called the space of states) evolving with time. More precisely, a dynamical system is a triple $(\mathbb{X}, \Phi, G)$ where $\mathbb{X}$ denotes the *state space* usually given by a topological space, Φ is the *flow* of the system, that is, the rule of evolution, given by a continuous map from $G \times \mathbb{R}$ into $\mathbb{X}$ and $G \subseteq \mathbb{R}$ a semigroup of times. When $G = \mathbb{Z}$ or $G = \mathbb{Z}^+ \cup 0$ the dynamical system is called *discrete* and it is denoted by the pair $(\mathbb{X}, f)$ where $\mathbb{X}$ is a nonempty metric space and the flow is $\Phi(n, x) = f^n(x)$ where f is a continuous map form $\mathbb{X}$ into itself. Given $\mathbb{X}$, we will denote by $C(\mathbb{X})$ the set of continuous maps from $\mathbb{X}$ into itself. For $f \in C(\mathbb{X})$ we define its nth*-iterate* by $f^n = f \circ f^{n-1}$, $n \geq 1$, $f^0 =$ identity, with $f \circ g$ denoting the composition of f and g. When $G = \mathbb{R}$ the system is called *continuous*.

The main goal when considering dynamical systems is to understand the long term behavior of states in evolving according with the flow. The systems often involve several variables and are usually nonlinear. In a variety

F. Balibrea: *Chaos, Periodicity and Complexity on Dynamical Systems*, StudFuzz **206**, 1–52 (2006)
www.springerlink.com

of settings, very complicated behavior is observed even though the equations themselves describing the system are not very complicated. Thus simple algebraic forms of the equations do not mean that the dynamical behavior is simple; in fact, it can be very complicated or even *chaotic*. One aspect of the chaotic nature of systems is described by the *sensitive dependence on initial conditions* which means that initial close states of the system evolve separately.

Definition 1.1. *The dynamical system* $(\mathbb{X}, f)$ *has sensitive dependence on initial conditions (s.d.i.c.) on* $Y \subseteq \mathbb{X}$ *if there exists an* $r > 0$ *(independent of the points of* Y*) such that for each point* $x \in Y$ *and for each* $\epsilon > 0$*, there exist* $y \in Y$ *with* $\rho(x, y) < \epsilon$ *and* $n \geq 0$ *such that* $\rho(f^n(x), f^n(y)) \geq r$.

One of the first situations where *s.d.i.c.* appeared was observed by E.Lorenz in his simplified well known system of three differential equation stated as a model for the prediction of the weather .

For such systems, if the initial conditions are only approximately specified, then the evolution of the state may be very different. This fact leads to important difficulties when using approximate, or even real, solutions to predict future states based on present knowledge. To develop an understanding of these aspects of chaotic dynamics, we want to find situations which exhibit this behavior and yet for which we can still understand the important features of how solutions evolves with time.

Sometimes we cannot follow a particular solution with complete certainty because there is round off error in the calculations or we are using some numerical scheme to find it. We are interested to know whether the approximate solution we calculate is related to a true solution of the exact equations. In some of the chaotic systems, we can understand how an ensemble of different initial conditions evolves, and prove that the approximate solution traced by a numerical scheme is shadowed by a true solution with some nearby initial conditions. One typical example of such behavior is given by the weather, see [61].

If the system models the weather, people may not be content with the range of possible outcomes of the weather that could develop from the known precision of the previous conditions, or to know that a small change of the previous conditions would have produced the weather which had been predicted. However, even for a subject like weather, for which quantitative as well as qualitative predictions are important, it is still useful to understand what factors can lead to instabilities in the evolution of the state of the system. It is now realized that no new better simulation of weather on more accurate computers of the future will be able to predict the weather more than about fourteen days ahead, because of the very nonlinear nature of the evolution of the state of weather. This type of knowledge can by itself be useful.

In recent years, dynamical systems has had many applications to science and engineering, some of which have gone under the related headings of chaos theory or nonlinear analysis. Behind these applications there lies a rich mathematical subject. The subject centers usually on the orbits of iteration of a nonlinear map or on the solutions of nonlinear ordinary differential equations. Nevertheless we restrict ourselves to the setting of discrete dynamical systems $(\mathbb{X}, f)$.

The sequence $(f^n(x))_{n=0}^{\infty}$ is called the *orbit* of the point $x \in \mathbb{X}$ under the action of f or simply the *orbit* of x. The simplest behavior for one orbit is when $f(x) = x$, we say that x is a *fixed point* of the system and if $f^m(x) = x$ for some m, we say that x is *periodic* of period m and the minimal number m is *the period* of x. Of course if x is periodic of period m then it is also periodic of periods km where k is any positive integer.

From now on we are dealing with topics connected with chaos and its measurement and with the effect of nonlinearity in the systems. These topics started to be developed in the 1960's and nowadays has become a fruitful field of research in mathematics, physics, experimental sciences and engineering.

In dynamical systems there are several notions of chaos which depend on the characteristic we would consider as the most important to indicate the disorder or the complexity on the system. Roughly speaking, a dynamical system is chaotic if its dynamics is complicated in an invariant set Y $(f(Y) \subseteq Y)$. The use of the word chaos in dynamical systems was introduced in [49] by L. Li and J. Yorke in 1975. In that paper they showed that if an interval or line map into itself has a periodic points of period three, then the map has periodic points of all periods. The existence of a lot of periodic points was considered as a chaotic situation for the system. They proved also that if there exists a periodic point of period three then there is an uncountable invariant set $S \subset \mathbb{X}$ where $\mathbb{X} = \mathbb{I} = [0,1]$ or $\mathbb{R}$ (called a *scrambled set*) such that for all $x, y \in S$ with $x \neq y$ we have

$$\limsup_{n\to\infty} \mid f^n(x) - f^n(y) \mid > 0 \qquad (1)$$

$$\liminf_{n\to\infty} \mid f^n(x) - f^n(y) \mid = 0 \qquad (2)$$

In fact for $\mathbb{X} = \mathbb{I}$, if there are two distinct points x and y whose orbits satisfy these two conditions, then there exists an uncountable invariant scrambled set containing the two points, [44]. Conditions (1) and (2) mean that the orbit of different points x and y separate and are close at infinite iterations. We see that such behavior is also connected with the sensitive dependence on initial conditions property.

Definition 1.2. *The dynamical system $(\mathbb{X}, f)$ where $\mathbb{X}$ is a metric space, is LY_2-chaotic (in the sense of Li and Yorke) if there are two different points x, y in the space (a L-Y pair) such that (1) and (2) hold.*

Since in the general setting of metric spaces is not true that if there is a L-Y pair then there is an uncountable scrambled set (see the examples of [31]) then it is taken as adequate the following

Definition 1.3. *The dynamical system* $(\mathbb{X}, f)$ *where* $\mathbb{X}$ *is a metric space is* LY_u*-chaotic (in the sense of Li and Yorke) if there is an uncountable scrambled set.*

The notion of (s.d.i.c.) is weaker than the closely related notion of *expansiveness.* Roughly speaking, a map is expansive if the distance between any two orbits become at least a constant.

Definition 1.4. *The dynamical system* $(\mathbb{X}, f)$ *is expansive on* $Y \subseteq \mathbb{X}$ *if there exists an* $r > 0$ *(independent of the points of* Y*) such that for each pair of points* $x, y \in Y$*, there exists* $n \geq 0$ *such that* $\rho(f^n(x), f^n(y)) \geq r$*. When* f *is an homeomorphism, then* n *can belong to* $\mathbb{Z}$*.*

It is easy to prove that when $\mathbb{X}$ is a perfect (without isolated points) metric space, if f is expansive, then f has (s.d.i.c.).

Considering the above notions and that of *transitivity* (for any two open neighborhoods U and V of $\mathbb{X}$ there is $n > 0$ such that $f^{-n}(U) \cap V \neq \emptyset$). When $\mathbb{X}$ has not isolated points, transitivity is equivalent to the existence of a dense orbit in $\mathbb{X}$). Devaney introduced in [26] the following notion of chaotic system, which will be called *D-chaotic*,

Definition 1.5. *The dynamical system* $(\mathbb{X}, f)$ *is D-chaotic on* $Y \subseteq \mathbb{X}$ *if:*

1. f *is transitive*
2. f *has in* Y *a dense set of periodic points*
3. f *has in* Y *(s.d.i.c.)*

In [14] was proved that the first two conditions imply the third and moreover, in line or interval maps, the first is equivalent to the second. That is, in such case D-chaotic is equivalent to the existence of a dense orbit, that is, to transitivity.

As a weaker definition, in [9] (see also [61]) it is introduced as chaotic system the following

Definition 1.6. *The dynamical system* $(\mathbb{X}, f)$ *is chaotic on* $Y \subseteq \mathbb{X}$ *if in* Y *are held:*

1. f *is transitive*
2. f *has (s.d.i.c.)*

Related to this Definition, in [9] are constructed examples of homeomorphisms which have not (s.d.i.c.) which proves hat in general, condition (1) is not equivalent to (2).

There are others properties which a chaotic system tends to have. One of them can be understood throughout the notion of *Lyapunov exponent* of an orbit. This notion generalizes that of eigenvalue for a periodic orbit and associates to an orbit a growth rate of the infinitesimal separation of nearby points. This quantity can be also defined even when the invariant set does not have a *hyperbolic structure* like the Cantor set for the *quadratic map* $f(x) = 4x(1-x)$. The Lyapunov exponents are often used to measure chaos since it is the most computable quantity in a computer. These Lyapunov exponents can be defined also in higher dimensions.

Definition 1.7. *Let $f : \mathbb{R} \to \mathbb{R}$ be a C^1 map. Given $x \in \mathbb{R}$ the Lyapunov exponent of x is*

$$\begin{aligned}\lambda(x) &= \limsup_n \frac{1}{n} \log(|(f^n)'(x)|) \\ &= \limsup_n \frac{1}{n} \sum_{j=0}^{n-1} \log(|f'(x_j|)\end{aligned}$$

where $x_j = f^j(x)$

Clearly the right hand side of the formula is an average on an orbit of the logarithm of the derivative. Applying the above formula it can be calculated the Lyapunov exponents in the following typical examples:

- Let
$$t(x) = \begin{cases} 2x, & 0 \leq x \leq \frac{1}{2}, \\ 2(1-x), & \frac{1}{2} \leq x \leq 1. \end{cases}$$
be the *tent map*. For any point $y \in [0,1]$ except those holding $t^j(y) = \frac{1}{2}$ where the derivative does not exist, the Lyapunov exponent is $\lambda(y) = \log 2$.
- Let $q(x) = 4x(1-x)$ the quadratic map. Then if $y \in [0,1]$ is a pre-image of $\frac{1}{2}$, is $\lambda(y) = -\infty$. For $y = 0, 1$ is $\lambda(y) = \log 4$. In the rest of points is $\lambda(y) = \log 2$.

When we want to give a quantitative measure of chaos, the notion of *topological entropy* is more adequate than the previously considered, since it gives a number between zero and infinite, describing and evaluating in some sense, the complexity of the system. It will be introduced after the section devoted to the *Kolmogorov-Sinai entropy.* Let us only say that zero topological entropy means that the system (or the corresponding map f) is not chaotic while positive topological entropy ($h(f) > 0$) describes various degrees of chaos.

There are other notions of chaos, some of them like *distributional chaos* are connected with a probabilistic and ergodic point of view (see [70] and [13]).

Among the notions of chaos there are several implications. Here we will present a few of them, first in the general setting of compact metric spaces and then in dynamical systems on $\mathbb{I} = [0,1]$.

Theorem 1.1. *Let $(\mathbb{X}, f)$ be a dynamical system with $\mathbb{X}$ a compact metric space. Then the following hold:*

1. *D-chaos $\Rightarrow$ LY_2-chaos (see [37])*

2. $h(f) > 0 \Rightarrow LY_2$*-chaos (see [17])*

Theorem 1.2. *Let* $(\mathbb{I}, f)$ *be a dynamical system. Then the following conditions hold:*

1. D-chaos $\Rightarrow h(f) > 0 \Rightarrow LY_u$*-chaos (see [35] and [5]*
2. L-Y chaos $\not\Rightarrow h(f) > 0$
3. $h(f) > 0 \not\Rightarrow$ *D-chaos*

In fact here are maps $f \in C(\mathbb{I})$ of type $\{2^\infty\}$ which are LY_u chaotic but with $h(f) = 0$ (see [69]) and also maps with $h(f) > 0$ which are not transitive (see [35]).

There is an interesting implication between the existence of positive Lyapunov exponents and the fact that topological entropy is positive. In fact in [38], it is proved that if a map preserves a non-atomic (continuous) Borel probability measure μ such that μ-almost all initial conditions have non-zero Lyapunov exponents, then the topological entropy is positive and therefore the system is chaotic in such sense.

We have pointed out some notion of chaos, but there are others not considered here (scattering chaos, space-temporal chaos, etc). In most cases there are connections among them difficult to see. One pending question is to clarify such connections and to state what are the ideas of complexity that those notions represent.

1.2 On the periodic structure of some dynamical systems

The existence of an infinite number of periodic orbits in a system must be interpreted to represent complicated dynamics. In some situations like in low-dimensional dynamical systems, the existence of an orbit of a determined period forces the existence of periodic orbits of other periods, even infinite many. The first result we are dealing with in this part was obtained by A.N. Sharkovsky in 1964 and it is concerned with the relations of forcing among the periods of all periodic orbits of the map which defines the discrete system. This theorem does not deal exactly with complicated dynamics and in principle it is not connected with any chaotic behavior, but according to it, if a map f has a periodic point of odd period, then f has infinitely many different periods and we can consider this is a chaotic situation.

The existence of infinitely many different periods is an indication of the complexity of such a map. After the publication in 1964 of the result, a new branch on the dynamics of discrete dynamical systems started which was called *Combinatorial Dynamics* in [5]. It deals with objects of a combinatorial nature, such as permutations, graphs, periodic points, etc. Complementary, in most papers written in the subject, a measure of complexity of the maps (the *topological entropy)* is also treated. In the excellent monograph [5] and in the tutorial papers [52] and [6], the current work in Combinatorial Dynamics is presented jointly with new trends and open problems. Currently, some of the problems presented continue open and in other cases some progress has been done, even new trends have appeared since then. To complete this view, see [7] for the state of art in what is called *minimal models* in *graph maps* and [48] for a complete treatment of the periodic structure of continuous maps on the **8** graph.

Since the theory has been developed in one-dimension systems and can be followed in the above cited references, we will concentrate in discrete dynamical systems of dimension greater than one, in order to present less known results.

From now on we will be concerned with continuous maps only. First of all it is natural to extend the Sharkovsky result on interval $\mathbb{I}$ or line maps to maps on the unit square $\mathbb{I}^2$, $\mathbb{R}^2$ or $\mathbb{T}^2$. There are good reasons for this. Nonlinear difference equations, models on population dynamics, models on economic theory, Poincaré transformation, etc, can be studied through dynamical systems of dimension two. In some cases the extensions can be made easily to $\mathbb{I}^n$, $\mathbb{R}^n$ or $\mathbb{T}^n$ with $n > 2$.

The aim of this section is to present some results on combinatorial dynamics on some particular dynamical systems of low-dimension and use them to interpret the dynamical complexity of such systems and to state the periodic structure of delayed difference equations of the form $x_n = f(x_{n-k})$, with $k \geq 2$, for $f \in C(\mathbb{I})$ and $f \in C(\mathbb{S}^1)$. We remark that these periodic structures are a consequence of the knowledge of the above general combinatorial properties and the particular periodic structure of interval and circle maps, respectively. Thus, the strategy followed in $\mathbb{I}^n$ and $\mathbb{T}^n$ to obtain the periodic structure of σ-permutation maps can be applied to more general and complicated spaces (see Remark 1.1). In subsection 2.2 we will link delayed difference equations of the form $x_n = f(x_{n-k})$ with appropriate σ-permutation maps. This allows us to deduce the periodic structure of such equations, when f is either an interval or a circle map. Finally, we obtain also other results concerning the existence of global attractors for the same type of equations.

1.2.1 Periodic structure of σ-permutation maps

Given a nonempty metric space $\mathbb{X}$, for $f \in C(\mathbb{X})$ we denote by $\mathrm{Per}(f)$, $\mathrm{P}(f)$ and $\mathrm{Fix}(f)$ the sets of periods, periodic points and fixed points of f, respectively. We denote by $\gcd(p,q)$ and $\mathrm{lcm}(p,q)$ the greatest common divisor and the lowest common multiple of two positive integers p, q, respectively, and we write $p \mid q$ if p divides q, and $p \nmid q$ if p does not divide q.

The well known Sharkovsky theorem ([67]) establishes the periodic structure of interval and line maps. By periodic structure we mean a type of forcing relation among all the periods which a map can have. When we try to extend this result to spaces of greater dimension and for general continuous maps, we find that it is not possible. Nevertheless for some classes of maps contained in the continuous, we obtain a result similar to Sharkovsky's for *σ-permutation maps* on appropriate spaces (see [10]), where with σ we denote the standard cyclic permutation of the set of numbers $\{1, 2, \cdots, n\}$, that is, $\sigma(i) = i + 1$ and $\sigma(n) = 1$. We prove also that without loss of generality we can use the σ transform as representative of any other permutation on the numbers $\{1, 2, ..., n\}$, since the resultant systems are conjugate and therefore similar from the dynamics point of view.

The map $F : \mathbb{X}^n \to \mathbb{X}^n$ is called a *σ-permutation* if

$$F(x_1, x_2, \cdots, x_n) = (f_{\sigma(1)}(x_{\sigma(1)}), f_{\sigma(2)}(x_{\sigma(2)}), \cdots, f_{\sigma(n)}(x_{\sigma(n)})),$$

where $f_i \in C(\mathbb{X})$ for $i = 1, 2, \cdots, n$. We denote them by $C_\sigma(\mathbb{X}^n)$.

This class of maps appear in connection with problems of economic theory, for example in the Cournot duopoly model (see [25]) where two firms are producing the same good and their productions depend on the production of the rival firm in the previous period. In this case, we have the dynamical system $(\mathbb{X}, F)$, where

$$F(x, y) = (g(y), f(x))$$

and $\mathbb{X}^2 = \mathbb{I}^2$.

The periodic structure of a σ-permutation map F is closely related with those of the maps $f_i^{(n)} : \mathbb{X} \to \mathbb{X}$ given by

$$f_i^{(n)} = f_{\sigma(i)} \circ f_{\sigma^2(i)} \circ \cdots \circ f_{\sigma^n(i)}, \qquad i \in \{1, \cdots, n\}.$$

In fact, the following results establish that relationship (see [11]).

Proposition 1.1. *Let $F : \mathbb{X}^n \to \mathbb{X}^n$ be a σ-permutation map. Then $\{p : p|n\} \subseteq \mathrm{Per}(F)$ if and only if $f_1^{(n)}$ has at least two different fixed points.*

Proposition 1.2. *Let $F : \mathbb{X}^n \to \mathbb{X}^n$ be a σ-permutation map. Suppose $t \nmid n$. If $1/\gcd(t, n) \in \mathrm{Per}(f_1^{(n)})$, then $t \in \mathrm{Per}(F)$.*

Observe that these results can be stated in the setting of a general set $\mathbb{X}$, so it is not necessary that $\mathbb{X}$ be endowed with a metric, even with a topological structure. Moreover, the Propositions work if the hypothesis of continuity of F and $f_i^{(n)}$ is removed. Therefore, the nature of Proposition 1.1 and 1.2 is merely combinatorial.

When $\mathbb{X}^n = \mathbb{I}^n$ or $\mathbb{R}^n$ for $n \geq 2$ it is possible to obtain the periodic structure of σ-permutation maps, including converse results.

The *Sharkovsky ordering* of $\mathbb{N} \cup 2^\infty = N^*$, denoted by $\geq_s$, is as follows

$$3 >_s 5 >_s 7 >_s \cdots >_s 3 \cdot 2 >_s 5 \cdot 2 >_s \cdots >_s 3 \cdot 2^2 >_s 5 \cdot 2^2 >_s \cdots$$
$$\cdots >_s 3 \cdot 2^n >_s 5 \cdot 2^n >_s \cdots >_s \{2^\infty\} >_s \cdots >_s 2^{n+1} >_s 2^n >_s \cdots > 1.$$

We denote by $S(m)$ the *initial segment* of the Sharkovsky ordering which ends at $m \in \mathbb{N}^*$, that is,

$$S(m) = \{n \in \mathbb{N}^* : n \leq_s m\} \text{ if } m \in \mathbb{N}, \text{ and}$$
$$S(2^\infty) = \{1, 2, 2^2, \cdots, 2^n, \cdots\}.$$

For any interval map f, Sharkovky's theorem establishes that there exists $m \in \mathbb{N}^*$ such that $\mathrm{Per}(f) = S(m)$. In the case of a line map g either $\mathrm{Per}(g) = S(m)$ or $\mathrm{Per}(g) = \emptyset$.

When $\mathbb{X} = \mathbb{I}^n$ the periodic structure of σ-permutation maps is known (see [12]). Next, we explain it. For $k \in \mathbb{N}$ and $m \in \mathbb{N}^*$, we introduce

$$S_k(m) = \left\{t \in \mathbb{N} : t \nmid k \quad \text{and} \quad \frac{t}{\gcd(t,k)} \in S(m)\right\} \bigcup \{1\}.$$

Obviously $t \geq_s m$ implies $S_k(m) \subseteq S_k(t)$. Moreover, it is easily seen that

$$S_k(m) = \{pt : p \mid k,\ t \in (S(m) \setminus \{1\}),\ \gcd(k/p, t) = 1\} \bigcup \{1\}.$$

Theorem 1.3. *(Periodic structure of σ-permutation maps on $\mathbb{I}^k$)*

1. *Let F be a σ-permutation map on $\mathbb{I}^k$, $k \geq 2$. Then there exists $m \in \mathbb{N}^*$ such that*
$$\mathrm{Per}(F) = S_k(m) \qquad \text{or}$$
$$\mathrm{Per}(F) = S_k(m) \bigcup \{p : p|k\}.$$
We have $\mathrm{Per}(f_i^{(k)}) = S(m)$ for every $i \in \{1, 2, \cdots, k\}$. Moreover, $\mathrm{Per}(F) \neq S_k(m)$ if and only if $f_i^{(k)}$ possesses at least two different fixed points.
2. *Suppose that $\mathcal{P} = S_k(m)$ or $\mathcal{P} = S_k(m) \cup \{p : p|k\}$ for some $m \in \mathbb{N}^*$. Then there exists a σ-permutation map $F : \mathbb{I}^k \to \mathbb{I}^k$ with $\mathrm{Per}(F) = \mathcal{P}$.*

The key of the proof is to use the periodic structure of the interval maps $f_i^{(k)}$ (see [12]). It can be also extended to the case $\mathbb{X}^k = \mathbb{R}^k$, obtaining a similar result to the case $\mathbb{I}^k$, except in the fact there is another possibility for the direct and converse statements, the case $\mathrm{Per}(F) = \emptyset$.

In order to visualize better the above periodic structure, we present the situations $k = 2, 3$. We are using the notation $a \Rightarrow b$ to indicate that the period a forces the presence of period b in the map, and $(a \Leftrightarrow b)$ indicates that periods a and b are mutually forced.

Theorem 1.4. *(Periodic structure on $\mathbb{I}^2$) For $k = 2$, the periodic structure of a σ-permutation map $\mathbb{I}^2 \to \mathbb{I}^2$ is described in the frame of forcing on $\mathbb{N} \setminus \{2\}$*

$$(3 \Leftrightarrow 2 \cdot 3) \Rightarrow (5 \Leftrightarrow 2 \cdot 5) \Rightarrow \cdots \Rightarrow (2n+1 \Leftrightarrow 2 \cdot (2n+1)) \Rightarrow \cdots$$
$$2^2 \cdot 3 \Rightarrow 2^2 \cdot 5 \Rightarrow \cdots \Rightarrow 2^2 \cdot (2n+1) \Rightarrow \cdots$$
$$2^3 \cdot 3 \Rightarrow 2^3 \cdot 5 \Rightarrow \cdots \Rightarrow 2^3 \cdot (2n+1) \Rightarrow \cdots$$
$$\cdots$$

$$2^k \cdot 3 \Rightarrow 2^k \cdot 5 \Rightarrow \cdots \Rightarrow 2^k \cdot (2n+1) \Rightarrow \cdots$$
$$\cdots$$
$$\cdots \Rightarrow 2^{m+1} \Rightarrow 2^m \Rightarrow \cdots \Rightarrow 2^3 \Rightarrow 2^2 \Rightarrow 1.$$

Moreover, $2 \in \mathrm{Per}(F)$ *if and only if* $f_i^{(2)}$ *has at least two different fixed points.*

Theorem 1.5. *(Periodic structure on* $\mathbb{I}^3$*) For* $k = 3$*, the periodic structure of a* σ*-permutation map* $\mathbb{I}^3 \to \mathbb{I}^3$ *is described in the frame of forcing on* $\mathbb{N} \setminus \{3\}$

$$3 \cdot 3 \Rightarrow (3 \cdot 5 \Leftrightarrow 5) \Rightarrow (3 \cdot 7 \Leftrightarrow 7) \Rightarrow 3 \cdot 9 \Rightarrow (3 \cdot 11 \Leftrightarrow 11) \Rightarrow \cdots$$
$$3 \cdot 2 \cdot 3 \Rightarrow (3 \cdot 2 \cdot 5 \Leftrightarrow 2 \cdot 5) \Rightarrow (3 \cdot 2 \cdot 7 \Leftrightarrow 2 \cdot 7) \Rightarrow 3 \cdot 2 \cdot 9 \Rightarrow \cdots$$
$$\cdots$$
$$3 \cdot 2^k \cdot 3 \Rightarrow (3 \cdot 2^k \cdot 5 \Leftrightarrow 2^k \cdot 5) \Rightarrow (3 \cdot 2^k \cdot 7 \Leftrightarrow 2^k \cdot 7) \Rightarrow 3 \cdot 2^k \cdot 9 \Rightarrow \cdots$$
$$\cdots$$
$$\cdots \Rightarrow (3 \cdot 2^m \Leftrightarrow 2^m) \Rightarrow \cdots \Rightarrow (3 \cdot 2^2 \Leftrightarrow 2^2) \Rightarrow (3 \cdot 2 \Leftrightarrow 2) \Rightarrow 1.$$

Moreover, $3 \in \mathrm{Per}(F)$ *if and only if* $f_i^{(3)}$ *has at least two different fixed points.*

Now we consider $\mathbb{X} = S^1$, where $\mathbb{S}^1 = \{z \in \mathbb{C} : |z| = 1\}$ is the unit circle. We wonder about the periodic structure of σ-permutations maps on the k-dimensional torus $\mathbb{X}^k\mathbb{T}^k = \mathbb{S}^1 \times \overset{k}{\ldots} \times \mathbb{S}^1$ with $k \geq 2$. The periodic structure will depend on the periodic structure of maps on $\mathbb{S}^1$ which in turn depends on the *degree* of the map. We need to use now results on periodic structure from circle maps. A complete and detailed treatment of the subject can be seen in [5]. The *degree* of a σ-permutation map F on $\mathbb{T}^k$, denoted by D, is defined as

$$D = \deg(F) = \deg(f_1^{(k)}) = \deg(f_2^{(k)}) = \cdots = \deg(f_k^{(k)}) = \prod_{i=1}^{k} \deg(f_i).$$

In a general sense, D plays an important role since the periodic structure of F depends on the periodic structure of $f_1^{(k)}$, and in turn it is influenced by the degree of the composition above. We separate the cases $D \neq 1$ and $D = 1$. In the first case we use the same notation with the same meaning for the initial segments $S_k(m)$ than in the case $\mathbb{I}^k$.

For $D \neq 1$ we obtain (see [11])

Proposition 1.3. *Let* $F : \mathbb{T}^k \to \mathbb{T}^k$ *be a* σ*-permutation map. Suppose* $t \nmid k$ *and* $D \neq 1$*. Then* $t \in \mathrm{Per}(F)$ *if and only if* $t/\gcd(t,k) \in \mathrm{Per}(f_1^{(k)})$.

This result and Proposition 1.1 allow us to obtain the main result for $D \neq 1$ ([11]).

Theorem 1.6. *Let* $F : \mathbb{T}^k \to \mathbb{T}^k$ *be a* σ*-permutation map.*

1. If $|D| > 2$*, then* $\mathrm{Per}(F) = \mathbb{N}$.

2. *If* $D=-2$, *then* $\mathrm{Per}(F)=\mathbb{N}$ *or* $\mathrm{Per}(F)=\mathbb{N}\setminus\{2p : p \mid k, 2p \nmid k\}$.
3. *If* $D=2$, *then* $\mathrm{Per}(F)=\mathbb{N}$ *or* $\mathrm{Per}(F)=\mathbb{N}\setminus\{p : p \mid k, p \geq 2\}$.
4. *If* $D=-1$ *then there exists* $m\in\mathbb{N}^*$ *such that* $\mathrm{Per}(F)=S_k(m)\cup\{p : p \mid k\}$.
5. *If* $D=0$, *then there exists* $m\in\mathbb{N}^*$ *such that* $\mathrm{Per}(F)=S_k(m)$ *or* $\mathrm{Per}(F)=S_k(m)\cup\{p : p \mid k\}$.

Now we establish the result on periodic structure for σ-permutation maps F with $D=1$. We need to introduce the following notation. For $\alpha,\beta\in\mathbb{R}$, $m_\alpha, m_\beta\in\mathbb{N}^*$ and $t\in\mathbb{N}$, we define

$$\widetilde{S_k}(t)=\{m\in\mathbb{N} : m\nmid k, \frac{m}{\gcd(m,k)}=t\},$$

$$\widetilde{S_k}(\alpha,\beta,m_\alpha,m_\beta)=\bigcup_{t\in S(\alpha,m_\alpha)\cup M(\alpha,\beta)\cup S(\beta,m_\beta)}\widetilde{S_k}(t),$$

where for $\rho\in\{\alpha,\beta\}$

$$S(\rho,m_\rho)=\begin{cases}\emptyset, & \text{if}\,\rho\notin\mathbb{Q}\\ \{n_\rho p : p\in S(m_\rho)\}, & \text{if}\,\rho=k_\rho/n_\rho,\ \gcd(k_\rho,n_\rho)=1,\end{cases}$$

$$M(\alpha,\beta)=\begin{cases}\emptyset, & \text{if}\,\alpha=\beta\\ \{n\in\mathbb{N} : \alpha<t/n<\beta\,\text{for some } t\in\mathbb{Z}\}, & \text{if}\,\alpha<\beta\end{cases}$$

As in the case of $S_k(m)$ we find that

$$\widetilde{S_k}(\alpha,\beta,m_\alpha,m_\beta)=$$
$$\{pt : p\mid k,\ t\in S(\alpha,m_\alpha)\bigcup M(\alpha,\beta)\bigcup S(\beta,m_\beta),\ t\neq 1,\quad \gcd(k/\rho,t)=1\}$$

When $D=1$ a new situation appears. In general, there exist some exceptions in order to establish the converse result of Proposition 1.2 (see [11]).

Let $F=(f_{\sigma(i)})_{i=1}^k$ be a σ-permutation map with $D=1$. We say that F is *exceptional* if the rotation interval of $f_1^{(k)}$ equals $[\alpha,\beta]$, where $\alpha,\beta\in\mathbb{Q}$, $\alpha=s_\alpha/n_\alpha$ with $\gcd(s_\alpha,n_\alpha)=1$, $\beta=s_\beta n_\beta$ with $\gcd(s_\beta,n_\beta)=1$, $n_\alpha\neq n_\beta$, $\mathrm{Per}(f_1^{(k)})=n_\alpha S(m_\alpha)\cup M(\alpha,\beta)\cup n_\beta S(m_\beta)$ for some $m_\alpha,m_\beta\in\mathbb{N}^*$, $\beta-\alpha=1/\mathrm{lcm}(n_\alpha,n_\beta)$, $\mathrm{lcm}(n_\alpha,n_\beta)\notin\mathrm{Per}(f_1^{(k)})$ and $1\notin\mathrm{Per}(f_1^{(k)})$. In this case we put

$$E_{\alpha,\beta}=\{(qe)\ \mathrm{lcm}(n_\alpha,n_\beta) : e=\gcd(k,\mathrm{lcm}(n_\alpha,n_\beta)),\ qe\mid k,\ qe\neq 1,\\ (qe)\ \mathrm{lcm}(n_\alpha,n_\beta)\nmid k\}$$

Now we can give the result for the case $D=1$ (see [11]).

Proposition 1.4. *Let* $F:\mathbb{T}^k\to\mathbb{T}^k$ *be a* σ*-permutation map. Suppose that* $D=1$ *and* F *is exceptional. Set* $e=\gcd(k,\mathrm{lcm}(n_\alpha,n_\beta))$. *Then* $qe\mid k$, $(qe)\mathrm{lcm}(n_\alpha,n_\beta)\nmid k$ *and* $(qe)\neq 1$ *imply* $(qe)\mathrm{lcm}(n_\alpha,n_\beta)\in\mathrm{Per}(F)$. *Moreover,* $\mathrm{lcm}(n_\alpha,n_\beta)\notin\mathrm{Per}(F)$ *if* $e=1$.

To complete Proposition 1.2, in the case $D = 1$ we distinguish the exceptional case from the rest and obtain (see [11])

Proposition 1.5. *Let $F : \mathbb{T}^k \to \mathbb{T}^k$ be a σ-permutation map with $D = 1$, and let $m \in \mathbb{N}$ with $m \nmid n$.*

1. *Suppose that F is not exceptional. Then $m \in \mathrm{Per}(F)$ if and only if $m/\gcd(m,k) \in \mathrm{Per}(f_1^{(k)})$.*
2. *Assume that F is exceptional. Then $m \in \mathrm{Per}(F)$ if and only if $m/\gcd(m,k) \in \mathrm{Per}(f_1^{(k)})$ or $m = (qe)\mathrm{lcm}(n_\alpha, n_\beta)$ for some $q \in \mathbb{N}$ with $qe \mid k$ and $qe \neq 1$, where $e = \gcd(k, \mathrm{lcm}(n_\alpha, n_\beta))$.*

Using this result we deduce ([11])

Theorem 1.7. *Let $F : \mathbb{T}^k \to \mathbb{T}^k$ be a σ-permutation map with $D = 1$. Then one of the following situations holds:*

1. $\mathrm{Per}(F) = \emptyset$.
2. $\mathrm{Per}(F) = \widetilde{S_k}(\alpha, \beta, m_\alpha, m_\beta) \bigcup E_{\alpha,\beta}$, *for some $\alpha, \beta \in \mathbb{Q}$ and for some $m_\alpha, m_\beta \in \mathbb{N}^*$, if F is exceptional.*
3. *Otherwise there exist $\alpha, \beta \in \mathbb{R}$ and $m_\alpha, m_\beta \in \mathbb{N}^*$ such that* $\mathrm{Per}(F) = \wp = \widetilde{S_k}(\alpha, \beta, m_\alpha, m_\beta)$, *or* $\mathrm{Per}(F) = \wp \cup \{1\}$, *or* $\mathrm{Per}(F) = \wp \cup \{p : p \mid k\}$.

The corresponding converse results can be also given (see [11]), that is, given one of the set of periods $\mathcal{P}$ appearing in Theorems 1.6 and 1.7, it is possible to construct a σ-permutation map $F : \mathbb{T}^k \to \mathbb{T}^k$ in such a way that $\mathrm{Per}(F) = \mathcal{P}$.

The following example is given to illustrate and clear the exceptional case mentioned in the statement of the previous result.

Example 1.1. Consider $t = 12$, $k = 2$. Therefore $\gamma := \gcd(t, n) = 2$ and $t/\gamma = 6$. We are going to choose a $\sigma-$permutation map $F : \mathbb{T}^2 \to \mathbb{T}^2$ such that $t = 12 \in \mathrm{Per}(F)$ but $k/\gamma = 6 \notin \mathrm{Per}(f_1^{(2)})$. According to [5, Section 3.10], for any given $\alpha, \beta \in \mathbb{R}$, and $m_\alpha, m_\beta \in \mathbb{N}^*$, we can find $f \in C(\mathbb{S}^1)$ such that $\deg(f) = 1$ and $\mathrm{Per}(f) = S(\alpha, m_\alpha) \cup M(\alpha, \beta) \cup S(\beta, m_\beta)$. In particular, if $\alpha = 1/3$, $\beta = 1/2$, $m_\alpha = m_\beta = 1$, $n_\alpha = 3$, $n_\beta = 2$, it is $\mathrm{Per}(f) = \{3\} \cup M(1/3, 1/2) \cup \{2\}$, with $M(1/3, 1/2) = \{5, 7, 8, 9, 10, 11, 12, 13, ...\}$. In this case, $t/\gamma = 6 = \mathrm{lcm}(n_\alpha, n_\beta)$.

$$\beta - \alpha = \frac{1}{2} - \frac{1}{3} = \frac{1}{6} = \frac{1}{\mathrm{lcm}(n_\alpha, n_\beta)}$$

and $1 \notin \mathrm{Per}(f)$. Now we define $F : \mathbb{T}^2 \to \mathbb{T}^2$ by

$$F(x_1, x_2) = (x_2, f(x_1))$$

We deduce that $f_i^{(2)} = f$ for $i = 1, 2$, and F is exceptional. Hence $k/\gamma = 6 \notin \text{Per}(f_1^{(2)})$. Finally we prove that $t = 12 \in \text{Per}(F)$. Since F is exceptional, according to Theorem 1.7 we have

$$\begin{aligned}
&\widetilde{S_k}(\alpha, \beta, m_\alpha, m_\beta) = \\
&\{pt : p \mid k,\ t \in S(\alpha, m_\alpha) \bigcup M(\alpha, \beta) \bigcup S(\beta, m_\beta),\ t \neq 1, \quad \gcd(k/\rho, t) = 1\} \\
&\qquad = 2\,\text{Per}(f) \bigcup \{m : m \neq 1,\ m \in \text{Per}(f), \quad m \text{ odd} = \mathbb{N} \setminus \{1, 2, 8, 12\}\}
\end{aligned}$$

and

$$\begin{aligned}
E_{\alpha,\beta} &= \{(qe)\ \text{lcm}(n_\alpha, n_\beta) : e = \gcd(k, \text{lcm}(n_\alpha, n_\beta)),\ qe \mid k,\ qe \neq 1, \\
&\qquad\qquad\qquad\qquad\qquad\qquad (qe)\ \text{lcm}(n_\alpha, n_\beta) \nmid k\} \\
&= \{2 \cdot q \cdot 6 : 2q \mid 2,\ 2q \neq 1,\ 2 \cdot q \cdot 6 \nmid 2\} = \{12\}
\end{aligned}$$

Therefore $\text{Per}(F) = \mathbb{N} \setminus \{1, 2, 8\}$. Observe that in this example we have $t = 12 \in \text{Per}(F)$ but $t/\gcd(t, k) = \text{lcm}(n_\alpha, n_\beta) = 6 \notin \text{Per}(f)$.

In Theorems 1.6 and 1.7 we include the finite sets of periods which a σ-permutation map on $\mathbb{T}^k$ can have. To be more precise, in the next result we state a simple characterization of these finite sets of periods, specially in the case $D = 1$ and $k = 2$. In general we have (see [11])

Proposition 1.6. *Let $F : \mathbb{T}^k \to \mathbb{T}^k$ be a σ-permutation map. Suppose that* $\text{Per}(F)$ *is a finite set. Then*

1. $D \in \{-1, 0, 1\}$.
2. *If $D = -1$ then there exists $0 \leq r < \infty$ such that* $\text{Per}(F) = S_k(2^r) \cup \{p : p \mid k\}$.
3. *If $D = 0$, then there exists $0 \leq r < \infty$ such that* $\text{Per}(F) = S_k(2^r)$ *or* $\text{Per}(F) = S_k(2^r) \cup \{p : p \mid k\}$.
4. *If $D = 1$, then* $\text{Per}(F) = \emptyset$ *or there exist integers m, r with $m \geq 1$ and $r \geq 0$ such that*

$$\text{Per}(F) = \left\{pm2^i : p \mid k, \quad \gcd\left(\frac{k}{p}, m2^i\right) = 1,\ i = 0, 1, \cdots, r\right\}$$

for $m > 1$, or

$$\text{Per}(F) = \left\{p2^i : p \mid k, \quad \gcd\left(\frac{k}{p}, 2^i\right) = 1,\ i = 1, 2, \cdots, r\right\} \bigcup \{p : p \mid k\}$$

for $m = 1$ and $r \geq 1$. In the case $m = 1$, $r = 0$, only one of the following situations can occur

$$\text{Per}(F) = \{p : p \mid k\},$$
$$\text{Per}(F) = \{1\}.$$

5. *If* $D = 1$ *and* $k = 2$, *then* $\mathrm{Per}(F) = \emptyset$ *or there exists* $0 \leq r < \infty$ *such that* $\mathrm{Per}(F) = \{m, 2m, \cdots, 2^{r+1}m\}$ *for some* m *odd*, $\mathrm{Per}(F) = \{2m, 2^2m, \cdots, 2^{r+1}m\}$ *for some* m *even or* $\mathrm{Per}(F) = \{1\}$.

Remark 1.1. Given a σ-permutation map $F : \mathbb{X}^k \to \mathbb{X}^k$, where $\mathbb{X}$ is simply a set, without specifying a particular topology or metric, Proposition 1.1 holds (see [11]). Hence we can know if the divisors of k belong or not to $\mathrm{Per}(F)$. This is obviously a combinatorial result. On the other hand, for nondivisors of k, Proposition 1.2 also holds for any set $\mathbb{X}$, (see [11]). Therefore in order to know completely the periodic structure of $F \in C_\sigma(X)$ we must study if there exist some exceptions to the converse result of this property. For $\mathbb{X} = \mathbb{I}$ there are no exceptions (see [12], and the same occurs for $\mathbb{X} = \mathbb{S}^1$ in the case $D \neq 1$ (see Proposition 1.3). However, in the case $\mathbb{X} = \mathbb{S}^1$, $D = 1$, we have seen that exceptional maps do not verify the converse result which we must separate from the rest.

In view of these results, if we would know the periodic structure of a map $f : \mathbb{X} \to \mathbb{X}$ (for example, in dimension one, the periodic structure of continuous maps on trees, graphs, dendrites, dendroids, etc., for definitions see [55]) we could establish the periodic structure of σ-permutation maps on $\mathbb{X}^k$ and describe in what cases is possible to obtain finite sets of them.

1.2.2 Periodic structure of delayed difference equations $x_n = f(x_{n-k})$, $k \geq 2$

Let

$$x_n = f(x_{n-k}) \tag{1.2.1}$$

be a delayed difference equation, where $f : \mathbb{X} \to \mathbb{X}$, and $\mathbb{X}$ is any set. Observe that for all $r \geq 0$, $s \geq 1$ we have

$$f^r(x_s) = x_{s+rk}, \tag{1.2.2}$$

and consider the dynamics generated by the initial condition

$$C_0 = (x_1, x_2, \cdots, x_k) \in \mathbb{X}^k.$$

In the simplest situation, the periodic case, we say that the sequence $\{x_n\}_{n=1}^{\infty}$ generated by (1.2.1) from the initial condition C_0 is periodic whenever $x_{n+p} = x_n$ for some $p \in \mathbb{N}$ and for all $n \in \mathbb{N}$. The smallest of these values p is called the *period* of $\{x_n\}_n$. We use $\mathrm{Per}_{DE}(f)$ to denote the set of periods of the equation. To study its periodic structure, consider the map

$$F(z_1, z_2, \cdots, z_k) = (z_2, z_3, \cdots, z_k, f(z_1))$$

defined from $\mathbb{X}^k$ into itself. The following result is concerned with the general expression of any iterate of F and it is straightforward to obtain.

Lemma 1.1. *Let* $F(z_1, z_2, \cdots, z_k) = (z_2, z_3, \cdots, z_k, f(z_1))$. *Then*

$$F^{n+k+j}((x_i)_{i=1}^k) = \\ (f^n(x_{j+1}), f^n(x_{j+2}), \cdots, f^n(x_k), f^{n+1}(x_1), f^{n+1}(x_2), \cdots, f^{n+1}(x_j))$$

for every $n \geq 0$ *and* $1 \leq j \leq k$.

Then we can study the periodic points of the equation with the periodic points of F.

Proposition 1.7. *We have*

$$\mathrm{Per}(F) = \mathrm{Per}_{DE}(f).$$

Proof. Let $(x_1, x_2, \cdots, x_k) \in \mathbb{X}^k$ be an initial condition which generates a periodic solution of (1.2.1) $\{x_n\}_{n=1}^{\infty}$ of period p. If $p = nk + j$, with $n \geq 0$ and $0 \leq j < k$, by (1.2.2) and Lemma 1.1 we have

$$\begin{aligned} F^p(x_1, x_2, \cdots, x_k) &= (f^n(x_{j+1}), f^n(x_{j+2}), \cdots, f^n(x_k), f^{n+1}(x_1), \cdots, f^{n+1}(x_j)) \\ &= (x_{j+1+nk}, x_{j+2+nk}, \cdots, x_{k+nk}, x_{1+(n+1)k}, \cdots, x_{j+(n+1)k}) \\ &= (x_{p+1}, x_{p+2}, \cdots, x_{k+p-j}, x_{k+1+p-j}, \cdots, x_{p+k}) \\ &= (x_1, x_2, \cdots, x_{k-j}, x_{k-j+1}, \cdots, x_k) \\ &= (x_1, x_2, \cdots, x_k). \end{aligned}$$

So $(x_1, x_2, \cdots, x_k)$ is a periodic point of F, and $q = \mathrm{ord}_F((x_i)_{i=1}^k)$ verifies $q|p$. We have to prove that in fact $q = p$. Suppose that $q < p$, $q|p$, and $q = mk + i$, with $0 \leq i < k$. According to Lemma 1.1, if $F^q((x_i)_{i=1}^k) = (x_i)_{i=1}^k$ we find

$$x_1 = f^m(x_{i+1}), \cdots, x_{k-i} = f^m(x_k), x_{k-i+1} = f^{m+1}(x_1), \cdots, x_k = f^{m+1}(x_i). \tag{1.2.3}$$

On the other hand, from (1.2.2) and (1.2.3) we obtain

$$x_s = x_{s+i+mk} = x_{s+q}$$

for $1 \leq s \leq k - i$, and

$$x_t = x_{t-k+i+(m+1)k} = x_{t-k+k+q} = x_{t+q}$$

for $k - i + 1 \leq t \leq k$.

If $r \geq 1$, with $r = a_r k + b_r$, $1 \leq b_r \leq k - 1$, $a_r \geq 0$, now

$$x_{k+r} = x_{(a_r+1)k+b_r} = f^{a_r+1}(x_{b_r}) = f^{a_r+1}(x_{b_r+q}) = x_{(a_r+1)k+b_r+q} = x_{k+r+q}.$$

Then $\{x_i\}_{i=1}^{\infty}$ is periodic for (1.2.1) and its period is smaller or equal than q, which is a contradiction since $q < p$. This proves that $\mathrm{Per}_{DE}(f) \subseteq \mathrm{Per}(F)$.

If $(x_1, x_2, \cdots, x_k)$ is a periodic point of F of order p, then it is easy to obtain that the sequence $x_{n}{}_{n=1}^{\infty}$ has also period p and therefore $\mathrm{Per}(F) \subseteq \mathrm{Per}_{DE}(f)$. □

From this result we deduce that the study of the periodic structure of (1.2.1) is equivalent to analyze the periodic structure of F. In the case of $\mathbb{X} = \mathbb{I}$ and $C(\mathbb{I})$ this periodic structure is known (see Theorem 1.3). Therefore, using Proposition 1.7 and Theorem 1.3, with $F(z_1, z_2, \cdots, z_k) = (z_2, z_3, \cdots, z_k, f(z_1))$, we obtain

Corollary 1.1. *(Periodic structure of difference equations of type $x_n = f(x_{n-k})$ on $\mathbb{I}$, $k \geq 2$)*

1. *Let $x_n = f(x_{n-k})$ be a delayed difference equation, with $f \in C(\mathbb{I})$ and $k \geq 2$. Then there exists $m \in \mathbb{N}^*$ such that*

$$\mathrm{Per}_{DE}(f) = S_k(m) \qquad \text{or}$$
$$\mathrm{Per}_{DE}(f) = S_k(m) \bigcup \{p : p|k\}.$$

2. *Given $\mathcal{P} = S_k(m)$ or $\mathcal{P} = S_k(m) \cup \{p : p|k\}$, with $m \in \mathbb{N}^*$, there exists a difference equation of type $x_n = f(x_{n-k})$, with $f \in C(I)$, such that $\mathrm{Per}_{DE}(f) = \mathcal{P}$.*

Now we will obtain the periodic structure of delayed difference equations of the type given in (1.2.1) for circle maps $f \in C(\mathbb{S}^1)$. It can be easily done knowing the periodic structure of σ-permutation maps on $\mathbb{T}^k$. If $\deg(f) \neq 1$, let us consider $F \in C(\mathbb{T}^k)$ given by $F(z_1, z_2, \cdots, z_k) = (z_2, z_3, \cdots, z_k, f(z_1))$, $k \geq 2$. Applying Theorem 1.6 and Proposition 1.7 we have

Corollary 1.2. *Let $x_n = f(x_{n-k})$ be a delayed difference equation defined on the circle $\mathbb{S}^1$, with $f \in C(\mathbb{S}^1)$ and $k \geq 2$. Let $d = \deg(f) \neq 1$.*

1. *If $|d| > 2$, then $\mathrm{Per}_{DE}(f) = \mathbb{N}$.*
2. *If $d = -2$, then $\mathrm{Per}_{DE}(f) = \mathbb{N}$ or $\mathrm{Per}_{DE}(f) = \mathbb{N} \setminus \{2p : p \mid k, 2p \nmid k\}$.*
3. *If $d = 2$, then $\mathrm{Per}_{DE}(f) = \mathbb{N}$ or $\mathrm{Per}_{DE}(f) = \mathbb{N} \setminus \{p : p \mid k, p \geq 2\}$.*
4. *If $d = -1$ then there exists $m \in \mathbb{N}^*$ such that $\mathrm{Per}_{DE}(f) = S_k(m) \cup \{p : p \mid k\}$.*
5. *If $d = 0$, then there exists $m \in \mathbb{N}^*$ such that $\mathrm{Per}_{DE}(f) = S_k(m)$ or $\mathrm{Per}_{DE}(f) = S_k(m) \cup \{p : p \mid k\}$.*

The converse result is also true.

Corollary 1.3. *Given one of the sets $\mathcal{P}$ of periods stated in Corollary (1.2) there exists a circle map $f \in C(\mathbb{S}^1)$ such that the associate delayed difference equation $x_n = f(x_{n-k})$ holds $\mathrm{Per}_{DE}(f) = \mathcal{P}$.*

Given one of the sets of periods $\mathcal{P}$ appearing in Theorem 1.7, in ([11]) we construct σ-permutation maps F on $\mathbb{T}^n$ with $\mathrm{Per}(F) = \mathcal{P}$, and with $F(x_1, x_2, \cdots, x_n) = (x_2, x_3, \cdots, x_n, f(x_1))$. Then as an immediate consequence of Proposition 1.7 we have

Corollary 1.4. *Let $x_n = f(x_{n-k})$ be a delayed difference equation defined on the circle $\mathbb{S}^1$, with $k \geq 2$. Suppose that $d = \deg(f) = 1$. Then one of the following situations holds:*

1. $\mathrm{Per}_{DE}(f) = \emptyset$.
2. $\mathrm{Per}_{DE}(f) = \widetilde{S_k}(\alpha, \beta, m_\alpha, m_\beta) \bigcup E_{\alpha,\beta}$, *for some* $\alpha, \beta \in \mathbb{Q}$ *and* $m_\alpha, m_\beta \in \mathbb{N}^*$.
3. *Otherwise there exist* $\alpha, \beta \in \mathbb{R}$ *and* $m_\alpha, m_\beta \in \mathbb{N}^*$ *such that* $\mathrm{Per}_{DE}(f) = \wp = \widetilde{S_k}(\alpha, \beta, m_\alpha, m_\beta)$, *or* $\mathrm{Per}_{DE}(f) = \wp \cup \{1\}$, *or* $\mathrm{Per}_{DE}(f) = \wp \cup \{p : p \mid k\}$.

Corollary 1.5. *Given one of the sets $\mathcal{P}$ of periods stated in Corollary 1.4 there exists a circle map $f \in C(\mathbb{S}^1)$ such that the associate delayed difference equation $x_n = f(x_{n-k})$ holds $\mathrm{Per}_{DE}(f) = \mathcal{P}$.*

Finally, according to Remark 1.1, if we know the periodic structure of continuous maps defined from a topological space into itself, we can know the periodic structure of delayed difference equations $x_n = f(x_{n-k})$. To do it, the divisors of the dimension k are analyzed in Proposition 1.3, and for non-divisors of k we must obtain the exceptional cases for which the property: $t \in \mathrm{Per}(F)$ implies $k/\gcd(k,n) \in \mathrm{Per}(f_1^{(n)})$ whenever $k \nmid n$ is not completely fulfilled.

1.2.3 Further results

Analogously to the periodic situation, we can consider other properties of the sequences defined in equation (1.2.1) by an initial condition $(x_1, x_2, \cdots, x_k)$. For example, we can study the existence of global attractors of (1.2.1) in terms of

$$F(x_1, x_2, \cdots, x_k) = (x_2, \cdots, x_k, f(x_1))$$

We say that $\overline{x}$ is a global attractor of the general difference equation $x_{n+1} = \Phi(x_n, x_{n-1}, \cdots, x_{n-k})$ defined from $\mathbb{R}^{k+1}$ into $\mathbb{R}$ if $\lim_{n\to\infty} x_n = \overline{x}$ for any initial conditions x_{-k}, x_{-k+1}, $\cdots$, x_0. We obtain the following result.

Proposition 1.8. *Let $f : \mathbb{R} \to \mathbb{R}$ be continuous.*

1. *Suppose that f has a unique fixed point $\overline{x}$, and assume that $\overline{x}$ is a global attractor of the difference equation $x_n = f(x_{n-1})$. Then $\overline{x}$ is a global attractor of the delayed difference equation $x_n = f(x_{n-k})$ for all $k \geq 2$.*
2. *Conversely, suppose that $\overline{x}$ is a global attractor of $x_n = f(x_{n-k})$ for some $k \geq 1$. Then $\mathrm{P}(f) = \mathrm{Fix}(f) = \{\overline{x}\}$ and $\overline{x}$ is a global attractor.*

In Corollary 2.4.1 of [43] it is stated the following result concerning the existence of global attractors of the difference equation $x_{n+1} = \alpha x_n + f(x_{n-k})$.

Proposition 1.9. *Let $\alpha \in [0,1)$ and let $k \in \mathbb{N}$. Let $f \in C([0,\infty),(0,\infty))$ be decreasing. Suppose that f has a unique fixed point $\overline{x} \in (0,\infty)$ and that $\overline{x}$ is a global attractor of all positive solutions of the first order difference equation*

$$y_{n+1} = \frac{f(y_n)}{1-\alpha}, \qquad n = 0, 1, \cdots,$$

where $y_0 \in (0,\infty)$. Then $\overline{x}$ is a global attractor of all positive solutions of

$$x_{n+1} = \alpha x_n + f(x_{n-k}). \tag{1.2.4}$$

Now we are going to prove that this result works even in the case in which f is not decreasing. Moreover, the hypotheses imply that $\alpha = 0$, so we can apply Proposition 1.8 in order to conclude that $\overline{x}$ is a global attractor of $x_{n+1} = f(x_{n-k})$.

Proposition 1.10. *Let $f \in C([0,\infty),(0,\infty))$. Suppose that f has a unique fixed point $\overline{x} \in (0,\infty)$ and that $\overline{x}$ is a global attractor of all positive solutions of the first order difference equation*

$$y_{n+1} = \frac{f(y_n)}{1-\alpha}, \qquad n = 0, 1, \cdots, \ \alpha \in [0,1), \ y_0 \in (0,\infty).$$

Then $\alpha = 0$ and $\overline{x}$ is a global attractor of

$$x_{n+1} = f(x_{n-k}),$$

for all $k \geq 1$.

Proof. Since $\overline{x}$ is a global attractor of $y_{n+1} = f(y_n/(1-\alpha)$, by continuity of the map $g(z) = f(z)/(1-\alpha)$ we find $\overline{x} = f(\overline{x})/(1-\alpha)$. This implies that $f(\overline{x}) = (1-\alpha)\overline{x}$. On the other hand, $\overline{x}$ is a fixed point of f, so $\overline{x} = (1-\alpha)\overline{x}$. Since $\overline{x} \neq 0$, we obtain $\alpha = 0$, and the difference equation (1.2.4) reduces to $x_{n+1} = f(x_{n-k})$. According to Proposition 1.8, $\overline{x}$ is a global attractor of $x_{n+1} = f(x_{n-k})$, for all $k \geq 1$.

Besides the results on periodic structure, in one dimensional dynamical systems coming from population dynamics is interesting the detection of such periodic points. In practical situations, such detection can be difficult due to irregular prediction of them ([34]). In the computation of periodic points their stability plays also a relevant role. Another relevant problem (in systems depending on parameters) is to state the robustness of such periodic points.

The above treated situations are related to one dimensional systems on the interval with or without delay and in dimension two with σ-permutation maps. Nevertheless in the applications are stated other problems. We present here some of them and add some examples:

- Nonlinear difference equations $x_{n+1} = f(x_n, x_{n-1})$ It is equivalent to consider the plane dynamical systems $(F, \mathbb{R}^2)$ where $F(x,y) = (y, f(x,y))$. It is not known if in this situation there are forcing relations among periods. The following example are dynamical systems equivalent to the corresponding difference equations. The Henón transformations
  1. $F(x,y) = (1 - ax^2 + by, x)$
  2. $F(x,y) = (y, (x + ay^2 - 1)/b)$

  or logistic two-dimensional transformations
  - $F(x,y) = (y, ay(1-x))$
  - $F(x,y) = (y, ay - bxy))$

  are good examples.
- Nonlinear systems of difference equations:
  1. $x_{n+1} = F(x_n, y_n)$
  2. $y_{n+1} = G(x_n, y_n)$

  In this more complicate case, we have positive and negative answers to the same question. For the *triangular systems*
  1. $x_{n+1} = f(x_n)$
  2. $y_{n+1} = g(x_n, y_n)$

  and for the *anti-triangular systems*
  1. $x_{n+1} = g(y_n)$
  2. $y_{n+1} = f(x_n)$

  the answer is positive, that is, there are forcing relations among the periods. In the triangular case it is held a Sharkovsky ordering [39] and the anti-triangular case has been treated before.

The *Lotka-Volterra* systems equivalent to the corresponding systems of difference equations, are the great interest in the applications.

$$F(x,y) = (x(a_1 + b_1 x + c_1 y), y(a_2 + b_2 x + c_2 y))$$

There are many examples in the literature, like

$$F(x,y) = ((1 + a - bx - cy)x, dxy)$$

with a, b, c and d real parameters, was introduced in 1968 by Maynard Smith in the setting of the population dynamics [50]. In the same setting, Scudo et *al* in [47] proposed to consider the transformation

$$F(x,y) = ((5 - 1.9x - 10y)x, xy)$$

From Electronics we have the transformation proposed by Sharkovsky ([71]).

$$F(x,y) = (x(4 - x - y), xy)$$

In all these cases, it is not known if there exists or not forcing relations.

When we are looking for periodic points, it is useful to consider some stability results like the following. For a continuous map f and a periodic point of period n, there exists a neighborhood of f in the $C(I,I)$ topology, such that every g belonging to this neighborhood keeps periodic points of all periods in the Sharkovsky ordering (see [18]). In this setting it is relevant to see when the periodic points disappear or not after small perturbation in the map of the system under consideration.

1.3 Metric entropy or Kolmogorov-Sinai entropy (KS)

When one deals with chaotic behavior of a dynamical system, one interesting problem is to find a measure of the complexity of the system. It is generally assumed that KS entropy means a scale of measure of such complexity. Since we move from zero to positive KS entropy, it establishes and describes the transition from regular to chaotic behavior. More precisely, the claim is that having positive KS entropy is a sufficient condition for a dynamical system to be chaotic (see [15]). The notion is important and has been introduced in several books and papers (see for example [73]). Nevertheless we have chosen a more general way using the approach introduced by J. Canovas in [24] in his Doctoral Dissertation.

In some settings, chaos is explained in terms of random behavior, and random behavior is explained in turn in terms of having positive KS entropy. The connection between this entropy and random behavior is justified in some papers (see for example [30]), proving that in Hamiltonian systems, KS es equivalent to a generalized version of Shannon's communication-theoretic entropy under certain assumptions.

By a probability space $(X, \beta(X), \mu)$ we will understand a set X, a σ-algebra on X, $\beta(X)$, and a probability measure μ defined on $\beta(X)$. Consider a sequence of measure preserving transformations $T_{1,\infty} = (T_i)_{i=1}^{\infty}$, that is, measurable transformations $T_i : X \to X$ satisfying the condition $\mu(T_i^{-1}A) = \mu(A)$ for any $A \in \beta(X)$ and for any $i \in \mathbb{N}$. By a *measure theoretical non-autonomous system* (in short *m.t.n.s.*) we understand the pair $(X, T_{1,\infty})$. If $x \in X$, then the *orbit* of x is given by the sequence

$$(x, T_1(x), T_2(T_1(x)), \cdots).$$

When $T_i = T$ for all $i \in \mathbb{N}$ where T is a measure preserving transformation T, the pair (X,T) is a classical *measure theoretical dynamical system* (shortly *m.t.d.s*) deeply studied in the literature (see eg. [73] or [28]).

Sequences of measure preserving transformations have been studied from the point of view of Ergodic Theory. In few words, it deals with the convergence of the sequence

$$\frac{1}{n} S_n(f,x) = \frac{1}{n} \sum_{i=0}^{n-1} f \circ T_i \circ \cdots \circ T_1(x),$$

where $f : X \to \mathbb{R}$ is a continuous map and $x \in X$ (see [33] [21], [22], [23], [16] or [20]).

A useful tool to study classical m.t.d.s. is the *metric entropy or Kolmogorov-Sinai entropy* of T, introduced by Kolmogorov and Sinai (KS) (see the definition below). In this chapter and motivated by a paper of Kolyada and Snoha [40], we extend this notion in the setting of sequences of measure preserving transformations and study its properties and make an special emphasis on the properties of metric sequence entropy. In order to give a complete account of the subject we have given proofs of some results when they are difficult to get mostly taken from [24]; otherwise, we introduce the results giving some references.

1.3.1 Metric entropy of a finite partition

First of all, we recall some necessary notions and notation. A set $\mathcal{A} = \{A_1, \cdots, A_k\}$ is called a *finite partition* of X if $A_i \in \beta(X)$ for $i = 1, 2, \cdots, k$, $A_i \cap A_j = \emptyset$ if $i \neq j$, and $\cup_{i=1}^k A_i = X$. Let $\mathcal{Z}$ denote the set containing all the finite partitions of X. If $\mathcal{B} = \{B_1, \cdots, B_l\}$ is also a finite partition of X, $\mathcal{A} \vee \mathcal{B} = \{A_i \cap B_j : (A_i \in \mathcal{A}), (B_j \in \mathcal{B})\}$. We write $\mathcal{A} \leq \mathcal{B}$ to mean that each element of $\mathcal{A}$ is a union of elements of $\mathcal{B}$. We also say that $\mathcal{B}$ is *finer* than $\mathcal{A}$. Under the convention that $0 \log 0 = 0$, we have (see for example [73]):

Definition 1.8. *Let $\mathcal{A} = \{A_1, \cdots, A_k\} \in \mathcal{Z}$. The metric entropy of the finite partition $\mathcal{A}$ is*

$$H_\mu(\mathcal{A}) = -\sum_{i=1}^{k} \mu(A_i) \log \mu(A_i)$$

We also introduce the metric conditional entropy as:

Definition 1.9. *Let $\mathcal{A} = \{A_1, \cdots, A_k\}$ and $\mathcal{B} = \{B_1, \cdots, B_l\}$ be finite partitions of X. The metric conditional entropy of $\mathcal{A}$ relative to $\mathcal{B}$ is*

$$\begin{aligned} H_\mu(\mathcal{A}/\mathcal{B}) &= \sum_{\mu(B_j)\neq 0} \mu(B_j) \sum_{i=1}^{k} \frac{\mu(A_i \bigcap B_j)}{\mu(B_j)} \log \frac{\mu(A_i \bigcap B_j)}{\mu(B_j)} \\ &= \sum_{\mu(B_j)\neq 0} \sum_{i=1}^{k} \mu(A_i \bigcap B_j) \log \frac{\mu(A_i \bigcap B_j)}{\mu(B_j)}. \end{aligned}$$

Let $\mathcal{A}, \mathcal{B}$ be finite partitions of X. We introduce a relation between these partitions by $\rho(\mathcal{A}, \mathcal{B}) = H_\mu(\mathcal{A}/\mathcal{B}) + H_\mu(\mathcal{B}/\mathcal{A})$. It can be easily checked that ρ is a distance and then $(\mathcal{Z}, \rho)$ is a metric space.

The main properties of metric entropy and metric conditional entropy of partitions are summarized in the following result (see Theorem 4.3 of [73]).

Theorem 1.8. *If $\mathcal{A}, \mathcal{B}, \mathcal{C} \in \mathcal{Z}$ then*

(a) $H_\mu(\mathcal{A} \vee \mathcal{B}/\mathcal{C}) = H_\mu(\mathcal{A}/\mathcal{C}) + H_\mu(\mathcal{B}/\mathcal{A} \vee \mathcal{C})$.
(b) $H_\mu(\mathcal{A} \vee \mathcal{B}) = H_\mu(\mathcal{A}) + H_\mu(\mathcal{B}/\mathcal{A})$.
(c) If $\mathcal{A} \leq \mathcal{B}$ *then* $H_\mu(\mathcal{A}/\mathcal{C}) \leq H_\mu(\mathcal{B}/\mathcal{C})$.
(d) If $\mathcal{A} \leq \mathcal{B}$ *then* $H_\mu(\mathcal{A}) \leq H_\mu(\mathcal{B})$.
(e) If $\mathcal{B} \leq \mathcal{C}$ *then* $H_\mu(\mathcal{A}/\mathcal{C}) \leq H_\mu(\mathcal{A}/\mathcal{B})$.
(f) $H_\mu(\mathcal{A}) \geq H_\mu(\mathcal{A}/\mathcal{B})$.
(g) $H_\mu(\mathcal{A} \vee \mathcal{B}/\mathcal{C}) \leq H_\mu(\mathcal{A}/\mathcal{C}) + H_\mu(\mathcal{B}/\mathcal{C})$.
(h) $H_\mu(\mathcal{A} \vee \mathcal{B}) \leq H_\mu(\mathcal{A}) + H_\mu(\mathcal{B})$.
(i) If $T : X \to X$ *is a measure preserving transformation then*

$$H_\mu(T^{-1}\mathcal{A}/T^{-1}\mathcal{B}) = H_\mu(\mathcal{A}/\mathcal{B}).$$

(j) If $T : X \to X$ *is a measure preserving transformation then*

$$H_\mu(T^{-1}\mathcal{A}) = H_\mu(\mathcal{A}).$$

1.3.2 Metric entropy of a sequence of measure preserving transformations

Let $T_{1,\infty} = (T_i)_{i=1}^\infty$ be a sequence of measure preserving transformations $T_i : X \to X$ with $i \in \mathbb{N}$. Denote by $T_1^n = T_n \circ \cdots \circ T_1$, and by $T_1^{-n}(A) = T_1^{-1}(\cdots(T_n^{-1}A))$ for any $A \in \beta(X)$.

Definition 1.10. *Let* $\mathcal{A}$ *be a finite partition of* X. *The metric entropy of* $T_{1,\infty}$ *relative to the partition* $\mathcal{A}$ *is*

$$h_\mu(T_{1,\infty}, \mathcal{A}) = \limsup_{n\to\infty} \frac{1}{n} H_\mu\left(\bigvee_{i=0}^{n-1} T_1^{-n}\mathcal{A}\right).$$

and the metric entropy of $T_{1,\infty}$

$$h_\mu(T_{1,\infty}) = \sup_{\mathcal{A}} h_\mu(T_{1,\infty}, \mathcal{A}).$$

We will denote the partition $\bigvee_{i=0}^{n-1} T_1^{-n}\mathcal{A}$ by $\mathcal{A}^n(T_{1,\infty})$ if it is necessary.

Let $B = (b_i)_{i=1}^\infty$ be a sequence of positive integers. The sequence $A = (a_i)_{i=1}^\infty$ defined by $a_i = \sum_{k=1}^i b_k$ is an increasing sequence of positive integers. Consider the sequence of measure preserving transformations $T_{1,\infty} = (T^{b_i})_{i=1}^\infty$, where $T : X \to X$ is a measure preserving transformation. Then its metric entropy $h_\mu(T_{1,\infty}) = h_{\mu,A}(T)$ coincides with the *metric sequence entropy* of T (see [45]). When the sequence B is the constant sequence $b_i = 1$, then $h_\mu(T_{1,\infty}) = h_\mu(T)$ is the *metric entropy* of T (see Chapter 4 from [73]).

The basic properties of the metric entropy of sequences of measure preserving transformations are summarized below.

Theorem 1.9. *Let* $T_{1,\infty} = (T_i)_{i=1}^\infty$ *be a sequence of measure preserving transformations. Let* $\mathcal{A}, \mathcal{B} \in \mathcal{Z}$. *Then:*

(a) $h_\mu(T_{1,\infty}, \mathcal{A}) \leq H_\mu(\mathcal{A})$.
(b) $h_\mu(T_{1,\infty}, \mathcal{A} \vee \mathcal{B}) \leq h_\mu(T_{1,\infty}, \mathcal{A}) + h_\mu(T_{1,\infty}, \mathcal{B})$.
(c) If $\mathcal{A} \leq \mathcal{B}$ *then* $h_\mu(T_{1,\infty}, \mathcal{A}) \leq h_\mu(T_{1,\infty}, \mathcal{B})$.
(d) $h_\mu(T_{1,\infty}, \mathcal{A}) \leq h_\mu(T_{1,\infty}, \mathcal{B}) + H_\mu(\mathcal{A}/\mathcal{B})$.
(e) $h_\mu(T_{1,\infty}, \mathcal{A}) = h_\mu(T_{2,\infty}, \mathcal{A})$ *where* $T_{2,\infty} = (T_2, T_3, \cdots)$.
(f) $|h_\mu(T_{1,\infty}, \mathcal{A}) - h_\mu(T_{1,\infty}, \mathcal{B})| \leq \rho(\mathcal{A}, \mathcal{B})$. *Hence the map* $h_\mu(T_{1,\infty}, \cdot) : \mathcal{Z} \to \mathbb{R}^+ \cup \{0\}$ *is continuous.*

Proof. By Theorem 1.8 (h) and (j) we have that

$$H_\mu\left(\bigvee_{i=1}^{n} T_1^{-i}\mathcal{A}\right) \leq \sum_{i=1}^{n} H_\mu(T_1^{-i}\mathcal{A}) \leq nH_\mu(\mathcal{A}),$$

and this gives (a).

Now, using Theorem 1.8 (h) it follows that

$$H_\mu\left(\bigvee_{i=1}^{n} T_1^{-i}(\mathcal{A} \vee \mathcal{B})\right) \leq H_\mu\left(\bigvee_{i=1}^{n} T_1^{-i}\mathcal{A}\right) + H_\mu\left(\bigvee_{i=1}^{n} T_1^{-i}\mathcal{B}\right),$$

and this proves (b).

Since $\mathcal{A} \leq \mathcal{B}$ we have that $\bigvee_{i=1}^{n} T_1^{-i}\mathcal{A} \leq \bigvee_{i=1}^{n} T_1^{-i}\mathcal{B}$, and so applying Theorem 1.8 (d) it holds

$$H_\mu\left(\bigvee_{i=1}^{n} T_1^{-i}\mathcal{B}\right) \leq H_\mu\left(\bigvee_{i=1}^{n} T_1^{-i}\mathcal{A}\right),$$

which gives (c).

To prove (d) we will use first Theorem 1.8 (c) and (b). We have then that

$$\begin{aligned} H_\mu\left(\bigvee_{i=0}^{n-1} T_1^{-i}\mathcal{A}\right) &\leq H_\mu\left(\left(\bigvee_{i=0}^{n-1} T_1^{-i}\mathcal{A}\right) \vee \left(\bigvee_{i=0}^{n-1} T_1^{-i}\mathcal{B}\right)\right) \\ &\leq H_\mu\left(\bigvee_{i=0}^{n-1} T_1^{-i}\mathcal{B}\right) + H_\mu\left(\bigvee_{i=0}^{n-1} T_1^{-i}\mathcal{A}/\bigvee_{i=0}^{n-1} T_1^{-i}\mathcal{B}\right). \end{aligned}$$

On the other hand, by Theorem 1.8 (c), (e) and (i) it follows that

$$\begin{aligned} H_\mu\left(\bigvee_{i=0}^{n-1} T_1^{-i}\mathcal{A}/\bigvee_{i=0}^{n-1} T_1^{-i}\mathcal{B}\right) &\leq \sum_{i=0}^{n-1} H_\mu\left(T_1^{-i}\mathcal{A}/\bigvee_{i=0}^{n-1} T_1^{-i}\mathcal{B}\right) \\ &\leq \sum_{i=0}^{n-1} H_\mu\left(T_1^{-i}\mathcal{A}/T_1^{-i}\mathcal{B}\right) \\ &= nH_\mu\left(\mathcal{A}/\mathcal{B}\right). \end{aligned}$$

Since there exists $\lim_{n\to\infty}(1/n)nH_\mu(\mathcal{A}/\mathcal{B}) = H_\mu(\mathcal{A}/\mathcal{B})$ it holds that

$$\begin{aligned}h_\mu(T_{1,\infty},\mathcal{A}) &= \limsup_{n\to\infty}\frac{1}{n}H_\mu\left(\bigvee_{i=0}^{n-1}T_1^{-i}\mathcal{A}\right)\\&\le \limsup_{n\to\infty}\frac{1}{n}\left(H_\mu\left(\bigvee_{i=0}^{n-1}T_1^{-i}\mathcal{B}\right)+nH_\mu(\mathcal{A}/\mathcal{B})\right)\\&= \limsup_{n\to\infty}\frac{1}{n}\left(H_\mu\bigvee_{i=0}^{n-1}T_1^{-i}\mathcal{B}\right)+H_\mu(\mathcal{A}/\mathcal{B})\\&= h_\mu(T_{1,\infty},\mathcal{B})+H_\mu(\mathcal{A}/\mathcal{B}).\end{aligned}$$

Now we will prove the property (e). Note that by Theorem 1.8 (b) and (d) we have that

$$\begin{aligned}H_\mu\left(\bigvee_{i=0}^{n-1}T_1^{-i}\mathcal{A}\right) &\le H_\mu(\mathcal{A})+H_\mu\left(\bigvee_{i=1}^{n-1}T_1^{-i}\mathcal{A}\right)\\&= H_\mu(\mathcal{A})+H_\mu\left(T_1^{-1}\left(\bigvee_{i=1}^{n-1}T_2^{-i}\mathcal{A}\right)\right)\\&= H_\mu(\mathcal{A})+H_\mu\left(\bigvee_{i=1}^{n-1}T_2^{-i}\mathcal{A}\right).\end{aligned}$$

Dividing by n and taking upper limits when n tends to infinite, we have that

$$\begin{aligned}h_\mu(T_{1,\infty},\mathcal{A}) &= \limsup_{n\to\infty}\frac{1}{n}H_\mu\left(\bigvee_{i=0}^{n-1}T_1^{-i}\mathcal{A}\right)\\&\le \limsup_{n\to\infty}\frac{1}{n}\left(H_\mu(\mathcal{A})+H_\mu\left(\bigvee_{i=1}^{n-1}T_2^{-i}\mathcal{A}\right)\right)\\&= h_\mu(T_{2,\infty},\mathcal{A}).\end{aligned}$$

On the other hand, by Theorem 1.8 (d), it follows that

$$\begin{aligned}H_\mu\left(\bigvee_{i=0}^{n-1}T_2^{-i}\mathcal{A}\right) &= H_\mu\left(T_1^{-1}\left(\bigvee_{i=0}^{n-1}T_2^{-i}\mathcal{A}\right)\right)\\&= H_\mu\left(\bigvee_{i=1}^{n}T_1^{-i}\mathcal{A}\right)\\&\le H_\mu\left(\bigvee_{i=0}^{n}T_1^{-i}\mathcal{A}\right).\end{aligned}$$

Dividing by n and taking upper limits when n tends to infinite, we conclude

$$h_\mu(T_{1,\infty},\mathcal{A}) = \limsup_{n\to\infty}\frac{1}{n}H_\mu(\bigvee_{i=0}^{n-1}T_1^{-i}\mathcal{A})$$

$$\geq \limsup_{n\to\infty} \frac{1}{n} H_\mu(\bigvee_{i=0}^{n-1} T_2^{-i}\mathcal{A})$$
$$= h_\mu(T_{2,\infty}, \mathcal{A}).$$

The proof of (f) is easy to see. □

In case of metric sequence entropy we have the following interesting property which will be useful later.

Proposition 1.11. *Let $T : X \to X$ be a measure preserving transformation and let $A = (a_i)_{i=1}^{\infty}$ be a sequence of positive integers. Then for all $\mathcal{A} \in \mathcal{Z}$ it holds*

$$h_{\mu,A}(T, \mathcal{A}) = h_{\mu,A}(T, T^{-1}\mathcal{A})$$

Proof. It follows from the equality $H_\mu(\mathcal{D}) = H_\mu(T^{-1}\mathcal{D})$ where $\mathcal{D} = \bigvee_{i=1}^{n} T^{-a_i}\mathcal{A}$ for some arbitrary partition $\mathcal{A}$ of X. □

Let $(\mathcal{A}_i)_{i=1}^{\infty}$ be a sequence of finite partitions of X. Denote by $\bigvee_{i=1}^{\infty}\mathcal{A}_i$ the biggest partition satisfying $\mathcal{A}_n \leq \bigvee_{i=1}^{\infty}\mathcal{A}_i$ for all $n \in \mathbb{N}$. The following lemma allows us to compute the metric entropy of measurable sequences in a simple way, and it is analogous to the same lemma in [45].

Lemma 1.2. *Let $T_{1,\infty} = (T_i)_{i=1}^{\infty}$ be a sequence of measure preserving transformations. Let $\mathcal{A}_1 \leq \mathcal{A}_2 \leq \cdots \leq \mathcal{A}_n \leq \cdots$ be a sequence of measurable partitions such that $\bigvee_{i=1}^{\infty} \mathcal{A}_i = \epsilon$, the partition into individual points. Then*

$$h_\mu(T_{1,\infty}) = \lim_{i\to\infty} h_\mu(T_{1,\infty}, \mathcal{A}_i).$$

Proof. Let $\mathcal{Z}_k$ be the set of finite partitions of X such that for any $\mathcal{B} \in \mathcal{Z}_k$ it holds that $\mathcal{B} \leq \mathcal{A}_k$. It follows from Lemma 2 in [45] that $\mathcal{M} = \cup_{k\geq 1}\mathcal{Z}_k$ is everywhere dense in $\mathcal{Z}$. Then

$$h_\mu(T_{1,\infty}) = \sup_{\mathcal{A}\in\mathcal{Z}} h_\mu(T_{1,\infty}, \mathcal{A}) = \sup_{\mathcal{A}\in\mathcal{M}} h_\mu(T_{1,\infty}, \mathcal{A})$$
$$= \sup_{\mathcal{A}_i} h_\mu(T_{1,\infty}, \mathcal{A}_i) = \lim_{i\to\infty} h_\mu(T_{1,\infty}, \mathcal{A}_i),$$

and the proof ends. □

The following technical lemmas can be found in [73].

Lemma 1.3. *Let $r \geq 1$ be an integer. Then for all $\epsilon > 0$ there exists $\delta > 0$ such that if $\mathcal{A} = \{A_1, \cdots, A_r\}$ and $\mathcal{B} = \{B_1, \cdots, B_r\}$ are two finite partitions with $\sum_{i=1}^{r} \mu(A_i \triangle B_i) < \delta$, it holds that $\rho(\mathcal{A}, \mathcal{B}) < \epsilon$.*

Lemma 1.4. *Let β_0 be an algebra such that the σ-algebra generated by β_0, which it is denoted $\sigma(\beta_0)$, is $\beta(X)$. Let $\mathcal{A}$ be a finite partition of X containing elements from $\beta(X)$. Then, for all $\epsilon > 0$ there exists a finite partition $\mathcal{B}$ containing elements from β_0 and holding $\rho(\mathcal{A}, \mathcal{B}) < \epsilon$.*

The above lemmas help us to prove the following result.

Proposition 1.12. *Let $T_{1,\infty} = (T_i)_{i=1}^{\infty}$ be a sequence of measure preserving transformations. Let β_0 be an algebra such that $\sigma(\beta_0) = \beta(X)$. Let $\mathcal{Z}_0$ be the set of finite partitions of X containing elements from β_0. Then*

$$h_\mu(T_{1,\infty}) = \sup\{h_\mu(T_{1,\infty}, \mathcal{A}) : \mathcal{A} \in \mathcal{Z}_0\}.$$

Proof. By Lemma 1.4, given an arbitrary real number $\epsilon > 0$ and $\mathcal{B} \in \mathcal{Z}$ there exists a finite partition $\mathcal{A}_\epsilon \in \mathcal{Z}_0$ such that $H_\mu(\mathcal{B}/\mathcal{A}_\epsilon) < \epsilon$. Then by Theorem 1.9 (d) we have that

$$\begin{aligned} h_\mu(T_{1,\infty}, \mathcal{B}) &\leq h_\mu(T_{1,\infty}, \mathcal{A}_\epsilon) + H_\mu(\mathcal{B}/\mathcal{A}_\epsilon) \\ &\leq h_\mu(T_{1,\infty}, \mathcal{A}_\epsilon) + \epsilon. \end{aligned}$$

So

$$h_\mu(T_{1,\infty}, \mathcal{B}) \leq \epsilon + \sup\{h_\mu(T_{1,\infty}, \mathcal{A}) : \mathcal{A} \in \mathcal{Z}_0\},$$

and since ε was arbitrary it follows that

$$h_\mu(T_{1,\infty}) \leq \sup\{h_\mu(T_{1,\infty}, \mathcal{A}) : \mathcal{A} \in \mathcal{Z}_0\}.$$

The reverse inequality is obvious, and so the proof ends.□

1.3.3 Isomorphisms of non-autonomous systems

Let $(X, \beta(X), \mu)$ and $(Y, \beta(Y), \nu)$ be two probability spaces and let $T_{1,\infty} = (T_i)_{i=1}^{\infty}$ and $S_{1,\infty} = (S_i)_{i=1}^{\infty}$ be two sequences of measure preserving transformations. We define the notion of isomorphism between $T_{1,\infty}$ and $S_{1,\infty}$ as follows.

Definition 1.11. *We say that $T_{1,\infty}$ is* isomorphic *to $S_{1,\infty}$ if there exist two measurable sets $M_1 \subset X$ and $M_2 \subset Y$ with $\mu(M_1) = 1 = \nu(M_2)$ such that*

(a) $T_i(M_1) \subset M_1$ and $S_i(M_2) \subset M_2$ for any $i \in \mathbb{N}$.
(b) There exists an invertible measure preserving transformation $\phi : M_1 \to M_2$ with $\phi \circ T_i = S_i \circ \phi$ for every integer i.

If the measure preserving transformation $\phi : M_1 \to M_2$ is surjective we will say that $S_{1,\infty}$ is a factor *of $T_{1,\infty}$. $T_{1,\infty}$ and $S_{1,\infty}$ are said* weakly isomorphic *if $S_{1,\infty}$ is a factor of $T_{1,\infty}$ and viceversa.*

In case of constant sequences of measure preserving transformations, we have the standard isomorphism of classical dynamical systems, and then it is well known that if $T_{1,\infty} = (T)$ and $S_{1,\infty} = (S)$ then $h_\mu(T) = h_\nu(S)$. If S is a factor of T we have $h_\mu(T) \geq h_\nu(S)$.

When we consider sequences of measure preserving transformations a similar result is obtained.

Theorem 1.10. *Let $T_{1,\infty}$ and $S_{1,\infty}$ be two sequences of measure preserving transformations. Then*

(a) If $T_{1,\infty}$ and $S_{1,\infty}$ are isomorphic, then $h_\mu(T_{1,\infty}) = h_\nu(S_{1,\infty})$.
(b) If $S_{1,\infty}$ is a factor of $T_{1,\infty}$, then $h_\mu(T_{1,\infty}) \geq h_\nu(S_{1,\infty})$.
(c) If $T_{1,\infty}$ and $S_{1,\infty}$ are weakly isomorphic, then $h_\mu(T_{1,\infty}) = h_\nu(S_{1,\infty})$.

Proof. Let $\mathcal{A}$ be a finite partition of Y. If $S_{1,\infty}$ is a factor of $T_{1,\infty}$, by Definition 1.11, there exists a measurable onto (modulo zero measure sets) $\phi: X \to Y$. Then

$$H_\nu\left(\bigvee_{i=1}^{n} S_1^{-i}\mathcal{A}\right) = H_\nu\left(\bigvee_{i=1}^{n} \pi^{-1}S_1^{-i}\pi\mathcal{A}\right) = H_\mu\left(\bigvee_{i=1}^{n} S_1^{-i}\pi\mathcal{A}\right),$$

which provides the inequality

$$h_\nu(S_{1,\infty}) \leq h_\mu(T_{1,\infty}),$$

and this proves (b).

Applying (b) two times we obtain (a) and (c) and this concludes the proof.
□

1.3.4 Examples

In this section, we compute the metric entropy of some sequences of measurable maps. We will make an special emphasis on zero metric entropy sequences.

Recall that given a measure preserving transformation $T : X \to X$ it can be defined a linear operator $U_T : L^2(X, \mu) \to L^2(X, \mu)$ given by $U_T(f) = f \circ T$ for all $f \in L^2(X, \mu)$. λ is said an eigenvalue of T if there exists a non zero map $f \in L^2(X, \mu)$ with $U_T(f) = \lambda f$; f is said an eigenvector of T associated to λ. We say that T has discrete spectrum if $L^2(X, \mu)$ has an orthonormal basis of eigenvectors.

Proposition 1.13. *Let $T_{1,\infty} = (T_i)_{i=1}^\infty$ be a sequence of measure preserving transformations. Suppose that for all positive integer i the measure preserving transformation T_i has discrete spectrum with the same orthonormal basis. Then $h_\mu(T_{1,\infty}) = 0$.*

Proof. Let $f \in L^2(X, \mu)$ be a characteristic function of a set of measure 1/2 and let M be the set of such functions contained in $L^2(X, \mu)$. Consider the map $\psi : M \to \mathcal{Z}$ such that $\psi(f) = \left\{f^{-1}(1), f^{-1}(0)\right\}$. Defining the metric ρ' in M as

$$\rho'(f, g) = \mu(f^{-1}(1) \triangle g^{-1}(1)),$$

it follows from [45] that ψ is a continuous map and if a set $K \subset M$ is compact, then $\psi(K)$ is also compact in $\psi(M)$. By Lemma 2 in [45], a closed set $L \subset \psi(M) \subset \mathcal{Z}$ is compact if and only if for any sequence $\{\xi_i\}_{i=1}^\infty \subset L$ it holds

$$\lim_{n\to\infty}\frac{1}{n}H_\mu\left(\bigvee_{i=1}^{n}\xi_i\right)=0.$$

Consider the sequence of unitary operators $U_{1,\infty}=\{U_{T_1},U_{T_2},\cdots\}$ where $U_{T_i}(g)=g\circ T_i$ for all $g\in L^2(X,\mu)$. If each transformation T_i has discrete spectrum with the same orthonormal basis $\{e_i\}_{i=0}^{\infty}$, then for any $f\in L^2(X,\mu)$ it follows that the closure of the set $\{U_{T_1}\circ\cdots\circ U_{T_n}(f)\}_{n=0}^{\infty}$, is compact. In order to see this, we know that $U_{T_i}(e_j)=\lambda_{ij}e_j$, $|\lambda_{ij}|=1$ for any pair $(i,j)\in\mathbb{N}^2$. For any $f\in L^2(X,\mu)$, $f=\sum_{i=0}^{\infty}a_ie_i$ with $\sum_{i=0}^{\infty}|a_i|^2<\infty$, it follows that:

$$\mathrm{Cl}\left(\{U_{T_1}\circ\cdots\circ U_{T_n}(f)\}_{n=0}^{\infty}\right)\subset\left\{g=\sum_{i=0}^{\infty}b_ie_i:|b_i|\le|a_i|\right\}=B.$$

Since B is compact, our set $\mathrm{Cl}(\{U_{T_1}\circ\cdots\circ U_{T_n}(f)\}_{n=0}^{\infty})$ is compact.

Let $f\in M\subset L^2(X,\mu)$. Since $\mathrm{Cl}(\{U_{T_1}\circ\cdots\circ U_{T_n}(f)\}_{n=0}^{\infty})$ is compact and the map ψ is continuous, the set

$$\psi\left(\mathrm{Cl}(\{U_{T_1}\circ\cdots\circ U_{T_n}(f)\}_{n=0}^{\infty})\right)=\mathrm{Cl}(\{T_1^{-n}\psi(f)\}_{n=0}^{\infty})$$

is compact in $\psi(M)\subset\mathcal{Z}$ and then

$$h_\mu(T_{1,\infty},\psi(f))=\limsup_{n\to\infty}\frac{1}{n}H_\mu\left(\bigvee_{i=0}^{n-1}T_1^{-n}\psi(f)\right)=0.$$

If we consider the finite partitions of X given by $\bigvee_{i=1}^{n}\xi_i$ with $\xi_i\in\psi(M)$, then by Theorem 1.9 we have

$$h_\mu(T_{1,\infty},\bigvee_{i=1}^{n}\xi_i)\le\sum_{i=1}^{n}h_\mu(T_{1,\infty},\xi_i)=0.$$

Since it follows from [45] that the set of finite partitions $\bigvee_{i=1}^{n}\xi_i$ with $\xi_i\in\psi(M)$ is everywhere dense in $\mathcal{Z}$, and the map $h_\mu(T_{1,\infty},\cdot):\mathcal{Z}\to R^+\cup\{0\}$ is continuous, by Theorem 1.9, $h_\mu(T_{1,\infty})=0$ and the proof concludes. □

Let $\mathbb{S}^1=\{z\in\mathbb{C}:|z|=1\}$. Consider a sequence $(\alpha_i)_{i=1}^{\infty}$, $\alpha_i\in\mathbb{C}$, and construct the sequence $R_{1,\infty}=(R_i)_{i=1}^{\infty}$ where $R_i:\mathbb{S}^1\to\mathbb{S}^1$ is given by $R_i(x)=\alpha_{nx}$, for all $x\in\mathbb{S}^1$. Then maps R_i are rotations on $\mathbb{S}^1$. Every rotation preserves the normalized Haar measure (see page 20 of [73]). Then we can prove the following proposition.

Proposition 1.14. *Under the above conditions, if μ is the normalized Haar measure in $\mathbb{S}^1$ we have that $h_m(R_{1,\infty})=0$.*

Proof. It is known that every rotation on $\mathbb{S}^1$ has discrete spectrum with the same orthonormal basis (see Section 3.3 from [73]). Applying Proposition 1.13 the proof ends. □

Now we apply Proposition 1.13 to the sequences of ergodic rotations on a compact group. Let G be a compact group and $a \in G$. Consider the measure preserving transformation $T_a : G \to G$ given by $T_a(g) = ag$ called rotation of *angle* a. It can be seen that each rotation preserves the Haar measure in G (see [73]). Then we can prove the following result.

Proposition 1.15. *Let $(a_i)_{i=1}^{\infty}$ be a sequence of elements of an abelian metrizable compact topological group G and let $T_{1,\infty} = (T_{a_i})_{i=1}^{\infty}$ be a sequence of ergodic rotations over G. If m denotes the Haar measure, then it follows that $h_m(T_{1,\infty}) = 0$.*

Proof. Let $\widehat{G}$ be the set of characters of G, that is, the set of continuous homomorphisms of G onto $\mathbb{S}^1$. It is know (see Section 3.3 of [73]) that $\widehat{G}$ is a discrete countable group, whose members are mutually orthogonal members of $L^2(G, m)$.

By Theorem 3.5 from [73], we know that every rotation T_{a_i} has discrete spectrum with basis $\widehat{G}$ for all $i \in \mathbb{N}$. Now apply Proposition 1.13 to obtain $h_m(T_{1,\infty}) = 0$. $\square$

There are sequences of measure preserving transformations with zero metric entropy $T_{1,\infty} = (T_i)_{i=1}^{\infty}$ satisfying that every map T_i, $i \in \mathbb{N}$, has no discrete spectrum . Let us see in the following example.

Example 1.2. Let $\Sigma^2 = \{0,1\}^{\mathbb{Z}}$. Denote by β the product σ-algebra of Σ^2 and by μ the product measure satisfying $\mu([i]) = 1/2$ where

$$[i] = \{\{x_i\}_{i=-\infty}^{\infty} : x_0 = i\} \qquad \text{with } i = 0, 1.$$

Let $\sigma : \Sigma^2 \to \Sigma^2$ be the shift map. Since $h_\mu(\sigma) = h_\mu(\sigma^{-1}) = \log 2$ it follows σ and σ^{-1} have not discrete spectrum (see [45]). Consider the periodic sequence $\sigma_{1,\infty} = (\sigma, \sigma^{-1}, \sigma, \sigma^{-1}, \cdots)$. By Theorem 1.18 we have $h_\mu(\sigma_{1,\infty}^2) = 2h_\mu(\sigma_{1,\infty})$. Since $\sigma_{1,\infty}^2 = (Id, Id, \cdots)$ with Id the identity map on Σ^2 we have $h_\mu(\sigma_{1,\infty}^2) = 0$ and then $h_\mu(\sigma_{1,\infty}) = 0$.

1.4 Properties of the metric entropy of measurable sequences

In this section we will prove some formulas to compute the metric entropy of sequences of measure preserving transformations.

The product formula

Proposition 1.12 helps us to prove the following formula. Let $(X_j, \beta(X_j), \mu_j)$ be two probability spaces with $j = 1, 2$ and consider two sequences of measure preserving transformations $T_{1,\infty} = (T_i)_{i=1}^{\infty}$ and $S_{1,\infty} = (S_i)_{i=1}^{\infty}$ where T_i :

$X_1 \to X_1$ and $S_i : X_2 \to X_2$ for $i \in \mathbb{N}$. Consider the product probability space $(X_1 \times X_2, \beta(X_1 \times X_2), \mu_1 \times \mu_2)$ and the sequence of measure preserving transformations $T_{1,\infty} \times S_{1,\infty} = (T_i \times S_i)_{i=1}^{\infty}$. Then we prove the following result.

Proposition 1.16. *Under the above conditions it follows that*

(a) $h_{\mu_1 \times \mu_2}(T_{1,\infty} \times S_{1,\infty}) \leq h_{\mu_1}(T_{1,\infty}) + h_{\mu_2}(S_{1,\infty})$.
(b) If $S_{1,\infty} = T_{1\infty}$, *it follows* $h_{\mu_1 \times \mu_2}(T_{1,\infty} \times T_{1,\infty}) = 2h_{\mu}(T_{1,\infty})$.

Proof. First we will prove (a). Consider two measurable partitions, $\mathcal{A}_1$ of X_1 and $\mathcal{A}_2$ of X_2. Then $\mathcal{A}_1 \times \mathcal{A}_2$ is a measurable partition of $X_1 \times X_2$. Hence

$$H_{\mu_1 \times \mu_2}(\bigvee_{i=0}^{n-1} (T_1^{-i} \times S_1^{-i})(\mathcal{A}_1 \times \mathcal{A}_2)) = H_{\mu_1}(\bigvee_{i=0}^{n-1} T_1^{-i}\mathcal{A}_1) + H_{\mu_2}(\bigvee_{i=0}^{n-1} S_1^{-i}\mathcal{A}_2).$$

So

$$\begin{aligned}
h_{\mu_1 \times \mu_2}(T_{1,\infty} \times S_{1,\infty}, \mathcal{A}_1 \times \mathcal{A}_2) &= \limsup_{n\to\infty} \frac{1}{n} H_{\mu_1 \times \mu_2}(\bigvee_{i=0}^{n-1}(T_1^{-i} \times S_1^{-i})(\mathcal{A}_1 \times \mathcal{A}_2)) \\
&\leq \limsup_{n\to\infty} \frac{1}{n} H_{\mu_1}(\bigvee_{i=0}^{n-1} T_1^{-i}\mathcal{A}_1) \\
&\quad + \limsup_{n\to\infty} \frac{1}{n} H_{\mu_2}(\bigvee_{i=0}^{n-1} S_1^{-i}\mathcal{A}_2) \\
&\leq h_{\mu_1}(T_{1,\infty}, \mathcal{A}_1) + h_{\mu_2}(S_{1,\infty}, \mathcal{A}_2) \\
&\leq h_{\mu_1}(T_{1,\infty}) + h_{\mu_2}(S_{1,\infty})
\end{aligned}$$

Since the algebra $\beta(X_1) \times \beta(X_2)$ generates the product σ-algebra $\beta(X_1 \times X_2)$, using Proposition 1.12, it follows

$$\begin{aligned}
h_{\mu_1 \times \mu_2}(T_{1,\infty} \times S_{1,\infty}) &= \sup_{\mathcal{A}_1 \times \mathcal{A}_2} h_{\mu_1 \times \mu_2}(T_{1,\infty} \times S_{1,\infty}, \mathcal{A}_1 \times \mathcal{A}_2) \\
&\leq h_{\mu_1}(T_{1,\infty}) + h_{\mu_2}(S_{1,\infty}),
\end{aligned}$$

and part (a) follows. In order to prove the part (b) consider $\mathcal{A}$ a finite partition of X, then

$$H_{\mu_1 \times \mu_1}(\bigvee_{i=0}^{n-1} (T_1^{-i} \times T_1^{-i})(\mathcal{A} \times \mathcal{A})) = 2H_{\mu_1}(\bigvee_{i=0}^{n-1} T_1^{-i}\mathcal{A})$$

Therefore

$$h_{\mu \times \mu}(T_{1,\infty} \times T_{1,\infty}, \mathcal{A} \times \mathcal{A}) = \limsup_{n\to\infty} \frac{1}{n} H_{\mu_1 \times \mu_1}(\bigvee_{i=0}^{n-1} (T_1^{-i} \times T_1^{-i})(\mathcal{A} \times \mathcal{A}))$$

$$= \limsup_{n\to\infty} \frac{1}{n} 2H_\mu(\bigvee_{i=0}^{n-1} T_1^{-i}\mathcal{A})$$
$$= 2h_\mu(T_{1,\infty}, \mathcal{A})$$

Taking the supremum over all the finite partitions

$$\sup_{\mathcal{A}} h_{\mu\times\mu}(T_{1,\infty}\times T_{1,\infty}, \mathcal{A}\times\mathcal{A}) = 2h_\mu(T_{1,\infty}).$$

Then

$$h_{\mu\times\mu}(T_{1,\infty}\times T_{1,\infty}) \geq 2h_\mu(T_{1,\infty}),$$

and using part (a), the proof ends. □

Iterated sequence formula

Let $T_{1,\infty}$ be a sequence of measure preserving maps $T_i : X \to X$. We can define a new sequence of maps $T_{1,\infty}^{[n]} = (S_i)_{i=1}^{\infty}$ where $S_i = T_{ni}\circ T_{ni-1}\circ\ldots\circ T_{(n-1)i+1}$ for all $i \in \mathbb{N}$. Then we obtain the following result.

Proposition 1.17. *Under the above conditions* $h_\mu(T_{1,\infty}^{[n]}) \leq nh_\mu(T_{1,\infty})$.

Proof. Let $\mathcal{A}$ be a finite partition of X. Then

$$\begin{aligned} h_\mu(T_{1,\infty}^{[n]}, \mathcal{A}) &= \limsup_{k\to\infty} \frac{1}{k} H_\mu(\mathcal{A}^k(T_{1,\infty}^{[n]})) \\ &\leq n \limsup_{k\to\infty} \frac{1}{kn} H_\mu(\mathcal{A}^{kn}(T_{1,\infty})) \\ &\leq n \limsup_{m\to\infty} \frac{1}{m} H_\mu(\mathcal{A}^m(T_{1,\infty})) \\ &= nh_\mu(T_{1,\infty}, \mathcal{A}). \end{aligned}$$

Taking the supremum over all the finite partitions the proof ends. □

The equality in Proposition 1.17 does not hold in general (see [27]). However it is true in some special cases. One of them is the following.

Definition 1.12. *Let* $T_{1,\infty} = (T_i)_{i=1}^{\infty}$ *be a sequence of measure preserving transformations. We say that* $T_{1,\infty} = (T_i)_{i=1}^{\infty}$ *is* periodic *if there exists a positive integer* n *such that* $T_{i+n} = T_i$ *for all* $i \in \mathbb{N}$. *The smallest integer satisfying this property is called the* period *of* $T_{1,\infty}$.

Then we prove the following result.

Proposition 1.18. *Let* $T_{1,\infty} = (T_i)_{i=1}^{\infty}$ *be a periodic sequence of measure preserving transformations. Then the formula* $h_\mu(T_{1,\infty}^m) = mh_\mu(T_{1,\infty})$ *holds for any positive integer* m.

Proof. First of all we prove the equality for n the period of $T_{1,\infty}$. Then $T_{1,\infty}^{[n]}$ is the constant sequence $(T_i^{[n]})_{i=1}^{\infty}$ with $T_i^{[n]} = T_1^n$ for all $i \in \mathbb{N}$. For any finite partition $\mathcal{A}$ and $t = mn + r$ we have:

$$\mathcal{A}^t(T_{1,\infty}) = (\mathcal{A} \vee T_1^{-1}\mathcal{A} \vee \cdots \vee T_1^{-(n-1)}\mathcal{A}) \bigvee (T_1^{-n}\mathcal{A} \vee T_1^{-1}\mathcal{A} \vee \cdots \vee T_1^{(2n-1}\mathcal{A})$$
$$\bigvee \cdots \bigvee (T_1^{-mn}\mathcal{A} \vee T_1^{-1}\mathcal{A} \vee \cdots \vee T_1^{-(mn-1)}\mathcal{A}) \bigvee (\bigvee_{i=0}^{r-1} T_1^{-i}\mathcal{A}).$$

Then

$$\frac{1}{m} H_\mu(\mathcal{A}^{mn+r}(T_{1,\infty})) \leq \frac{1}{m} H_\mu(\mathcal{A}^m(T_{1,\infty}^{[n]})) + \frac{1}{m} H_\mu(\bigvee_{i=0}^{r-1} T_1^{-i}\mathcal{A}),$$

and therefore

$$\frac{1}{m} H_\mu(\mathcal{A}^m(T_{1,\infty}^{[n]})) \geq \frac{1}{m} H_\mu(\mathcal{A}^{mn+r}(T_{1,\infty})) - \frac{1}{m} H_\mu(\bigvee_{i=0}^{r-1} T_1^{-i}\mathcal{A}).$$

Since $\limsup_{m\to\infty}(1/m)H_\mu(\bigvee_{i=0}^{r-1} T_1^{-i}\mathcal{A}) = 0$, we have

$$\begin{aligned} h_\mu(T_{1,\infty}^{[n]}, \bigvee_{i=0}^{n-1} T_1^{-i}\mathcal{A}) &= \limsup_{m\to\infty} \frac{1}{m} H_\mu(\mathcal{A}^m(T_{1,\infty}^{[n]})) \\ &\geq \limsup_{m\to\infty} \frac{1}{m} H_\mu(\mathcal{A}^{mn+r}(T_{1,\infty})) \\ &= n \limsup_{m\to\infty} \frac{1}{nm+r} H_\mu(\mathcal{A}^{mn+r}(T_{1,\infty})) \\ &= n(h_\mu T_{1,\infty}, \mathcal{A}) \end{aligned}$$

and then

$$h_\mu(T_{1,\infty}^{[n]}) \geq n h_\mu(T_{1,\infty}).$$

Applying Proposition 1.17 we get the equality.

Now, suppose that $m = kn$ with $k \in \mathbb{N}$. Then it holds $T_{1,\infty}^{[m]}$ is the constant sequence $(T_i^{[m]})_{i=1}^{\infty}$ where $T_i^{[m]} = (T_n \circ \cdots \circ T_1)^k$ and since it is periodic it holds

$$\begin{aligned} h_\mu(T_{1,\infty}^{[m]}) &= h_\mu(T_n \circ \cdots \circ T_1)^k) \\ &= k h_\mu(T_n \circ \cdots \circ T_1) \\ &= m h_\mu(T_{1,\infty}). \end{aligned}$$

Finally, suppose that n does not divide m. Then the sequence $T_{1,\infty}^{[m]}$ is periodic with the same period as $T_{1,\infty}$. Then

$$h_\mu(T^{nm}_{1,\infty}) = nh_\mu(T^m_{1,\infty}),$$

and on the other hand we have

$$h_\mu(T^{nm}_{1,\infty}) = nmh_\mu(T_{1,\infty}).$$

Combining both equalities we have that $h_\mu(T^m_{1,\infty}) = mh_\mu(T_{1,\infty})$ and the proof ends. □

The equality in Proposition 1.17 also can be attained in the following case. First of all we introduce a definition.

Definition 1.13. *Let $T_{1,\infty} = (T_i)_{i=1}^\infty$ be a sequence of measure preserving transformations. We say that $T_{1,\infty} = (T_i)_{i=1}^\infty$ is* eventually periodic *if there exists a positive integer k such that $T_{k,\infty}$ is periodic.*

Proposition 1.19. *Let $T_{1,\infty} = (T_i)_{i=1}^\infty$ be an eventually periodic sequence of measure preserving transformations. Then for any $m \in \mathbb{N}$ it follows that $h_\mu(T^{[m]}_{1,\infty}) = mh_\mu(T_{1,\infty})$.*

Proof. Let k be the first positive integer such that $T_{k,\infty}$ is periodic. Consider the sequence $T^{[m]}_{1,\infty} = (T_1^{m-1}, T_m^{2m-1}, \dots T_{(i-1)m}^{im-1}, \cdots)$. This sequence is also eventually periodic and $(T^{[m]})_{k,\infty} = (T_{km}^{2km-1}, \cdots, T_{(i-1)m}^{im-1}, \cdots)$ is periodic.

By Theorem 1.9 (e) and Proposition 1.18 it follows that

$$h_\mu(T^{[m]}_{1,\infty}) = h_\mu((T^{[m]})_{k,\infty}) = mh_\mu(T_{k,\infty}) = mh_\mu(T_{1,\infty}),$$

which concludes the proof. □

1.4.1 Some particular formulas for metric entropy

Now we consider periodic sequences of measure preserving transformations of period 1, that is, $T_{1,\infty} = (T)$. So the metric entropy of these sequences is called metric entropy of the transformation T.

The computation of the metric entropy is in general difficult. However, there exist several formulas which allow us to compute it when the measure preserving transformations have an special form. The main formulas that we will consider are the following.

Proposition 1.20. *Let $T : X \to X$ be a measure preserving transformation. Then*

(a) If T is invertible then $h_\mu(T^k) = |k|h_\mu(T)$ for all $k \in \mathbb{Z}$.
(b) If T is not invertible then $h_\mu(T^k) = kh_\mu(T)$ for all $k \in \mathbb{N}$.

Let $(X_i, \beta(X_i), \mu_i)$ be two probability spaces for $i = 1, 2$. Suppose that $T_i : X_i \to X_i$ are measure preserving transformations for $i = 1, 2$. Consider the product probability space $(X_1 \times X_2, \beta(X_1 \times X_2), \mu_1 \times \mu_2)$ and the measure preserving transformation $T_1 \times T_2 : X_1 \times X_2 \to X_1 \times X_2$ given by $T_1 \times T_2(x_1, x_2) = (T_1(x_1), T_2(x_2))$ for all $(x_1, x_2) \in X_1 \times X_2$. Then the metric entropy of this transformation can be computed with the following proposition.

Proposition 1.21. *Under the above conditions*

$$h_{\mu_1 \times \mu_2}(T_1 \times T_2) = h_{\mu_1}(T_1) + h_{\mu_2}(T_2).$$

Consider again two probability spaces $(X_i, \beta(X_i), \mu_i)$ as before. Let $g : X_1 \to X_1$ be a measure preserving transformation and let $f_{x_1} : X_2 \to X_2$ be a family of measure preserving transformations with $x_1 \in X_1$. Consider the product space $(X_1 \times X_2, \beta(X_1 \times X_2), \mu_1 \times \mu_2)$ and the triangular measure preserving transformation $T : X_1 \times X_2 \to X_1 \times X_2$ defined by

$$T(x_1, x_2) = (g(x_1), f(x_1, x_2)) = (g(x_1), f_{x_1}(x_2))$$

for all $(x_1, x_2) \in X_1 \times X_2$. When g is the identity the following proposition allows us to compute the metric entropy of T.

Proposition 1.22. *Under the above conditions*

$$h_{\mu_1 \times \mu_2}(T) = \int_{X_1} h_{\mu_2}(f_{x_1}) d\mu_1.$$

When g is not the identity, there exists a formula proved at the same time by Abramov and Rokhlin in [2] and by Adler in [3], which allows to compute the metric entropy of T. Let $\mathcal{A}$ be a finite partition of X_2 and let $x_1 \in X_1$. Define

$$\mathcal{A}_{x_1}^n = \bigvee_{k=0}^{n-1} f_{x_1}^{-1} f_{g(x_1)}^{-1} \cdots f_{g^{k-1}(x_1)}^{-1} \mathcal{A},$$

$$h_{\mu_2,g}(f, \mathcal{A}) = \lim_{n \to \infty} \frac{1}{n} \int_{X_1} H_{\mu_2}(\mathcal{A}_{x_1}^n) d\mu_1(x_1),$$

$$h_{\mu_2,g}(f) = \sup \{h_{\mu_2,g}(f, \mathcal{A}) : \mathcal{A} \in \mathcal{Z}_2\},$$

where $\mathcal{Z}_2$ is the space of finite partitions of X_2. Then

Proposition 1.23. *Under the above conditions*

$$h_{\mu_1 \times \mu_2}(T) = h_{\mu_1}(g) + h_{\mu_2,g}(f).$$

1.4.2 Some particular formulas for metric sequence entropy

It is natural to wonder if the formulas presented in Section 4.01 are or not true in the setting of metric sequence entropy. In this section we will consider the same formulas and we will show that the answers are not always positive.

Iterated map formula

Let $T : X \to X$ be a measure preserving transformation and let $n \in \mathbb{N}$. The formula $h_\mu(T^n) = nh_\mu(T)$ is based on an arithmetic property of the sequence $(i)_{i=0}^\infty$. Take the subsequences $(ni)_{i=0}^\infty$. Then clearly $ni + k$ is an element of the sequence $(i)_{i=0}^\infty$ for all $k \in \mathbb{N}$. This property does not hold in general by arbitrary sequences of positive integers. For example, if we take the sequence $(2^i)_{i=0}^\infty$ the above property does not hold. Using similar properties we can prove some formulas in the case of metric sequence entropy. To this end we need a previous result. Let $\mathcal{I}$ be the set of increasing sequences of positive integers and consider the shift map $\sigma : \mathcal{I} \to \mathcal{I}$ given by $\sigma((a_i)_{i=1}^\infty) = (a_{i+1})_{i=1}^\infty$.

Lemma 1.5. *Let $T : X \to X$ a measure preserving transformation and let $A = (a_i)_{i=1}^\infty$ be a sequence of positive integers. Let $k \in \mathbb{N}$. Then:*

$$h_{\mu,A}(T) = h_{\mu,\sigma^k(A)}(T).$$

Proof. Let $\mathcal{A} \in \mathcal{Z}$. By Theorem 1.8 (h) we have

$$H_\mu(\bigvee_{i=1}^n T^{-a_i}\mathcal{A}) \leq H_\mu(\bigvee_{i=1}^k T^{-a_i}\mathcal{A}) + H_\mu(\bigvee_{i=k+1}^n T^{-a_i}\mathcal{A}).$$

Since $\limsup_{n\to\infty}(1/n)H_\mu(\bigvee_{i=1}^k T^{-a_i}\mathcal{A}) = 0$, we have that

$$\begin{aligned} h_{\mu,A}(T,\mathcal{A}) &= \limsup_{n\to\infty} \frac{1}{n} H_\mu(\bigvee_{i=1}^n T^{-a_i}\mathcal{A}) \\ &\leq \limsup_{n\to\infty} \frac{1}{n} H_\mu(\bigvee_{i=k+1}^n T^{-a_i}\mathcal{A}) \\ &= h_{\mu,\sigma^k(A)}(T,\mathcal{A}). \end{aligned}$$

Then

$$h_{\mu,A}(T) = \sup_{\mathcal{A}} h_{\mu,A}(T,\mathcal{A}) \leq \sup_{\mathcal{A}} h_{\mu,\sigma^k(A)}(T,\mathcal{A}) = h_{\mu,\sigma^k(A)}(T).$$

On the other hand, since $\bigvee_{i=1}^n T^{-a_i}\mathcal{A}$ is finer than $\bigvee_{i=k+1}^n T^{-a_i}\mathcal{A}$ it follows from Theorem 1.8 (d) that

$$H_\mu(\bigvee_{i=1}^n T^{-a_i}\mathcal{A}) \geq H_\mu \bigvee_{i=k+1}^n T^{-a_i}\mathcal{A}.)$$

Therefore

$$h_{\mu,A}(T,\mathcal{A}) = \limsup_{n\to\infty} \frac{1}{n} H_\mu\left(\bigvee_{i=1}^n T^{-a_i}\mathcal{A}\right)$$

$$\geq \limsup_{n\to\infty} \frac{1}{n} H_\mu\left(\bigvee_{i=k+1}^{n} T^{-a_i}\mathcal{A}\right)$$
$$= h_{\mu,\sigma^k(A)}(T,\mathcal{A}).$$

Then

$$h_{\mu,A}(T) = \sup_{\mathcal{A}} h_{\mu,A}(T,\mathcal{A}) \geq \sup_{\mathcal{A}} h_{\mu,\sigma^k(A)}(T,\mathcal{A}) = h_{\mu,\sigma^k(A)}(T),$$

and we get the equality.□

Lemma 1.5 means that when we try to compute the metric sequence entropy of a measurable transformation, we can remove the first elements of the sequence A. This fact seems reasonable because the metric sequence entropy is an asymptotic value. Lemma 1.5 allows us to prove the following formula.

Proposition 1.24. *Let $T : X \to X$ be a measure preserving transformation and let $A = (m^i)_{i=0}^\infty$ with m a positive integer. Let $k \in \mathbb{N}$. Then*

$$h_{\mu,A}(T) = h_{\mu,A}(T^{m^k}).$$

Proof. Let $\mathcal{A} \in \mathcal{Z}$. Given an arbitrary sequence of positive integer $B = (b_i)_{i=1}^\infty$ it follows that

$$h_{\mu,B}(T^k,\mathcal{A}) = \limsup_{n\to\infty} \frac{1}{n} H_\mu(\bigvee_{i=1}^{n} T^{-kb_i}\mathcal{A}) = h_{\mu,kB}(T,\mathcal{A}).$$

where kB denotes the sequence of positive integers $(kb_i)_{i=1}^\infty$. Take the sequence $A = (m^i)_{i=0}^\infty$. The above formula gives

$$h_{\mu,A}(T^{m^k},\mathcal{A}) = h_{\mu,m^k A}(T,\mathcal{A}),$$

where $m^k A = (m^{i+k})_{i=0}^\infty = (m^i)_{i=k}^\infty$. So $\sigma^k(A) = m^k A$, and by Lemma 1.5 we have that

$$h_{\mu,A}(T^{m^k},\mathcal{A}) = h_{\mu,m^k A}(T,\mathcal{A}) = h_{\mu,A}(T,\mathcal{A}).$$

Then

$$h_{\mu,A}(T^{m^k}) = \sup_{\mathcal{A}} h_{\mu,A}(T^{m^k},\mathcal{A}) = \sup_{\mathcal{A}} h_{\mu,A}(T,\mathcal{A}) = h_{\mu,A}(T),$$

which concludes the proof. □

Proposition 1.24 can be used to find examples of measure preserving transformations for which the formula $h_{\mu,A}(T^n) = nh_{\mu,A}(T)$ does not hold. For that it is enough to find transformations with finite metric sequence entropy with respect to the sequence $(m^i)_{i=0}^\infty$ for any positive integer m. These examples can be taken from [27], [46] and [45]. We guess if it is possible to find similar formulas for other type of sequences like for example $(i^m)_{i=0}^\infty$ with $m \in \mathbb{N}$.

The product formula

Given an arbitrary sequence A, the product formula is not true (see [46]), but it can be replaced by the inequality (see [32])

$$h_{\mu_1\times\mu_2,A}(T_1\times T_2)\leq h_{\mu_1,A}(T_1)+h_{\mu_2,A}(T_2).$$

When $T_1 = T_2$ the formula is true, and the same occurs if $h_{\mu_1,A}(T_1)$ or $h_{\mu_2,A}(T_2)$ are zero.

The triangular formula

Although the formula $h_{\mu_1\times\mu_2,A}(T_1\times T_2)=h_{\mu_1,A}(T_1)+h_{\mu_2,A}(T_2)$ is not true in general, it holds when T_1 or T_2 are the identity map. Therefore, it is reasonable to think that the formula $h_{\mu_1\times\mu_2,A}(T)=\int_{X_1}h_{\mu_2,A}(f_{x_1})d\mu_1$ could be true. However, it is easy to provide an example showing that it is false in general. To show this we will consider the following example studied by Kusnhirenko in [45]. Let us denote by $\beta(\mathbb{S}^1)$ the Borel σ-algebra and by μ the Haar measure. Consider $\mathbb{T}^2=\mathbb{S}^1\times\mathbb{S}^1$ the two dimensional torus, and let $\beta(\mathbb{T}^2)$ and $\mu\times\mu$ be the product σ-algebra and the product measure respectively. Let $T:\mathbb{T}^2\to\mathbb{T}^2$ be the triangular map defined by

$$T(x,y)=(x,xy),$$

for all $(x,y)\in\mathbb{T}^2$. We call this map the *K-map* and was introduced by Kusnhirenko ([45]). Here it is also proved that $h_{\mu\times\mu,A}(T)=\log 2$ where $A=(2^i)_{i=0}^{\infty}$.

Note that the transformation T is triangular and its coordinate maps are rotations. In fact $T(x,y)=(x,R_x(x_2))$, where $R_x:\mathbb{S}^1\to\mathbb{S}^1$ are rotations for all $x\in\mathbb{S}^1$. It can also be seen in [73] that every rotation has discrete spectrum, and therefore for any sequence of positive integers B we have $h_{\mu,B}(R_x)=0$. Then for any sequence of positive integers B it follows

$$\int_{\mathbb{S}^1}h_{\mu,B}(R_x)d(\mu)=0.$$

Taking the sequence $A=(2^i)_{i=0}^{\infty}$ we obtain

$$\log 2=h_{\mu\times\mu,A}(T)>\int_{S^1}h_{\mu,A}(R_x)d\mu=0$$

and Proposition 1.4.2 does not hold for metric sequence entropy.

The metric entropy of sequences of measurable transformations can be used to compute the metric entropy of a triangular maps for which f^n is the identity map for some $n\in\mathbb{N}$.

Theorem 1.11. *Under the above conditions we have*

$$h_{\mu_1\times\mu_2}(T) = \int_{X_1} h_{\mu_2}(g^x_{1,\infty})d\mu_1(x),$$

where $g^x_{1,\infty}$ *is the periodic sequence* $g^x_{1,\infty} = (g^x_i)_{i=1}^{\infty}$ *with* $g_i = g_{f^i(x)}$ *for all* $i \in \mathbb{N}$ *and* $x \in X_1$.

Proof. Since f^n is the identity, we get by Rokhlin's formula that

$$h_{\mu_1\times\mu_2}(T^n) = \int_{X_1} h_{\mu_2}(g_{f^{n-1}(x)} \circ \cdots \circ g_{f(x)} \circ g_x)d\mu_1(x).$$

Since the sequence $g^x_{1,\infty}$ is periodic, by Proposition 1.18 we have

$$h_\nu(g_{f^{n-1}(x)} \circ \cdots \circ g_{f(x)} \circ g_x) = nh_\nu(g^x_{1,\infty}),$$

and therefore

$$h_{\mu_1\times\mu_2}(T^n) = nh_{\mu_1\times\mu_2}(T)$$

.□

Some special cases

In some special conditions, the computation of the metric sequence entropy can be made knowing the metric entropy of the transformation. First we need a definition (see [57]).

Definition 1.14. *Let* $A = (a_i)_{i=1}^n$ *be a sequence of integers. Define*

$$U_A(n,k) = \bigcup_{i=1}^{n} \{a_i, 1+a_i, \cdots, k+a_i\},$$

$$S_A(n,k) = \text{Card} \bigcup_{i=1}^{n} \{a_i, 1+a_i, \cdots, k+a_i\},$$

and

$$K(A) = \lim_{k\to\infty} \left(\limsup_{n\to\infty} \frac{S_A(n,k)}{n} \right).$$

Recall that T is *ergodic* if the condition $T^{-1}(A) = A$ implies $\mu(A) = 0$ or 1. Then

Theorem 1.12. *Let* $T : X \to X$ *be an invertible ergodic transformation. Then*

$$h_{\mu,A}(T) = \begin{cases} 0 & \text{if } \ K(A) = 0 \\ K(A)h_\mu(T) & \text{if } \ 0 < h_\mu(T) < \infty \\ 0 & \text{if } \ 0 < K(A) < \infty, h_\mu(T) = 0 \\ \infty & \text{if } \ 0 < K(A) \le \infty, h_\mu(T) = \infty \end{cases}$$

The hypothesis of Theorem 1.12 can be relaxed and it is true for arbitrary measure preserving transformations. First of all we need the following result.

Lemma 1.6. *Let $T : X \to X$ be a measure preserving transformation and let $A = (a_i)_{i=1}^{\infty}$ be an increasing sequence of positive integers. Then*

$$h_{\mu,A}(T) \leq \begin{cases} 0 & \text{if} \quad K(A) = 0 \\ K(A)h_\mu(T) & \text{if} \quad 0 < K(A) < \infty \end{cases}$$

Proof. Let $\mathcal{A} \in \mathcal{Z}$, and put $b_n = \frac{1}{n} H_\mu(\bigvee_{i=0}^{n-1} T^{-i}\mathcal{A})$. Then b_n decreases to $h_\mu(T,\mathcal{A})$ and the sequence $\epsilon_n = \sup_{j\geq n}(b_r - h_\mu(T,\mathcal{A}))$ is decreasing and converges to zero.

Let k,n be positive integers and define $U_A(n,k)$ and $S_A(n,k)$ like in Definition 1.14. Divide $U_A(n,k)$ into connected segments as follows. For $i,j \in U_A(n,k)$ with $i \leq j$ we say that $i \sim j$ if $i \leq l \leq j$ implies $l \in U_A(n,k)$. Clearly $\sim$ is an equivalence relation, and denote by U_i for $i = 1,..,r$ the equivalence classes. Moreover Card $U_i = s_i \geq k$ for any $i = 1,\cdots,r$. Then

$$\bigvee_{i=1}^{n} T^{-a_i}\left(\bigvee_{j=0}^{k} T^{-j}\mathcal{A}\right) \leq \bigvee_{i\in U_A(n,k)} T^{-i}\mathcal{A}.$$

So

$$\begin{aligned} H_\mu\left(\bigvee_{i=1}^{n} T^{-a_i}\left(\bigvee_{j=0}^{k} T^{-j}\mathcal{A}\right)\right) &\leq H_\mu\left(\bigvee_{i\in U_A(n,k)} T^{-i}\mathcal{A}\right) \\ &\leq \sum_{j=1}^{r} H_\mu\left(\bigvee_{i\in U_j} T^{-i}\mathcal{A}\right) \\ &= \sum_{j=1}^{r} H_\mu\left(\bigvee_{i=0}^{s_i-1} T^{-i}\mathcal{A}\right). \end{aligned}$$

Since $(1/s_i)H_\mu\left(\bigvee_{i=0}^{s_i-1} T^{-i}\mathcal{A}\right) \leq h_\mu(T,\mathcal{A}) + \epsilon_k$ for $i = 1,\cdots,r$

$$\begin{aligned} H_\mu\left(\bigvee_{i=1}^{n} T^{-a_i}\left(\bigvee_{j=0}^{k} T^{-j}\mathcal{A}\right)\right) &\leq \sum_{j=1}^{r} s_j(h_\mu(T,\mathcal{A}) + \epsilon_k) \\ &= S_A(n,k)(h_\mu(T,\mathcal{A}) + \epsilon_k). \end{aligned}$$

Thus

$$h_{\mu,A}\left(T, \bigvee_{j=0}^{k-1} T^{-j}\mathcal{A}\right) = \limsup_{n\to\infty} \frac{1}{n} H_\mu\left(\bigvee_{i=1}^{n} T^{-a_i}\left(\bigvee_{j=0}^{k} T^{-j}\mathcal{A}\right)\right)$$

$$\leq \limsup_{n\to\infty} \frac{1}{n} S_A(n,k)(h_\mu(T,\mathcal{A})+\epsilon_k)$$
$$\leq K(A)(h_\mu(T,\mathcal{A})+\epsilon_k).$$

Since $h_{\mu,A}\left(T, \bigvee_{j=0}^{k-1} T^{-j}\mathcal{A}\right) \geq h_{\mu,A}(T,\mathcal{A})$, and taking the limit when $k \to \infty$ we have

$$h_{\mu,A}(T,\mathcal{A}) \leq K(A)(h_\mu(T,\mathcal{A})+\epsilon_k)$$
$$\leq K(A)h_\mu(T,\mathcal{A}).$$

Taking the supremum over $\mathcal{Z}$, it follows

$$h_{\mu,A}(T) \leq K(A)h_\mu(T).$$

Since $h_\mu(T,\mathcal{A}) \leq H_\mu(\mathcal{A}) < \infty$, if $K(A)=0$ then $h_{\mu,A}(T)=0$. If $K(A)>0$, we get $h_{\mu,A}(T) \leq K(A)h_\mu(T)$ and the proof ends. □

The following Lemma can be seen in [32].

Lemma 1.7. *Let $T: X \to X$ be a measure preserving transformation. Let $A=(a_i)_{i=1}^\infty$ be an increasing sequence of positive integers. Then*

$$h_{\mu,A}(T) \geq \begin{cases} K(A)h_\mu(T) & \text{if } \ 0 < h_\mu(T) < \infty \\ \infty & \text{if } \ K(A)>0 \text{ and } h_\mu(T)=\infty \end{cases}$$

Lemmas 1.6 and 1.7 allow us to prove the following result, which is a refinement of Theorem 1.12.

Theorem 1.13. *Let $T: X \to X$ be a measure preserving transformation, and let $A=(a_i)_{i=1}^\infty$ be an increasing sequence of integers. Then*

$$h_{\mu,A}(T) = \begin{cases} 0 & \text{if } \ K(A)=0 \\ K(A)h_\mu(T) & \text{if } \ 0<h_\mu(T)<\infty \\ 0 & \text{if } \ 0<K(A)<\infty, h_\mu(T)=0 \\ \infty & \text{if } \ 0<K(A)\leq\infty, h_\mu(T)=\infty \end{cases}$$

Finally we prove the following result

Theorem 1.14. *Let $T: X \to X$ be a measure preserving transformation, and let $A=(a_i)_{i=1}^\infty$ be an increasing sequence of integers. If A and T satisfy one of the cases of Theorem 1.13, then Propositions 1.20, 1.21, 1.4.2 and 1.23 hold.*

Note that the case $K(A)=\infty$ and $h_\mu(T)=0$ is indeterminate in Theorem 1.13. In this case Propositions 1.20, 1.21, 22 and 1.23 do not hold in general.

1.4.3 The metric sequence entropy map

This section is devoted to the study of the map $h_{\mu,(\cdot)}(T) : \mathcal{I} \to \mathbb{R}^+ \bigcup\{0,\infty\}$, where $\mathcal{I}$ denotes the set of all the increasing sequences of positive integers. Some properties of this map have been studied by Saleski in [64] and Pickel in [60]. We wonder if this map is surjective, and on the relation between $h_{\mu,A}(T)$ and $h_{\mu,B}(T)$, where B is a subsequence of A.

Lemma 1.8. *Let $T : X \to X$ be a measure preserving transformation and let A be an increasing sequence of positive integers such that $\infty > h_{\mu,A}(T) > 0$. Then for all $b \in [0, h_{\mu,A}(T)]$ there exists a sequence $B \in \mathcal{I}$ such that $h_{\mu,B}(T) = b$.*

Proof. If $b \in \{0, h_{\mu,A}(T)\}$ the sequences $\mathbf{0} = (0,0,\cdots)$ and A give us zero metric sequence entropy and $h_{\mu,A}(T)$ respectively. Suppose that $b \in (0, h_{\mu,A}(T))$, and let $r \in (0,1)$ such that $b = rh_{\mu,A}(T)$.

Construct a new increasing sequence $B \in \mathcal{I}$ by repeating elements in A such that if

$$r_n = \mathrm{Card}\{b_i \neq 0 : 1 \leq i \leq n\},$$

then $\lim_{n\to\infty}(r_n/n) = r$. Let $\mathcal{A}$ be a finite partition of X, and then

$$\begin{aligned} h_{\mu,B}(T,\mathcal{A}) &= \limsup_{n\to\infty} \frac{1}{n} H_\mu\left(\bigvee_{i=1}^{n} T^{-b_i}\mathcal{A}\right) \\ &= \limsup_{n\to\infty} \frac{r_n}{n}\frac{1}{r_n} H_\mu\left(\bigvee_{i=1}^{r_n} T^{-a_i}\mathcal{A}\right) \\ &= \lim_{n\to\infty} \frac{r_n}{n} \limsup_{n\to\infty} \frac{1}{r_n} H_\mu\left(\bigvee_{i=1}^{r_n} T^{-a_i}\mathcal{A}\right) \\ &= rh_{\mu,A}(T,\mathcal{A}). \end{aligned}$$

Taking the supremum over $\mathcal{Z}$ we have that $h_{\mu,B}(T) = rh_{\mu,A}(T)$, and the proof ends. □

Now we study when the map $h_{\mu,(\cdot)}(T) : \mathcal{I} \to \mathbb{R} \cup \{0,\infty\}$ is surjective.

Proposition 1.25. *Let $T : X \to X$ be a measure preserving transformation with $\infty > h_\mu(T) > 0$. Then the map $h_{\mu,\cdot}(T)$ is surjective.*

Proof. Let $\infty > h_\mu(T) > 0$. Since $h_\mu(T)$ is finite, by Lemma 1.8, there exists a sequence $B \in \mathcal{I}$ such that $h_{\mu,B}(T) = b$ for all $b \in [0, h_\mu(T)]$.

Let $b > h_\mu(T)$. Since $h_\mu(T)$ is positive and finite, there exists a positive integer n_b such that $h_\mu(T^{n_b}) = h_{\mu,A}(T) \geq b$ where $A = (n_b i)_{i=0}^{\infty}$. Applying Lemma 1.8 we find a sequence B for which $h_{\mu,B}(T) = b$.

Now we consider a sequence B satisfying

$$\lim_{k\to\infty}\left(\limsup_{n\to\infty}\frac{S_B(n,k)}{n}\right)=\infty,$$

and use Theorem 1.13 to prove that $h_{\mu,B}(T)=\infty$. □

Remark 1.2. It is possible to construct an example of a measure preserving transformation T satisfying $h_{\mu,A}(T)\in\{0,\infty\}$ for all $A\in\mathcal{I}$. To this end consider the transformation of the example in Section 4.8 from [73]. For this map $h_\mu(T)=\infty$ and hence, applying Theorem 1.13, our assertion is proved.

We distinguish three basic types of measure preserving transformations from the point of view of metric sequence entropy. They are summarized in the following definition.

Definition 1.15. *A measure preserving transformation $T:X\to X$ is said*

(a) Null *if* $\sup\{h_{\mu,A}(T):A\in\mathcal{I}\}=0$.
(b) Bounded *if* $\infty>\sup\{h_{\mu,A}(T):A\in\mathcal{I}\}>0$.
(c) Unbounded *if* $\sup\{h_{\mu,A}(T):A\in\mathcal{I}\}=\infty$.

So we have the following result due to Kusnhirenko (see [45]).

Theorem 1.15. *A measure preserving transformation T is null if and only if it has discrete spectrum.*

On the other hand, by Proposition 1.25, any measure preserving transformation T with positive metric entropy is unbounded. This condition is not necessary as Hulse pointed out in [36]. On the other hand, by a result due to Pickel (see [60]) we know that

$$\sup\{h_{\mu,A}(T):A\in\mathcal{I}\}\in\{\log k:k\in\mathbb{N}\cup\infty\}.$$

Let $A=(a_i)_{i=1}^\infty\in\mathcal{I}$, and let B be a subsequence of A. We will study the relation between $h_{\mu,A}(T)$ and $h_{\mu,B}(T)$. To this end define the increasing sequence of positive integers $(S(B,i))_{i=1}^\infty$ given by

$$S(B,i)=\mathrm{Card}\{B\bigcap\{a_1,\cdots,a_i\}\},$$

and let

$$S(B)=\limsup_{n\to\infty}\frac{S(B,n)}{n}.$$

Proposition 1.26. *Let $T:X\to X$ be a measure preserving transformation. Let $A\in\mathcal{I}$ and let B,C be two subsequences of A satisfying $B\cap C=\emptyset$. Then*

$$h_{\mu,A}(T)\le S(B)h_{\mu,B}(T)+S(C)h_{\mu,C}(T).$$

Proof. Let $A = (a_i)_{i=1}^{\infty}$, $B = (b_i)_{i=1}^{\infty}$ and $C = (c_i)_{i=1}^{\infty}$. Let $\mathcal{A} \in \mathcal{Z}$. By Theorem 1.8 (h) it follows that

$$\frac{1}{n} H_\mu\left(\bigvee_{i=1}^{n} T^{-a_i}\mathcal{A}\right) \leq \frac{S(B,n)}{n}\,\frac{1}{S(B,n)}\,H_\mu\left(\bigvee_{i=1}^{S(B,n)} T^{-b_i}\mathcal{A}\right) + \frac{S(C,n)}{n}\,\frac{1}{S(C,n)}\,H_\mu\left(\bigvee_{i=1}^{S(C,n)} T^{-c_i}\mathcal{A}\right)$$

Taking upper limits when $n \to \infty$ we have

$$\begin{aligned} h_{\mu,A}(T,\mathcal{A}) &\leq \limsup_{n\to\infty} \frac{S(B,n)}{n}\,\frac{1}{S(B,n)}\,H_\mu\left(\bigvee_{i=1}^{S(B,n)} T^{-b_i}\mathcal{A}\right) \\ &\quad + \limsup_{n\to\infty} \frac{S(C,n)}{n}\,\frac{1}{S(C,n)}\,H_\mu\left(\bigvee_{i=1}^{S(C,n)} T^{-c_i}\mathcal{A}\right) \\ &\leq \limsup_{n\to\infty} \frac{S(B,n)}{n}\,\limsup_{n\to\infty} \frac{1}{S(B,n)}\,H_\mu\left(\bigvee_{i=1}^{S(B,n)} T^{-b_i}\mathcal{A}\right) \\ &\quad + \limsup_{n\to\infty} \frac{S(C,n)}{n}\,\limsup_{n\to\infty} \frac{1}{S(C,n)}\,H_\mu\left(\bigvee_{i=1}^{S(C,n)} T^{-c_i}\mathcal{A}\right) \\ &\leq S(B)h_{\mu,B}(T,\mathcal{A}) + S(C)h_{\mu,C}(T,\mathcal{A}). \end{aligned}$$

Taking the supremum over $\mathcal{Z}$ we have

$$h_{\mu,A}(T) \leq S(B)h_{\mu,B}(T) + S(C)h_{\mu,C}(T)$$

and the proof ends. □

Remark 1.3. Proposition 1.26 can be generalized to an arbitrary number of subsequences of A in the following way. Let $B_1, B_2, ...B_k$ be pairwise disjoint subsequences of A, Define $S(B_m, i) = \text{Card}\{B_m \cap \{(a_1, \cdots, a_k)\}$ and consider

$$\limsup_{n\to\infty} \frac{S(B_m, n)}{n} = S(B_m).$$

Following as in the former proof it can be seen that

$$h_{\mu,A}(T) \leq \sum_{i=1}^{n} S(B_i)h_{\mu,B_i}(T).$$

Proposition 1.27. *Let $T : X \to X$ be a measure preserving transformation and let A, B and C be increasing sequences of positive integers like in Proposition 1.26, and such that $S(C) = \lim_{n\to\infty} S(C,n)/n = 0$. Then $h_{\mu,A}(T) = h_{\mu,B}(T)$.*

Proof. Since $S(C) = 0$ it follows

$$S(B) = \limsup_{n\to\infty} \frac{S(B,n)}{n} = \limsup_{n\to\infty} \frac{n - S(C,n)}{n} = \lim_{n\to\infty} \frac{n - S(C,n)}{n} = 1.$$

Let $\mathcal{A} \in \mathcal{Z}$. By Theorem 1.8 (d) it follows that

$$\frac{1}{S(B,n)} H_\mu \left(\bigvee_{i=1}^{S(B,n)} T^{-b_i} \mathcal{A} \right) \le \frac{n}{S(B,n)} \frac{1}{n} H_\mu \left(\bigvee_{i=1}^{n} T^{-a_i} \mathcal{A} \right).$$

Taking the upper limit when $n \to \infty$ we have

$$\begin{aligned} h_{\mu,B}(T,\mathcal{A}) &\le \limsup_{n\to\infty} \frac{n}{S(B,n)} \frac{1}{n} H_\mu \left(\bigvee_{i=1}^{n} T^{-a_i} \mathcal{A} \right) \\ &= \lim_{n\to\infty} \frac{n}{S(B,n)} \limsup_{n\to\infty} \frac{1}{n} H_\mu \left(\bigvee_{i=1}^{n} T^{-a_i} \mathcal{A} \right) \\ &= h_{\mu,A}(T,\mathcal{A}). \end{aligned}$$

Taking the supremum over $\mathcal{Z}$ we get $h_{\mu,B}(T) \le h_{\mu,A}(T)$.

On the other hand, applying Proposition 1.26 we have that

$$h_{\mu,A}(T) \le S(B) h_{\mu,B}(T) + S(C) h_{\mu,C}(T) = h_{\mu,B}(T),$$

and this concludes the proof. □

We complete our study of metric sequence of subsequences of A with the following result.

Proposition 1.28. *Let $T : X \to X$ be a measure preserving transformation and let $A \in \mathcal{I}$. Let B be a subsequence of A such that $\limsup_{n\to\infty} n/S(B,n) = s(B)$. If $s(B) < \infty$ or $h_{\mu,A}(T) > 0$ then it follows that $h_{\mu,B}(T) \le s(B) h_{\mu,A}(T)$.*

Proof. Let $\mathcal{A} \in \mathcal{Z}$. By Theorem 1.8 (h)

$$\frac{1}{S(B,n)} H_\mu \left(\bigvee_{i=1}^{S(B,n)} T^{-b_i} \mathcal{A} \right) \le \frac{n}{S(B,n)} \frac{1}{n} H_\mu \left(\bigvee_{i=1}^{n} T^{-a_i} \mathcal{A} \right).$$

Taking the upper limit when $n \to \infty$

$$\begin{aligned} h_{\mu,B}(T,\mathcal{A}) &\le \limsup_{n\to\infty} \frac{n}{S(B,n)} \frac{1}{n} H_\mu \left(\bigvee_{i=1}^{n} T^{-a_i} \mathcal{A} \right) \\ &\le \limsup_{n\to\infty} \frac{n}{S(B,n)} \limsup_{n\to\infty} \frac{1}{n} H_\mu \left(\bigvee_{i=1}^{n} T^{-a_i} \mathcal{A} \right) \\ &= s(B) h_{\mu,A}(T,\mathcal{A}). \end{aligned}$$

Taking the supremum over the set of finite partitions the proof ends. □

The values $s(B) = \infty$ and $h_{\mu,A}(T) = 0$ give us an indeterminate case. In fact there exist examples with the sequences $A = (i)_{i=0}^{\infty}$ and $B = (2^i)_{i=0}^{\infty}$ satisfying $h_{\mu,A}(T) = 0$ and $h_{\mu,B}(T) = \log 2$ (see [46]) while for discrete spectrum transformations, $h_{\mu,B}(T) = 0$. Therefore there is not a direct relation, only based on the sequences, between $h_{\mu,A}(T)$ and $h_{\mu,B}(T)$.

Moreover, when $h_{\mu,A}(T) = 0$ using Proposition 1.28 we characterize a subset of subsequences of A for which the metric sequence entropy is also zero. These sequences do not add new information about the dynamical behavior of the transformation.

Now, we will study the structure of the set $\mathcal{S}_\alpha = \{A : h_{\mu,A}(T) = \alpha\} \subset \mathcal{I}$. Note that when a measure preserving transformation is null $\mathcal{S}_0 = \mathcal{I}$. The next result was proved by A. Saleski in [64] for invertible measure preserving transformations.

Theorem 1.16. *Let $A = (a_i)_{i=1}^{\infty}$ and $B = (b_i)_{i=1}^{\infty}$ be two increasing sequences of positive integers such that $\sup\{|a_i - b_i| : i \in \mathbb{N}\} = k \in \mathbb{N}$. If T is an invertible measure preserving transformation, then $h_{\mu,A}(T) = 0$ if and only if $h_{\mu,B}(T) = 0$.*

We will prove Theorem 1.16 in the case of non-invertible measure preserving transformations. Previously, we need the following result.

Lemma 1.9. *Let $A = (a_i)_{i=1}^{\infty}$ and $B = (b_i)_{i=1}^{\infty}$ be two sequences of positive integers such that for any $i \in \mathbb{N}$, either $b_i = a_i$ or $b_i = a_i + 1$. Then for any measure preserving transformation T it holds that $h_{\mu,A}(T) = 0$ if and only if $h_{\mu,B}(T) = 0$.*

Proof. Consider the sequence

$$A \wedge B = (a_1, b_1, a_2, b_2, \cdots, a_i, b_i, \cdots).$$

Hence, by Propositions 1.26 and 1.28 and Theorem 1.9 (e), we have

$$\frac{1}{2} h_{\mu,B}(T) \le h_{\mu,A\wedge B}(T) \le \frac{1}{2} h_{\mu,A}(T) + \frac{1}{2} h_{\mu,A+1}(T) = h_{\mu,A}(T).$$

Then, if $h_{\mu,A}(T) = 0$, then $h_{\mu,B}(T) = 0$. Similarly, we obtain that if $h_{\mu,B}(T) = 0$, then $h_{\mu,A}(T) = 0$ and the proof is over. □

Let $A = (a_i)_{i=1}^{\infty}$ and $B = (b_i)_{i=1}^{\infty}$ be two increasing sequences of positive integers such that $\sup\{|a_i - b_i| : i \in \mathbb{N}\} = k \in \mathbb{N}$. Let $A_j = \{a_{n_i} : a_{n_i} = b_{n_i} + j\}$ and let $B_j = \{b_{n_i} : b_{n_i} = a_{n_i} + j\}$ for $-k \le j \le k$. Then we can prove the following result, whose proof follows by induction using Lemma 1.9

Proposition 1.29. *Under the above conditions, $h_{\mu,A}(T) = 0$ if and only if $h_{\mu,B}(T) = 0$.*

On a question of D. Newton

Let $T : X \to X$ be an invertible ergodic transformation and let $A = (a_i)_{i=1}^{\infty}$ be an increasing sequence of positive integers such that

$$\limsup_{n\to\infty} \frac{a_n}{n} = d(A) < \infty.$$

D. Newton in [57] proved that $h_\mu(T) \le h_{\mu,A}(T) \le d(A)h_\mu(T)$, and wondered if the equality

$$h_{\mu,A}(T) = d(A)h_\mu(T) \tag{1.4.1}$$

was possible. Newton showed that this equality holds for bounded gap sequences, i.e. those sequences for which there exists a positive integer k with the property $a_{n+1} - a_n \le k$, for all $n \in \mathbb{N}$.

We will construct an example satisfying $h_{\mu,A}(T) < d(A)h_\mu(T)$. To this end consider $(n_i)_{i=1}^{\infty}$ a strictly increasing sequence of positive integers such that $n_1 = 1$ and with the following property: if

$$k_n = \mathrm{Card}\{n_j \in \{1, 2, \cdots, n\} : 1 \le j \le n\},$$

then $\lim_{n\to\infty} k_n/n = 0$. For example, take the sequence $n_j = 2^{j-1}$. Now let $A = (a_i)_{i=1}^{\infty}$ be the sequence of positive integers defined by $a_{n_j} = 2n_j$, and $a_{n_j+l} = 2n_j + l$ if $l < n_{j+1} - n_j$. It is easy to see that $d(A) = 2$. On the other hand consider $U_A(n,k)$, $S_A(n,k)$ and $K(A)$ like in Definition 1.14. Then $S_A(n,k) \le n + k \cdot k_n$ and then $K(A) = 1$.

Consider the measure preserving transformation of Example 1.2. It can be seen in Theorem 1.12 from [73] that $\sigma : \{0,1\}^{\mathbb{Z}} \to \{0,1\}^{\mathbb{Z}}$ is ergodic and $h_\mu(\sigma) = \log 2$. Applying Theorem 1.12 it follows

$$h_{\mu,A}(\sigma) = K(A)h_\mu(\sigma) = h_\mu(\sigma) = \log 2 < \log 4 = d(A)h_\mu(\sigma),$$

and then the equality (1.4.1) does not hold in general.

1.5 Topological entropy

The metric entropy measures how complicated is a system from the measure theoretical point of view. In particular, KS-entropy can not appreciate the dynamics concentrated in sets of measure zero. This problem is solved by the topological entropy which appreciates the dynamics concentrated on small sets. Nevertheless there are some relations between the KS-entropy and topological entropy. It is well known that the topological entropy of a map f is the supremum of the metric entropies of f over all probabilistic f-invariant measures (see [28] and [73]).

Topological entropy was introduced in 1965 by Adler, Konheim and McAndrew in [4] in the setting of dynamical systems $(\mathbb{X}, f)$ where $\mathbb{X}$ is a compact metric space and $f \in C(\mathbb{X})$ adapting ideas from Ergodic Theory. Later on, Bowen in [19] introduced the notion of *topological entropy for uniformly continuous maps* defined in arbitrary metric spaces. It is necessary to remark that Bowen's definition shows a good dynamical interpretation of topological entropy. When $\mathbb{X}$ is a compact metric space, both definitions are equivalent, and this is the general assumption in the rest of the paper.

Let us introduce this Bowen's notion. Assume that $(\mathbb{X}, d)$ is a metric space. Then it is easy to prove that for each $n \geq 1$ the function

$$d_n(x, y) = \max_{0 \leq i \leq \leq n-1} d(f^i x), f^i(y))$$

is a distance which is equivalent to d. A finite set $E \subset \mathbb{X}$ is called $(n, \varepsilon) -$ *separated* if for all $x, y \in E$. The set E is called $(n, \varepsilon) -$ *spanning* if for every $x \in \mathbb{X}$ there exists $y \in E$ such that $d_n(x, y) \leq \varepsilon$.

Now we denote by $s_n(f, \varepsilon)$ the maximal cardinality of an $(n, \varepsilon) -$ *separated* set and by $t_n(f, \varepsilon)$ the minimal cardinality of an $(n, \varepsilon) -$ *spanning* set. Then we state

$$\overline{s}(f) = \lim_{\varepsilon \to 0} \limsup_{n \to \infty} \frac{1}{n} \log s_n(f, \varepsilon),$$

$$\underline{s}(f) = \lim_{\varepsilon \to 0} \liminf_{n \to \infty} \frac{1}{n} \log s_n(f, \varepsilon),$$

$$\overline{t}(f) = \lim_{\varepsilon \to 0} \limsup_{n \to \infty} \frac{1}{n} \log t_n(f, \varepsilon),$$

$$\underline{t}(f) = \lim_{\varepsilon \to 0} \liminf_{n \to \infty} \frac{1}{n} \log t_n(f, \varepsilon).$$

In a not difficult way it can be proved that the former quantities all are the topological entropy introduce in [4]. The advantage is that throughout $(n, \varepsilon) -$ *separated* set can be given a clear interpretation of what topological entropy means.

Given two dynamical systems $(\mathbb{X}, f)$ and $(\mathbb{Y}, g)$, we say that the two systems are conjugate if there is a homeomorphism $h : \mathbb{X} \to \mathbb{Y}$ such that

$$h \circ f = g \circ h$$

Topological entropy is preserved by topological conjugacy, that is, two conjugate systems have the same topological entropy which means that the two systems have the same dynamics. Also was remarked that topological entropy can be taken as a measure of the chaoticity or complexity of a system. It is related with the existence of *Bernouilli shifts*([73]) which play a special role in *Smale's horseshoe* ([68]) and *Moser's theorem*([54]). Bernouilli shifts are complicated systems which have been used to prove the dynamical complexity of other systems. Roughly speaking, a usual technique to show a dynamical system is complicated is to find an invariant closed set which conjugate to a Bernouilli shift. Additionally, Bernouilli shifts have positive topological entropy. For this reason, positive topological entropy has been taken as a criterion to decide whether a continuous map is chaotic or not.

Topological entropy is in general difficult to compute. But there are several ways to do using some formulas held by it

$$h(f^n) = nh(f)$$

When f and g are conjugate, then

$$h(f) = h(g)$$

If f, g are continuous maps on the same state space then

$$h(f \circ g) = h(g \circ f)$$

For piecewise interval maps, let c_n the number of pieces of monotonicity of f^n, then for such maps

$$\lim_{n \to \infty} \frac{1}{n} \log c_n = h(f)$$

([53]). One easy application of this formula is computation of the topological entropy of the tent map (already considered in the Introduction) to see that it is $\log 2$. The same value is reached in the quadratic map $f(x) = 4x(1-x)$.Let f be an interval homeomorphism, then

$$h(f) = 0$$

In [5] an example is constructed of an interval maps with $h(f) = \infty$ which proves that $0 \leq h(f) \leq \infty$. But for C^1-interval maps, the topological entropy does not reach ∞ (see for example [73]).

Concerning the possibility of computing the topological or KS- entropy of a given system with a maximum error of ε in a reasonable term, it is interesting to find out the Milnor's opinion ([51]) in the sense that the answer to this question is in general cases negative which means in most cases, the two entropies are not effectively computable.

Finally, we remark that topological entropy can be introduced in an axiomatic way in the case of continuous interval maps. It has the advantage of underlying the importance of the different properties that has this quantity, (see [8]).

One open problem is try to introduce all the entropies in an axiomatic way in the most general settings trying to point out their properties. Also could be of interest try to give a topological version of the *Tsallis entropy* (see [72]) for discrete dynamical systems and establish the connections with the other entropies.

References

[1] Abramov, L. (1959) The Entropy of Derived Automorphism. *Dokl. Akad. Nauk. SSSR* **128**, 647-650. (Russian)

[2] Abramov, L. and Rokhlin, V.A. (1961) Entropy of a Skew Product of Mappings with Invariant Measure. *Vestnik Leningrad. Univ.* **17**, 5-13. (Russian)

[3] Adler, R. (1963) A Note on the Entropy of Skew Product Transformations. *Proc. Amer. Math. Soc.* **14**,665-669.

[4] Adler R., Konheim A. and McAndrew. (1965) Topological Entropy. *Trans. Amer. Math. Soc.* **114**, 309-319.

[5] Alsedà L., Llibre J. and Misiurewicz M. (2000) *Combinatorial Dynamics and Entropy in Dimension One*, second edition, World Scientific, Singapore.

[6] Alsedà L., Llibre J. and Misiurewicz M. (1999) Low-dimensional Combinatorial Dynamics. *Int.J. Bifurcation and Chaos* **9**, 1687-1704.

[7] Alsedá L., Juher D. and Mumbrú P. (2000) On the Minimal Models for Graph Maps. *Int.J.Bifurcation and Chaos* **13**, 1991-1996.

[8] Alsedà L., Kolyada S., Llibre L. and Snoha L. (2003) Axiomatic of the Topological Entropy on the Interval. *Aequationes Math.* **65**, 113-132.

[9] Auslander J. and Yorke J. (1980) Interval maps, Factors of Maps and Chaos. *Tôhoku Math. J.* **32**, 117-188.

[10] Balibrea F. and Linero A. (2001) Periodic Structure of σ-permutation Maps. *Aequationes Math.* **62**, 265-279.

[11] Balibrea F. and Linero A. (2002) Periodic Structure of σ-permutation Maps II. The Case $\mathbb{T}^n$. *Aequationes Math.* **64**, 34-52.

[12] Balibrea F. and Linero A. (2003) On the Periodic Structure of Delayed Difference Equations of the Form $x_n = f(x_{n-k})$. *Journal of Difference Equations and Applications* **9**(3/4), 359-371.

[13] Balibrea F. and Smítal J. (1993) A Chaotic Continuous Function Generates all Probability Distributions. *J. Math. Anal. Appl.* **180**, 587-598.

[14] Banks J., Brooks J., Cairns G., Davis G. and Stacey P. (1992) On Devaney's Definition of Chaos. *Amer. Math. Monthly* **99**, 332-334.

[15] Belot G. and Earman J. (1997) Chaos Out of Order: Quantum Mechanics, the Correspondence Principle and Chaos. *Studies in the History and Philosophy of Modern Physics* **28**, 147-182.

[16] Below A. and Losert V. (1984) On Sequences of Density Zero in Ergodic Theory. *Contemp. Math.* **26**, 49-60.

[17] Blanchard F., Glasner E., Kolyada S. and Mass A. (2002) On Li-Yorke Pairs. *J. Reine Angew. Math.* **547**, 51-68.

[18] Block L. and Coppel W.A. (1992) *Dynamics in One Dimension*, Lecture Notes in Math. **1513**, Springer, Berlin.

[19] Bowen R. (1971) Entropy for Group Endomorphisms and Homogeneous Spaces. *Trans. Amer. Math. Soc.* **153**, 401-414; erratum: (1973) *Trans. Amer. Math. Soc.* **181**, 509-510.

[20] Blum J.R. and Hanson D.L. (1960) On the Mean Ergodic Theorem for Subsequences. *Bull. Amer. Math. Soc.* **66**, 308-311.

[21] Bourgain J. (1988) On the Maximal Ergodic Theorem for certain Subsets of the Integers. *Israel J. Math.* **61**, 39-72.

[22] Bourgain J. (1988) On the Maximal Ergodic Theorem on L^p for Arithmetic Sets. *Israel J. Math.* **61**, 73-84.

[23] Bourgain J. (1988) Almost Sure Convergence and Bounded Entropy. *Israel J. Math.* **63**, 79-97.

[24] Cánovas J. (2000) *New Results on Entropy, Sequence Entropy and Related Topics* Memoir of Ph. Degree, Universidad de Murcia (Spain).

[25] Cournot A. (1838)*Recherches sur les Principes Mathematiques de la Théorie des Richesses*, Hachette, Paris.

[26] Devaney R.L. (1986) *An Introduction to Chaotic Dynamical Systems,* Benjamin/Cummings, Menlo Park.

[27] Dekking F.M. (1980) Some Examples of Sequence Entropy as an Isomorphism Invariant. *Trans. Amer. Mat. Soc.* **259**, 167-183.

[28] Denker M., Grillenberger C. and Sigmund K. (1976) *Ergodic Theory on Compact Spaces,* Lecture Notes in Mathematics, Springer,Berlin.

[29] Ellis M.H. and Friedman N.A. (1978) Subsequence Generators for Ergodic Group Translations. *Israel J. Math.* **31**, 115-121.

[30] Frigg R. (2004) In what Sense is the Kolmogorov-Sinai Entropy a Measure for Chaotic Behaviour?. Bridging the Gap between Dynamical Systems Theory and Communication Theory. *Brit. J. Phil. Sci.* **55**, 411-434.

[31] García Guirao J.L. and Lampart M. (2005)Li and Yorke Chaos with respect to the Cardinality of the Scrambled Sets. *Chaos, Solitons and Fractals.* **24**, 1203-1206.

[32] Goodman T.N. (1974) Topological Sequence Entropy. *Proc. London Math. Soc.* **29**, 331-350.

[33] Halmos P.R. (1956) *Lectures on Ergodic Theory,* The Mathematical Society of Japan, Tokyo.

[34] Henson S.M., Reilly J.R., Robertson S.L., Schub M.C., Rozier E.W. and Cushing J.M. (2003) Predicting Irregularities in Population Cycles. *SIAM J. Applied Systems* **2**, No2, 238-253.

[35] Hsu C. and Li M. (2002) Transitivity Implies Period Six: A Simple Proof. *Amer. Math. Monthly* **109**, 840-843.

[36] Hulse P. (1982) Sequence Entropy and Subsequence Generators. *J. London Math. Soc.* **26**, 441-450.

[37] Huang W. and Ye X. (2002) Devaney's Chaos or 2-scattering Implies Li-Yorke Chaos. *Topol. Appl.* **117**, 259-272.

[38] Katok A. (1980) Lyapunov Exponents, Entropy and Periodic Orbits for Diffeomorphisms. *Publ. Math. I.H.E.S.* **51**, 137-173.

[39] Kloeden P.E. (1979) On Sharkovsky Cycle Coexistence Orderings. *Bull. Austral. Math. Soc.* **20**, 171-177.

[40] Kolyada S. and Snoha L. (1996) Topological Entropy of Nonautonomous Dynamical Systems. *Random and Comp.Dynamics* **4**, 205-233.

[41] Krieger W. (1970) On Entropy and Generators of Measure Preserving Transformations. *Trans. Amer. Math. Soc.* **149**, 453-464.

[42] Kolmogorov A.N. (1958) A New Metric Invariant of Transient Dynamical Systems and Automorphisms in Lebesgue Spaces. *Dokl. Akad. Nauk. SSSR.* **119**, 861-864. (Russian)

[43] Kocic V.L. and Ladas G. (1993) *Global Behaviour of Nonlinear Difference Equations of Higher Order with Applications, Mathematics and its Applications*, Kluwer Academic Publishers, Netherlands, **256**.

[44] Kuchta M. and Smítal J. (1989) Two Points Scrambled Set Implies Chaos. *Proc. Europ.Conf. Iteration Theory, Spain 1987 (World Scientific, Singapore)*, 427-430.

[45] A. G. Kushnirenko A.G., (1967) On Metric Invariants of Entropy Type. *Russian Math. Surveys* **22**, 53-61.

[46] Lemanczycz M. (1985) The Sequence Entropy for Morse Shifts and Some Counterexamples. *Studia Mathematica* **82**, 221-241.

[47] Levine S.H., Scudo F.M. and Plunkett D.J. (1977) Persistence and Convergence of Ecosystems: an Analysis of some Order Difference Equations. *J. Math. Biol.* **4**no2, 171-182.

[48] Llibre J., Paraños J. and Rodríguez J.A. (2002) Periods for Continuous Self Maps of the Figure Eight Space. *Int. J. Bifurcation and Chaos* **13**No7, 1743-1754.

[49] Li T.Y. and Yorke J. (1975) Period Three Implies Chaos. *Amer. Math. Monthly* **82**, 985-992.

[50] Maynard Smit J. (1968) *Mathematical Ideas in Biology*, Cambridge University Press.

[51] Milnor J. (2002) Is Entropy Effectively Computable?. *Stony Brook*, preprint.

[52] Misiurewicz M. (1995) Thirty Years after Sharkovsky Theorem. *Int.J.Bifurcation and Chaos* **5**, 1275-1281.

[53] Misiurewicz M. and Szlenk W. (1980) Entropy of Piecewise Monotone Mappings. *Studia Math.* **67**, 45-63.

[54] Moser J. (1973) *Stable and Random Motions in Dynamical Systems* Princeton Uni.Press.

[55] Nadler S.B. (1992) *Continuum Theory. An Introduction.* Monographs and Textbooks in Pure and Applied Mathematics, **158**, Marcel Dekker, Inc. New York.

[56] Nagata J. (1965) *Modern Dimension Theory,* Interscience-Wiley, New York.

[57] Newton D. (1970) On Sequence Entropy I and II. *Math. Systems Theory* **4**, 119-125.

[58] Ornstein D. (1970) Two Bernuilli Shift with the Same Entropy are Isomorphic. *Advances in Math.* **4**, 337-352.

[59] Ornstein D. (1973) An Example of a Kolmogorov Automorphism that is not a Bernouilli Shift. *Advances in Math.* **10**, 49-62.

[60] Pickel B.S. (1969) Some Properties of A-entropy. *Mat. Zametki* **5**, 327-334 (Russian).

[61] Robinson C. (2001) *Dynamical Systems,Stability, Symbolic Dynamics and Chaos*, CRC Press, Inc.

[62] Rokhlin V.A. (1959) On the Entropy of a Metric Automorphism. *Dokl. Akad. Nauk. SSSR* **124** 980-983.

[63] V.A. Rokhlin V.A. (1967) Lectures on Entropy Theory of Transformations with Invariant Measure. *Russian Math. Surveys* **22**, 1-52.

[64] Saleski A. (1977) Sequence Entropy and Mixing. *J. Math. Anal.Appl.* **60**, 58-66.

[65] Sinai Y. (1959) On the Notion of Entropy of a Dynamical System. *Dokl.Akad.Nauk. SSSR* **124**, 768-771. (Russian)

[66] Shannon C.E. and Weaver W. (1962) *The Mathematical Theory of Communication,* University of Illinois Press, Urbana.

[67] Sharkovsky A.N. (1964) Coexistence of Cycles of a Continuous Map of the Line into itself. *Ukrain.Math.J.* **16**, 61-71 (in Russian). (1995) Proceedings of the Conference Thirty Years after Sharkovsky Theorem. New Perspectives. Murcia,1994. *Internat.J.Bifu.Chaos App.Sci.Engrg.* **5**, 1263-1273.

[68] Smale S. (1965) Diffeomorphisms with many Periodic Points. *Differential and Combinatorial Topology,* Princeton Univ. Press., 63-80.

[69] Smítal J. (1986) Chaotic Functions with Zero Topological Entropy, *Trans,Amer.Math.Soc.* **297**, 269-282.

[70] Schweizer A. and Smítal J. (1994) Measures of Chaos and a Spectral Decomposition of Dynamical Systems on the Interval. *Trans.Amer.Math.Soc.* **344**, 737-854.

[71] Swirszcz G. (1998) On a certain Map of a Triangle. *Fundamenta Mathematicae* **155**, 45-57.

[72] Tsallis C., Plastino A.R. and Zheng W.M. (1997) Power-law Sensivity to Initial Conditions- New Entropic Representation. *Chaos, Solitons and Fractals* **8**No6, 885-891.

[73] Walters P. (1982) *An Introduction to Ergodic Theory* Springer,Berlin.

2

Foundations of Nonextensive Statistical Mechanics

Sumiyoshi Abe

Institute of Physics, University of Tsukuba, Ibaraki 305-8571, JAPAN
E-mail: suabe@sf6.so-net.ne.jp

Summary. The fundamental aspects of nonextensive statistical mechanics based on the Tsallis entropy are surveyed. It is shown how the method of steepest descents, the counting algorithm and the evaluation of the density of states can appropriately be generalized for describing the power-law distributions. The generalized Boltzmann equation and the associated H-theorem are also considered for the Tsallis-type functional and the maximum Tsallis entropy distribution.

2.1 Introduction

Boltzmann-Gibbs statistical mechanics characterized by the exponential-type distributions has been playing a central role for more than a century in explaining/understanding physical properties of systems in thermal equilibria. Such systems are *simple* in the sense that they are ergodic and their energies are extensive, that is, are proportional to the numbers of elements contained. These two concepts, i.e., ergodicity and extensivity, are in fact essential premises in Boltzmann-Gibbs statistical mechanics. Another feature is that the long-time limit (of a macroscopic physical quantity) and the thermodynamic limit commute in those simple systems. This is trivial since the systems considered there are in equilibrium states and so the time does not play any role.

During the last quarter of the 20th century, researchers have come to notice that there exist ample examples of statistical systems in nature, which may not naively be described by ordinary Boltzmann-Gibbs statistical mechanics. Among others, of extreme interest and importance today are the so-called complex systems. They are often subjected to the power-law distributions, in marked contrast to the exponential distributions, indicating criticality. From the viewpoint of nonlinear dynamics, these systems are thought of as being prepared at the edge of chaos and therefore are nonergodic, in general. More precisely, they are characterized by the vanishing Lyapunov exponents. In addition, the phase space structure of a complex system at the edge of chaos

S. Abe: *Foundations of Nonextensive Statistical Mechanics*, StudFuzz **206**, 53–71 (2006)
www.springerlink.com

is quite different from that of a fully chaotic system: only a tiny part of it is occupied by the system. Accordingly, the number of microscopically accessible states is considered to be very small.

One of the most important points regarding a complex system is that its macroscopic properties cannot be understood in terms of the accumulation of the properties of each element contained in the system. This is largely due to strong correlation between the elements, and accordingly the separability condition commonly assumed in thermodynamics is violated. Thus, understanding complex systems requires a holistic approach.

Violation of separability in thermodynamics is a signal of the relevance of nonextensivity. Inseparability may result from both correlation and interaction. Consider a long-range interacting system, for example. Its internal energy does not scale with the number of particles, N. It has been known that such a system cannot naively be described by Boltzmann-Gibbs statistical mechanics. In this respect, it is of importance to notice that the long-range interacting systems may reside at the edge of chaos since their largest Lyapunov exponents tend to vanish in the thermodynamic limit [1, 2]. A crucial point regarding nonextensive systems such as long-range interacting ones (they are nonextensive since the internal energies do not scale with their numbers of particles) is that the thermodynamic limit and the long-time limit do not commute, in general [3]. Of particular interest is the order, in which the long-time limit is taken after the thermodynamic limit, since in this case nonequilibrium stationary states, which survive for long times (much longer than typical microscopic dynamical time scales), may be observed for a wide class of initial conditions [4, 5, 6, 7, 8]. Nonextensive statistical mechanics [3, 9, 10, 11, 12] is considered to offer a unified and consistent statistical description of such intriguing states of complex systems. Therefore, it is basically unconcerned with strict equilibrium.

Nonextensive statistical mechanics is a generalization of Boltzmann-Gibbs theory and is characterized by the power-law distributions often observed in complex systems. It is customarily formulated by the maximum entropy principle for a generalized entropy termed the Tsallis entropy

In this article, we present a review of the theoretical aspects of nonextensive statistical mechanics. This theory has widely been studied from various viewpoints, both theoretically and phenomenologically. Even limited to the theoretical investigations, there are a number of works done on both macroscopic (thermodynamic) and microscopic (statistical mechanical) foundations. Therefore, here we select only the specific discussions about the statistical foundations of the physics described by the power-law distributions in conformity with nonextensive statistical mechanics (in particular, we shall not discuss applications of the theory). The interested reader can visit the URL(http://tsallis.cat.cbpf.br/TEMUCO.pdf), from where a comprehensive list of references can be obtained about both the theoretical discussions and applications. Here, we shall discuss the bases for the power-law distributions in nonextensive statistical mechanics. We show how the ordinary maximum entropy principle, the method of steepest descents, the counting algorithm and the evaluation of the density of states can naturally be generalized for describing/understanding a class of complex systems characterized by the power-law distributions. We also see that an appropriate generalization of the Boltzmann equation can lead to the H-theorem for the Tsallis-type functional and accordingly the maximum Tsallis entropy distribution is obtained as the stationary solution of the generalized Boltzmann equation.

2.2 Maximum Tsallis entropy principle

This is the most widely discussed approach to nonextensive statistical mechanics. A central role is played by the Tsallis entropy [13] defined by

$$S_q[p] = \frac{1}{1-q}\left[\sum_{i=1}^{W}(p_i)^q - 1\right] \tag{2.2.1a}$$

$$= -\sum_{i=1}^{W}(p_i)^q \ln_q(p_i). \tag{2.2.1b}$$

Here, $\{p_i\}_{i=1,2,\ldots,W}$ is a normalized probability distribution of a system under consideration with W microscopically accessible states and q is the positive entropic index. The Boltzmann constant is set equal to unity for the sake of notational simplicity. The symbol, $\ln_q(x)$ means the q-logarithmic function defined by

$$\ln_q(x) = \frac{1}{1-q}(x^{1-q} - 1), \tag{2.2.2a}$$

which is the inverse function of the q-exponential function

$$e_q(x) = [1 + (1-q)x]_+^{1/(1-q)}, \tag{2.2.2b}$$

with the notation, $[a]_+ = \max\{0, a\}$. It is obvious that these functions tend respectively to the ordinary logarithmic and exponential functions in the limit $q \to 1$. Therefore, the Tsallis entropy converges to the familiar Boltzmann-Gibbs-Shannon entropy, $S[p] = -\sum_{i=1}^{W} p_i \ln p_i$, in the limit $q \to 1$.

A stationary state, which is regarded to be a stationary nonequilibrium state of a nonextensive system under consideration, may be obtained by maximizing the Tsallis entropy under appropriate constraints. In analogy with a microcanonical situation, the constraint is imposed only on the normalization condition

$$\sum_{i=1}^{W} p_i = 1. \tag{2.2.3}$$

Accordingly, the maximum Tsallis entropy principle reads

$$\delta_p\left\{S_q[p] - \alpha\left(\sum\nolimits_{i=1}^{W} p_i - 1\right)\right\} = 0, \tag{2.2.4}$$

where α stands for the Lagrange multiplier and δ_p denotes the variation with respect to the distribution. The solution to this problem is given by the equal *a priori* probability

$$\tilde{p}_i = \frac{1}{W} \qquad (i = 1, 2, \ldots, W), \tag{2.2.5}$$

where W depends on the energy, the number of particles and the volume of the system. Then, the maximum value of the Tsallis entropy is

$$S_q^{\max} = \ln_q W. \tag{2.2.6}$$

Eq. (2.2.6) has an important meaning. In Boltzmann-Gibbs statistical mechanics, where ergodicity is supposed to be valid, W is a huge number with the exponential dependence on the number of particles, N. The corresponding extensive entropy proportional to N is $S_q^{\max}$ with $q \to 1$, i.e. $S = \ln W$, which is the Boltzmann relation. However, as mentioned in Sec. 2.1, in the situation where ergodicity is broken, W is considered to be much smaller and depends more mildly on N, due to strong correlation between the particles. Only a tiny portion of the phase space (e.g., a multifractal subset) is occupied by the system, and the distribution in Eq. (2.2.5) is defined on it. Thus, Eq. (2.2.5) describes the principle of equal *a priori* probability on such a portion. Eq. (2.2.6) suggests that the Tsallis entropy with $q \neq 1$ can be a relevant extensive entropy in a case when W slowly grows as a power law with respect to N, for example. Necessity of entropy being extensive is actually a profound thermodynamic requirement, but this point will not be discussed here.

Next, let us look at a case when the constraints are imposed on the averages of the physical quantities. This case is in analogy with a (grand)canonical situation. To simplify the discussion, here we consider only the system energy as a physical quantity, i.e., the Hamiltonian, H, with its ith value, ε_i. The definition of the expectation value in nonextensive statistical mechanics is the so-called q-expectation value given as follows:

$$U_q = \langle H \rangle_q = \sum_{i=1}^{W} P_i^{(q)} \varepsilon_i, \tag{2.2.7a}$$

where $P_i^{(q)}$ is the escort distribution [14]

$$P_i^{(q)} = \frac{(p_i)^q}{\sum_{j=1}^{W} (p_j)^q}. \tag{2.2.7b}$$

Necessity of such a generalized expectation value has been clarified in [15] based on the consistency between the maximum entropy principle and the minimum cross entropy principle. Introducing the Lagrange multiplier, β, for the constraint on the q-expectation value of the energy, the corresponding maximum Tsallis entropy principle now reads [16]

$$\delta_p \left\{ S_q[p] - \alpha \left(\sum_{i=1}^{W} p_i - 1 \right) - \beta \left(\sum_{i=1}^{W} P_i^{(q)} \varepsilon_i - U_q \right) \right\} = 0. \tag{2.2.8}$$

The normalized solution to this problem is given by the so-called q-exponential distribution

$$\tilde{p}_i = \frac{1}{Z_q(\beta)} e_q(-\beta^*(\varepsilon_i - \tilde{U}_q)), \tag{2.2.9a}$$

$$Z_q(\beta) = \sum_{i=1}^{W} e_q(-\beta^*(\varepsilon_i - \tilde{U}_q)), \tag{2.2.9b}$$

where $\beta^* = \beta / c_q$ with $c_q = \sum_{i=1}^{W} (\tilde{p}_i)^q$ and $\tilde{U}_q$ is the internal energy in Eq. (2.2.7a) calculated in terms of $\tilde{p}_i$ in a self-referential manner. The distribution in Eq. (2.2.7a) is supposed to describe a nonequilibrium stationary state of the system of our interest. In the limit $q \to 1$, it becomes the familiar exponential distribution in Boltzmann-Gibbs statistical mechanics. In the case when $0 < q < 1$, the distribution has a cut-off at $\varepsilon_i^{\max} = \tilde{U}_q + c_q / [(1-q)\beta]$. On the other hand, if $q > 1$, it is a power-law distribution of the Zipf-Mandelbrot type. The q-exponential distributions have been observed in many complex systems and phenomena (see the URL given in Sec. 1).

It is known that nonextensive statistical mechanics is consistent with the most of the fundamental principles of thermodynamics. For the first and second laws, the proofs can be found in Ref. [17]. The third law also holds since the Tsallis entropy vanishes for a completely ordered state. The zeroth law, however, still remains somewhat unclear, though there are some numerical evidences for its validity [18]. This is because the zeroth law defines the concept of equilibrium, whereas nonextensive statistical mechanics is concerned with the nonequilibrium stationary situation. The thermodynamic Legendre transform structure can also be established as follows. Let $\tilde{S}_q$ be the Tsallis entropy calculated in terms of the stationary distribution in Eq. (2.2.9a). Then, it is straightforward to obtain

$$\frac{\partial \tilde{S}_q}{\partial \tilde{U}_q} = \beta. \tag{2.2.10}$$

Therefore, for example, the free energy may be defined in the usual way: $F_q = \tilde{U}_q - \tilde{S}_q / \beta$. For further discussions about the thermodynamic properties, see a recent work [19].

In ordinary Boltzmann-Gibbs statistical mechanics, there is a mathematical result called the Gibbs theorem, which states that a subsystem of a microcanonical ensemble with large degrees of freedom is uniquely characterized by the ordinary canonical distribution of the exponential type. This theorem has repeatedly been proved in various ways, such as the method of steepest descents, the counting algorithm and the evaluation of the density of states (nice discussions about the first two can be found in Ref. [20] and the last one in Ref. [21]). There, starting from the microcanonical basis with the principle of equal *a priori* probability, the Boltzmann-Gibbs distribution is derived for the canonical ensemble. In the subsequent three sections, we shall see how these three main methods can naturally be generalized to the power-law distribution in nonextensive statistical mechanics [22, 23, 24]. Once again, we emphasize that the equal *a priori* probability in Eq. (2.2.5) is defined on a tiny subset of the phase space (not on the full constant-energy hypersurface), and the number of states grows slowly (presumably as a power law) with respect to the number of particles.

2.3 Method of steepest descents

Following the discussions in Refs. [20, 22], let us begin with considering a classical system, s, and make its M replicas, $s_1, s_2, \cdots, s_M$. The collection $\mathbf{S} = \{s_\alpha\}_{\alpha=1,2,\cdots,M}$ is called a supersystem. Let A_α be a physical quantity of interest (such as the energy) of the system s_α, which should be bounded from below, in general. This random variable takes a value $a(m_\alpha)$, where m_α labels the allowed configurations of s_α. We are interested in a macroscopic quantity, which may be given by the arithmetic mean of $\{A_\alpha\}_{\alpha=1,2,\cdots,M}$ over the supersystem: $(1/M)\sum_{\alpha=1}^{M} A_\alpha$. We require that the probability of finding $\mathbf{S}$ in the configurations in which the value of this macroscopic quantity lies around a certain value $\bar{a}$, that is,

$$\left| \frac{1}{M}\sum_{\alpha=1}^{M} a(m_\alpha) - \bar{a} \right| < \frac{1}{M}\sum_{\alpha=1}^{M} |a(m_\alpha) - \bar{a}| < \varepsilon \tag{2.3.1}$$

is uniform, in conformity with the principle of equal *a priori* probability.

In Boltzmann-Gibbs statistical mechanics, ε is supposed to be of $O(1/\sqrt{M})$, which comes from the ordinary law of large numbers in the central limit theorem putting a basis for the universality of the Gaussian distributions. However, here, we are concerned with the power-law distributions. Therefore, what is relevant is the Lévy-Gnedenko generalized central limit theorem (in the half space) [25]. Accordingly, ε is assumed to be

$$\varepsilon \sim O\left(M^{-(1+\delta)}\right) \qquad \delta > 0. \tag{2.3.2}$$

The equiprobability, $p(m_1, m_2, \ldots, m_M)$, associated with Eq.(2.3.1) is given by

$$p(m_1, m_2, \cdots, m_M) \propto \theta\left(\varepsilon - |L|\right), \tag{2.3.3a}$$

$$L \equiv \frac{1}{M}\sum_{\alpha=1}^{M} a(m_\alpha) - \bar{a} = \frac{1}{M}\sum_{\alpha=1}^{M} [a(m_\alpha) - \bar{a}], \tag{2.3.3b}$$

where $\theta(x)$ is the Heaviside unit-step function. To obtain canonical ensemble theory, we select an objective system, say s_1, and eliminate the others. Then, the probability of finding such an objective system in the configuration $m_1 = m$ is

$$\tilde{p}(m) = \sum_{m_2,\cdots,m_M} p(m, m_2, \cdots, m_M). \tag{2.3.4}$$

Upon proving the Gibbs theorem, it is an ordinary way to employ the (inverse) Laplace transformation of the step function

$$\theta(x) = \frac{1}{2\pi i} \int_{\beta-i\infty}^{\beta+i\infty} \frac{d\phi}{\phi} e^{\phi x} \tag{2.3.5}$$

in Eq. (2.3.3a), where β is an arbitrary positive constant. Then, the method of steepest descents is applied to Eq. (2.3.4) in the large-M limit. The Boltzmann-Gibbs exponential distribution is seen to come from the exponential factor in the Laplace transformation [20].

Here, we are, however, interested in the power-law distribution, and so we need look for another representation of the step function. A point is that the step function takes the values of discrete topology, which may remain unchanged by a continuous deformation of the Laplace transformation. Accordingly, we examine the q-exponential function with $q > 1$. An analysis [22] shows that the equality

$$\theta\,(x) = \frac{1}{2\,\pi\,i} \int_{\beta-i\infty}^{\beta+i\infty} \frac{d\phi}{\phi}\, e_q(\phi x) \tag{2.3.6}$$

still holds if β satisfies

$$1 - (q-1)\beta x_{\max} > 0, \tag{2.3.7}$$

where $x_{\max}$ is the fixed maximum value of x in its range of interest (but later it turns out to be possible for it to be arbitrary large in the subsequent discussion of the steepest-descent approximation).

Let us evaluate the integral

$$\theta\,(\varepsilon - L) = \frac{1}{2\,\pi\,i} \int_{\beta-i\infty}^{\beta+i\infty} \frac{d\phi}{\phi}\, e_q((\varepsilon - L)\phi). \tag{2.3.8}$$

Recall that the q-exponential function has the property

$$e_q(a) e_q(b) = e_q(a + b + (1-q)ab). \tag{2.3.9}$$

However, taking Eqs. (2.3.1) and (2.3.2) into account, we see that in the large-M limit we can approximately realize the factorization to obtain

$$\theta(\varepsilon - L) \cong \frac{1}{2\pi i} \int_{\beta-i\infty}^{\beta+i\infty} \frac{d\phi}{\phi}\, e_q(\varepsilon\phi) \prod_{\alpha=1}^{M} e_q\left(-\phi\, \frac{1}{M}\left[a(m_\alpha) - \bar{a}\right]\right). \tag{2.3.10a}$$

Using $\theta(\varepsilon - |L|) = \theta(\varepsilon - L) - \theta(-\varepsilon - L)$ and changing the integration variable as $\phi \to M\phi$, we have

$$\theta(\varepsilon - |L|) \cong \frac{1}{\pi i} \int_{\beta^*-i\infty}^{\beta^*+i\infty} \frac{d\phi}{\phi} \sinh_q(M\phi\,\varepsilon) \prod_{\alpha=1}^{M} e_q(-\phi[a(m_\alpha) - \bar{a}]), \tag{2.3.10b}$$

where $\sinh_q(x) \equiv [e_q(x) - e_q(-x)]/2$ and

$$\beta^* = \frac{\beta}{M}. \tag{2.3.11}$$

Let us examine Eq. (2.3.7), as promised. It is now rewritten as follows:

$$1 - (q-1)\beta^* M\, |\pm\varepsilon - L|_{\max} > 0. \tag{2.3.12}$$

The equiprobability distribution (of the rectangular shape) we are considering has a very narrow support with the width 2ε. $|\pm\varepsilon - L|_{\max}$ is of $O(M^{-(1+\delta)})$ with $\delta > 0$, as mentioned in Eqs. (2.3.1) and (2.3.2). Therefore, in the large-M limit, β^* can be taken to be an arbitrary positive constant. Now, using the method of steepest descents in the large-M limit, we can make the following evaluation:

$$\begin{aligned}
\tilde{p}(m) &= \sum_{m_2,\cdots,m_M} p(m, m_2, \cdots, m_M) \\
&\cong \frac{1}{W}\frac{1}{\pi i} \int_{\beta^*-i\infty}^{\beta^*+i\infty} \frac{d\phi}{\phi} \sinh_q(M\phi\,\varepsilon) e_q(-\phi[a(m) - \bar{a}]) \\
&\qquad \times \sum_{m_2,\cdots,m_M} \prod_{\alpha=2}^{M} e_q(-\phi[a(m_\alpha) - \bar{a}]) \\
&= \frac{1}{W}\frac{1}{\pi i} \int_{\beta^*-i\infty}^{\beta^*+i\infty} \frac{d\phi}{\phi} \sinh_q(M\phi\,\varepsilon) \frac{e_q(-\phi[a(m) - \bar{a}])}{Z_q(\phi)} e^{M \ln Z_q(\phi)}.
\end{aligned} \tag{2.3.13}$$

In the above,

$$Z_q(\phi) = \sum_m e_q(-\phi[a(m) - \bar{a}]), \tag{2.3.14}$$

and W is the number of possible configurations satisfying Eq. (2.3.1) (i.e., essentially the normalization factor)

$$W = \frac{1}{\pi i} \int_{\beta^*-i\infty}^{\beta^*+i\infty} \frac{d\phi}{\phi} \sinh_q(M\phi\,\varepsilon) e^{M \ln Z_q(\phi)}. \tag{2.3.15}$$

Finally, using the real part β^* of ϕ, the steepest-descent condition is given by

$$\frac{\partial Z_q}{\partial \beta^*} = 0, \tag{2.3.16}$$

which consequently yields

$$\tilde{p}(m) = \frac{1}{Z_q} e_q(-\beta^* [a(m) - \bar{a}]), \tag{2.3.17}$$

$$\bar{a} = \sum_m \tilde{P}^{(q)}(m) a(m), \tag{2.3.18}$$

simultaneously, where $\tilde{P}^{(q)}(m)$ is the escort distribution

$$\tilde{P}^{(q)}(m) = \frac{[\tilde{p}(m)]^q}{\sum_n [\tilde{p}(n)]^q}, \tag{2.3.19}$$

the origin of which is in a very mathematical fact that $de_q(x)/dx = [e_q(x)]^q$.

Thus, we have seen how the power-law distribution (with the entropic index $q > 1$) in nonextensivity statistical mechanics can be derived from the principle of equal *a priori* probability by appropriately generalizing the ordinary discussion based on the method of steepest descents, as in Eqs. (2.3.1), (2.3.2) and (2.3.6). A point to be noticed is that entropy does not appear in the discussion.

2.4 Counting algorithm

A counting rule is an algorithm, which connects each other the entropy, a macroscopic quantity and the number of microscopic configurations of a system. Boltzmann's algorithm uses the multinomial counting of the microscopic configurations. There is, however, no *a priori* reason to assume such a special counting rule to be universal. In fact, there may yet exist a class of systems, which seems to prefer alternate kinds of counting rules.

Consider a supersystem defined in Sec. 2.3. The probability $\tilde{p}(m)$ of finding the system $s \equiv s_1$ in its mth configuration is given by $\tilde{p}(m) = W(m)/W$, where W is the total number of configurations satisfying Eq. (2.3.1) and $W(m)$ is the number of configurations of the objective system with $\bar{a}$ appearing in Eq. (2.3.1). To calculate this quantity, we rewrite Eq. (2.3.1) in the following form:

$$\left| \frac{1}{M} [a(m) - \bar{a}] + \frac{1}{M} \sum_{\alpha=2}^{M} a(m_\alpha) - \frac{M-1}{M} \bar{a} \right| < \varepsilon. \tag{2.4.1}$$

The number of configurations Y_M satisfying

$$\left| \frac{1}{M} \sum_{\alpha=2}^{M} a(m_\alpha) - \frac{M-1}{M} \bar{a} \right| < \varepsilon \tag{2.4.2}$$

is, according to Boltzmann's algorithm, counted in the large-M limit as follows [20]:

$$\ln\left[Y_M\left(\frac{M-1}{M}\bar{a}\right)\right]^{1/M} \cong \ln\left[Y_M(\bar{a})\right]^{1/M} \to \widetilde{S}(a). \tag{2.4.3}$$

Here, $\tilde{S}(\bar{a})$ is a certain function of $\bar{a}$ to be identified with the entropy. From Eqs. (2.4.1)-(2.4.3), it follows that

$$\begin{aligned}\ln W(m) &= \ln Y_M\left(\frac{M-1}{M}\bar{a} - \frac{1}{M}[a(m)-\bar{a}]\right)\\ &\cong \ln Y_M(\bar{a}) - [a(m)-\bar{a}]\frac{\partial \tilde{S}}{\partial \bar{a}}.\end{aligned} \tag{2.4.4}$$

Defining β as

$$\beta = \frac{\partial \tilde{S}(\bar{a})}{\partial \bar{a}} \tag{2.4.5}$$

and setting

$$Z(\beta) = \lim_{M\to\infty}\frac{W}{Y_M(\bar{a})}, \tag{2.4.6}$$

obtained is the ordinary canonical distribution

$$p(m) = \frac{1}{Z(\beta)}\exp\{-\beta[a(m)-\bar{a}]\}, \tag{2.4.7}$$

where $Z(\beta) = Z'(\beta)\exp(\beta\bar{a})$ with the ordinary partition function $Z(\beta) = \sum_m \exp[-\beta a(m)]$.

The counting rule in Eq. (2.4.3) is essential in the above derivation. Now, let us examine another kind of counting rule for another type of configurations, which is supposed to be realized by nonextensive systems. Specifically, we examine to replace Eq. (2.4.3) by the following rule [23]:

$$\ln_q\left[Y_M\left(\frac{M-1}{M}\bar{a}\right)\right]^{1/M} \cong \ln_q\left[Y_M(\bar{a})\right]^{1/M} \to \tilde{S}_q(\bar{a}). \tag{2.4.8}$$

$\tilde{S}_q(\bar{a})$ in Eq. (2.4.8) is a certain quantity dependent on $\bar{a}$ and later will be shown to be the Tsallis entropy. Then, instead of Eq. (2.4.4), this algorithm leads to

$$\begin{aligned}\ln_q W(m) &= \ln_q Y_M\left(\frac{M-1}{M}\overline{a} - \frac{1}{M}[a(m)-\overline{a}]\right)\\ &\cong \ln_q Y_M(\overline{a}) - \frac{1}{M}[a(m)-\overline{a}]\frac{\partial}{\partial\overline{a}}\ln_q Y_M(\overline{a}),\end{aligned} \tag{2.4.9}$$

provided that M may be large but finite. Defining

$$\frac{\partial \tilde{S}_q(\bar{a})}{\partial \bar{a}} = \beta \tag{2.4.10}$$

and using Eq. (2.4.8), we have

$$\beta = \frac{1}{M} \left[Y_M(\bar{a})\right]^{(1-q)/M-1} \frac{\partial Y_M(\bar{a})}{\partial \bar{a}}. \tag{2.4.11}$$

Therefore, we obtain

$$\frac{\partial}{\partial \bar{a}} \ln_q Y_M(\bar{a}) = \beta M \left[Y_M(\bar{a})\right]^{(1-q)(1-(1/M))}. \tag{2.4.12}$$

Substituting Eq. (2.4.12) into Eq. (2.4.9), we have

$$\ln_q W(m) - \ln_q Y_M(\bar{a}) = -\beta \left[Y_M(\overline{a})\right]^{(1-q)(1-(1/M))} [a(m) - \bar{a}]. \tag{2.4.13}$$

Making use of the property of the q-logarithmic function, that is,

$$\ln_q \left(\frac{x}{y}\right) = y^{q-1} [\ln_q(x) - \ln_q(y)], \tag{2.4.14}$$

we obtain

$$\frac{W(m)}{Y_M(\bar{a})} \cong e_q \left(-\beta^* \left[a(m) - \bar{a}\right]\right), \tag{2.4.15}$$

where $\beta^* = \beta / c_q$ with

$$c_q \equiv \left[Y_M(\bar{a})\right]^{(1-q)/M} = 1 + (1-q)\tilde{S}(\bar{a}). \tag{2.4.16}$$

From Eq. (2.4.15) and the total number of configurations, $W = \sum_m W(m)$, we also obtain

$$Z_q(\beta) \equiv \frac{W}{Y_M(\bar{a})} \cong \sum_m e_q(-\beta^* [a(m) - \bar{a}]). \tag{2.4.17}$$

Therefore, we consequently find the probability, $\tilde{p}(m) = W(m)/W$, to be the q-exponential distribution in nonextensive statistical mechanics [23]:

$$\tilde{p}(m) = \frac{1}{Z_q(\beta} e_q(-\beta^* [a(m) - \bar{a}]). \tag{2.4.18}$$

Actually, we have full agreement with nonextensive statistical mechanics, which is shown below. If we impose the condition

$$\left.\frac{\partial Z_q(\beta)}{\partial \beta}\right|_{\bar{a}} = 0, \tag{2.4.19}$$

then we have [23]

$$\bar{a} = \sum_m \tilde{P}^{(q)}(m) a(m), \tag{2.4.20}$$

where $\tilde{P}^{(q)}(m)$ is the escort distribution, as in Eq. (2.3.19). Now, if the entropy is given by

$$\tilde{S}_q = \ln_q Z_q(\beta), \tag{2.4.21}$$

then Eq. (2.4.19) is seen to be the maximum entropy condition. Eqs. (2.4.21) and (2.4.8) lead to the identification

$$Z_q(\beta) \cong [Y_M(\bar{a})]^{1/M}. \tag{2.4.22}$$

Furthermore, from Eq. (2.4.17), W is seen to be

$$W \cong [Y_M(\bar{a})]^{1/M+1}. \tag{2.4.23}$$

Recall that the identical relation

$$\sum_m [\tilde{p}(m)]^q = [Z_q(\beta)]^{1-q} \tag{2.4.24}$$

holds for the distribution in Eq. (2.4.18) with Eq. (2.4.17). Combining this equation with Eq. (2.4.22), we also have

$$[Y_M(\bar{a})]^{(1-q)/M} = \sum_m \tilde{p}(m)]^q. \tag{2.4.25}$$

Therefore, c_q in Eq. (2.4.16) is found to be given by

$$c_q = \sum_m [\tilde{p}(m)]^q. \tag{2.4.26}$$

Finally, using Eqs. (2.4.21), (2.4.24) and (2.4.26), we ascertain that both Eqs. (2.4.16) and (2.4.21) consistently lead to [23]

$$\tilde{S}_q = \frac{1}{1-q}\left\{\sum_m [\tilde{p}(m)]^q - 1\right\}, \tag{2.4.27}$$

which is precisely the Tsallis entropy of the nonequilibrium stationary state, $\tilde{p}(m)$.

2.5 Evaluation of the density of states

In this section, we present a macroscopic description of systems obeying the power-law distributions by reconsideration of contact with the heat bath, showing how the structure of nonextensive statistical mechanics can arise.

Let us start our discussion by examining a partition of energy between two nonextensive systems, I and II, in contact, whose entropies may not satisfy additivity. These systems have energies E_{I} and E_{II}, and the total energy is assumed to be fixed as $E = E_{\mathrm{I}} + E_{\mathrm{II}}$. If the state densities of I and II are denoted by $\Omega_{\rm I}$ and $\Omega_{\rm II}$, respectively and that of the total system I+II by Ω, then we have

$$\Omega(E)\Delta E = \int\limits_{E<E_{\rm I}+E_{\rm II}<E+\Delta E}\int \Omega_{\rm I}(E_{\rm I})\Omega_{\rm II}(E_{\rm II})dE_{\rm I}dE_{\rm II}$$

$$= \Delta E \int\limits_{0}^{E} \Omega_{\rm I}(E_{\rm I})\Omega_{\rm II}(E - E_{\rm I})\, dE_{\rm I}, \tag{2.5.1}$$

where ΔE is the thickness of the shell of constant energy in phase space of the total system. We should notice that this ignorance of the interaction part is a drastic approximation for nonextensive systems and may not be justified, in general. However, our purpose here is to examine how far we can proceed along with the ordinary settings for statistical mechanics. The probability of finding the system I in the range $(E_{\rm I}, E_{\rm I} + dE_{\rm I})$ is given by

$$p(E_{\rm I})dE_{\rm I} = \frac{\Omega_{\rm I}(E_{\rm I})\Omega_{\rm II}(E - E_{\rm I})\Delta E}{\Omega\,(E)\,\Delta E} dE_{\rm I}. \tag{2.5.2}$$

We are interested in the most probable partition of energy in conformity with the principle of equal *a priori* probability. It is determined by maximizing the quantity

$$\Omega_{\rm I}(E_{\rm I})\Omega_{\rm II}(E - E_{\rm I})\Delta E\, dE_{\rm I}. \tag{2.5.3}$$

In the traditional discussions of identifying the temperature of the system I and deriving its probability distribution, maximization is performed by taking the logarithm of the quantity in Eq. (2.5.3) [21]. The assumption underlying such a treatment is the logarithmic form (additivity) of entropy. Here, we relax this assumption in order to accommodate the power-law distributions. In particular, we consider the q-logarithmic evaluation since the q-logarithmic function is also a monotonically increasing one. In addition to Eq. (2.4.14), the q-logarithmic function possesses the following property:

$$\ln_q(xy) = \ln_q(x) + \ln_q(y) + (1-q)\ln_q(x)\ln_q(y). \tag{2.5.4}$$

Accordingly, the maximization condition can be written as follows [24]:

$$\ln_q(\Omega_{\rm I}(E_{\rm I})\Omega_{\rm II}(E - E_{\rm I})) = \max. \tag{2.5.5}$$

Though we do not have to specify the values of q of the composite system at this stage, we may assume that I and II have their own values of q, in general. Since the q-exponential and q-logarithmic functions are inverse to each other, we have

$$\Omega_i = e_{q_i}(\ln_{q_i}(\Omega_i)) \qquad i = {\rm I}, {\rm II}. \tag{2.5.6}$$

Here and hereafter, ΔE is set equal to unity for the sake of simplicity. Maximization (i.e., vanishing of the first derivative of Eq. (2.5.5) with respect to $E_{\rm I}$) leads to the expression

$$
\begin{aligned}
0 &= \frac{\partial}{\partial E_{\mathrm{I}}} \ln_q(\Omega_{\mathrm{I}}\,\Omega_{\mathrm{II}}) \\
&= (\Omega_{\mathrm{I}}\,\Omega_{\mathrm{II}})^{1-q} \left(\frac{1}{\Omega_{\mathrm{I}}} \frac{\partial \Omega_{\mathrm{I}}}{\partial E_{\mathrm{I}}} + \frac{1}{\Omega_{\mathrm{II}}} \frac{\partial \Omega_{\mathrm{II}}}{\partial E_{\mathrm{I}}} \right).
\end{aligned}
\tag{2.5.7}
$$

From this, we have

$$
\begin{aligned}
&\frac{1}{1+(1-q_{\mathrm{I}})\ln_{q_{\mathrm{I}}}(\Omega_{\mathrm{I}}(E_{\mathrm{I}}))} \frac{\partial \ln_{q_{\mathrm{I}}}(\Omega_{\mathrm{I}}(E_{\mathrm{I}}))}{\partial E_{\mathrm{I}}} \\
&\qquad = \frac{1}{1+(1-q_{\mathrm{II}})\ln_{q_{\mathrm{II}}}(\Omega_{\mathrm{II}}(E_{\mathrm{II}}))} \frac{\partial \ln_{q_{\mathrm{II}}}(\Omega_{\mathrm{II}}(E_{\mathrm{II}}))}{\partial E_{\mathrm{II}}},
\end{aligned}
\tag{2.5.8}
$$

where the assumption, $E_{\mathrm{I}} + E_{\mathrm{II}} = E$, has been used. Now, if we identify

$$
\tilde{S}_{iq_i} = \tilde{S}_{iq_i}(\Omega_i) \equiv \ln_{q_i}(\Omega_i) \tag{2.5.9}
$$

with a generalized entropy of the nonextensive system "i" characterized by the index $q_i, (i = \mathrm{I}, \mathrm{II})$ and define a parameter β_i by

$$
\beta_i = \frac{\partial S_{iq_i}(\Omega_i)}{\partial E_i}, \tag{2.5.10}
$$

then we obtain

$$
\frac{\beta_{\mathrm{I}}}{c_{\mathrm{I}q_{\mathrm{I}}}} = \frac{\beta_{\mathrm{II}}}{c_{\mathrm{II}q_{\mathrm{II}}}} \equiv \beta^*, \tag{2.5.11}
$$

where we have introduced the notation

$$
c_{iq_i} \equiv 1 + (1-q_i) S_{iq_i} = \Omega_i^{1-q_i}. \tag{2.5.12}
$$

To check the above state to be indeed the maximum, we need verify that the second derivative is negative. Carrying out such a calculation using Eqs. (2.5.7), (2.5.9) and (2.5.12), we have

$$
\begin{aligned}
&\frac{d^2}{dE_{\mathrm{I}}^2} \ln_q(\Omega_{\mathrm{I}}(E_{\mathrm{I}})\Omega_{\mathrm{II}}(E-E_{\mathrm{I}})) \\
&\quad = \frac{(\Omega_{\mathrm{I}}\Omega_{\mathrm{II}})^{1-q}}{1-q} \left\{ \frac{d^2}{dE_{\mathrm{I}}^2} \ln\left[1+(1-q)S_{\mathrm{I}q_{\mathrm{I}}}\right] + \frac{d^2}{dE_{\mathrm{I}}^2} \ln\left[1+(1-q)S_{\mathrm{II}q_{\mathrm{II}}}\right] \right\} \\
&\quad = (\Omega_{\mathrm{I}}\Omega_{\mathrm{II}})^{1-q} \left\{ \frac{1}{1-q_{\mathrm{I}}} \frac{d^2}{dE_{\mathrm{I}}^2} \ln\left[1+(1-q_{\mathrm{I}})S_{\mathrm{I}q_{\mathrm{I}}}\right] + \right. \\
&\qquad\qquad \left. + \frac{1}{1-q_{\mathrm{II}}} \frac{d^2}{dE_{\mathrm{I}}^2} \ln\left[1+(1-q_{\mathrm{II}})S_{\mathrm{II}q_{\mathrm{II}}}\right] \right\}.
\end{aligned}
\tag{2.5.13}
$$

In the above, we have given two equivalent expressions to exhibit the actual equivalence of seemingly different expressions. As in the case of the ordinary discussion of Boltzmann-Gibbs theory, we also assume that the second-order derivative of the generalized entropy in Eq. (2.5.9) is negative. Then, we see that each term appearing in Eq. (2.5.13) is negative for all positive values of q_i's, due to the monotonicity of the ordinary logarithmic function.

Now, let us study the necessary modification of the Gibbs theorem for the canonical measure. For this purpose, we regard the nonextensive system II as the heat bath

$$E_{\mathrm{I}} \ll E_{\mathrm{II}}. \tag{2.5.14}$$

The probability of finding the system I in its kth state of energy $E_{\mathrm{I}} = \varepsilon_k$ is proportional to the number of microscopic states of the system II as follows:

$$f(\varepsilon_k) \propto \frac{\Omega_{\mathrm{II}}(E - \varepsilon_k)}{\Omega_{\mathrm{II}}(E)}, \tag{2.5.15a}$$

which can obviously be rewritten in the form

$$f(\varepsilon_k) \propto e_{q_{\mathrm{II}}}\left(\ln_{q_{\mathrm{II}}}\left[\frac{\Omega_{\mathrm{II}}(E - \varepsilon_k)}{\Omega_{\mathrm{II}}(E)}\right]\right). \tag{2.5.15b}$$

From Eq. (2.4.14), it follows that

$$f(\varepsilon_k) \propto e_{q_{\mathrm{II}}}\left(\frac{1}{\Omega_{\mathrm{II}}^{1-q_{\mathrm{II}}}(E)}\left[\ln_{q_{\mathrm{II}}}(\Omega_{\mathrm{II}}(E - \varepsilon_k)) - \ln_{q_{\mathrm{II}}}(\Omega_{\mathrm{II}}(E))\right]\right). \tag{2.5.15c}$$

Eq. (2.5.14) with $E_{\mathrm{I}} = \varepsilon_k$ allows us to perform the following expansion to the leading order of ε_k:

$$\begin{aligned} f(\varepsilon_k) \propto e_{q_{\mathrm{II}}}&\left(\frac{1}{\Omega_{\mathrm{II}}^{1-q_{\mathrm{II}}}(E)}\left[\ln_{q_{\mathrm{II}}}(\Omega_{\mathrm{II}}(E))\right.\right. \\ &\left.\left. - \varepsilon_k \frac{\partial \ln_{q_{\mathrm{II}}}(\Omega_{\mathrm{II}}(E))}{\partial E} + \cdots - \ln_{q_{\mathrm{II}}}(\Omega_{\mathrm{II}}(E))\right]\right) \cong e_{q_{\mathrm{II}}}(-\beta^* \varepsilon_k), \end{aligned} \tag{2.5.16}$$

where Eqs. (2.5.10) and (2.5.11) have been used. Therefore, we obtain [24]

$$\Omega_{\mathrm{II}}(E - \varepsilon_k) \propto \Omega_{\mathrm{II}}(E)\, e_{q_{\mathrm{II}}}(-\beta^* \varepsilon_k). \tag{2.5.17}$$

This is the canonical measure for the nonextensive system. Since the heat bath is assumed to be large, it is appropriate to consider the relative probability of finding the system I as the ratio of the probability in the state with energy ε_k relative to the fixed value of energy ε_l:

$$\pi(\varepsilon_k; \varepsilon_l) = \frac{f(\varepsilon_k)}{f(\varepsilon_l)} = \frac{\Omega_{\mathrm{II}}(E - \varepsilon_k)}{\Omega_{\mathrm{II}}(E - \varepsilon_l)}, \tag{2.5.18}$$

which is rewritten as

$$\pi(\varepsilon_k; \varepsilon_l) = e_{q_{\mathrm{II}}}\left(\ln_{q_{\mathrm{II}}}\left(\frac{\Omega_{\mathrm{II}}(E - \varepsilon_k)}{\Omega_{\mathrm{II}}(E - \varepsilon_l)}\right)\right)$$

$$= e_{q_{\mathrm{II}}} \left(\frac{1}{\Omega_{\mathrm{II}}^{1-q_{\mathrm{II}}}(E-\varepsilon_l)} \left[\ln_{q_{\mathrm{II}}}(\Omega_{\mathrm{II}}(E-\varepsilon_k)) - \ln_{q_{\mathrm{II}}}(\Omega_{\mathrm{II}}(E-\varepsilon_l)) \right] \right). \tag{2.5.19}$$

Expanding the factors inside the q-exponential with respect to ε_k and ε_l and keeping the leading order terms, we find [24]

$$\pi(\varepsilon_k; \varepsilon_l) = e_{q_{\mathrm{II}}}(-\beta^*(\varepsilon_k - \varepsilon_l)). \tag{2.5.20}$$

We notice that, to derive the relative probability of this form, we had to resort to the first principle calculation shown in Eq. (2.5.19) and not to simply substitute the expression given in Eq. (2.5.16) into the first part in Eq. (2.5.18). This may be regarded as a feature of the nonadditive structure of the Tsallis entropy.

2.6 Generalized Boltzmann equation

In this section, we discuss a kinetic-theoretical foundation of nonextensive statistical mechanics developed in [26]. In particular, the celebrated hypothesis of molecular chaos, i.e., *Stosszahlansatz*, and accordingly the Boltzmann equation will appropriately be generalized. We shall see that the Tsallis-type functional satisfies the generalized H-theorem and the stationary solution to the generalized Boltzmann equation is given by the q-exponential distribution.

The number of particles whose positions and velocities are found in the intervals $\mathbf{r} \sim \mathbf{r}+d\mathbf{r}$ and $\mathbf{v} \sim \mathbf{v}+d\mathbf{v}$ at time t is denoted by $f(\mathbf{r},\mathbf{v},t)d^3\mathbf{r}d^3\mathbf{v}$. The total number of particles, N, is given by the integral, $N = \int d^3\mathbf{r}d^3\mathbf{v} f(\mathbf{r},\mathbf{v},t)$. The distribution, $f(\mathbf{r},\mathbf{v},t)$, satisfies the equation of the form

$$\frac{\partial f}{\partial t} + \mathbf{v} \cdot \frac{\partial f}{\partial \mathbf{r}} + \frac{\mathbf{F}}{m} \cdot \frac{\partial f}{\partial \mathbf{v}} = C(f), \tag{2.6.1}$$

where, $\mathbf{F}$ and m are a force assumed to be independent of $\mathbf{v}$ and the mass of the particle, respectively, and the right-hand side represents the collision term. As in the ordinary discussion, we consider only two-body collisions (consistent with symmetries and conservation laws): $(\mathbf{v},\mathbf{v}_1) \to (\mathbf{v}',\mathbf{v}_1')$ with any $\mathbf{v}_1$. Let us write $C(f)$ as follows:

$$C(f) = \int d\omega d^3\mathbf{v}_1 \sigma V_r R(f, f_1; f', f_1'), \tag{2.6.2}$$

where $V_r = |\mathbf{v} - \mathbf{v}_1|$ is the magnitude of the relative velocity before collision, σ the scattering cross section, and ω the solid angle familiar in geometry of collision kinematics. $R(f, f_1; f', f_1')$ describes the correlation difference in the system before and after collision in terms of the distributions, $f(\mathbf{r},\mathbf{v},t)$, $f_1 = f(\mathbf{r},\mathbf{v}_1,t)$, $f' = f(\mathbf{r},\mathbf{v}',t)$, and $f_1' = f(\mathbf{r},\mathbf{v}_1',t)$. Nontrivial physics is contained in this quantity. In the ordinary Stosszahlansatz, two colliding particles have no correlation

$$R(f, f; f', f_1') = f' f_1' - f f_1. \tag{2.6.3}$$

In a nonextensive system, however, the colliding particles are always strongly correlated, and therefore factorization in the each term in Eq. (2.6.3) is not realized. Thus, the following specific Stosszahlansatz may be examined [26]:

$$\begin{aligned} R_q(f, f_1; f', f_1') = e_q[(f')^{q-1} \ln_q(f') + (f_1')^{q-1} \ln_q(f_1')] - \\ - e_q[(f)^{q-1} \ln_q(f) + (f_1)^{q-1} \ln_q(f_1)], \quad (2.6.4) \end{aligned}$$

which describes complex correlation between colliding particles. In the limit $q \to 1$, this quantity tends back to that in Eq. (2.6.3) with vanishing correlation.

Now, let us consider the Tsallis-type H-function

$$H_q(\mathbf{r}, t) = \int d^3\mathbf{v}[f(\mathbf{r}, \mathbf{v}, t)]^q \ln_q[f(\mathbf{r}, \mathbf{v}, t)], \tag{2.6.5}$$

which becomes the Boltzmann H-function, $H(\mathbf{r}, t) = \int d^3\mathbf{v} f(\mathbf{r}, \mathbf{v}, t) \ln f(\mathbf{r}, \mathbf{v}, t)$, in the limit $q \to 1$. Taking the time derivative of H_q and using Eqs. (2.6.1) and (2.6.2) with R_q in Eq. (2.6.4) for R, we have

$$\frac{\partial H_q}{\partial t} = \int d^3\mathbf{v}[1 + q(f)^{q-1} \ln_q(f)] \left(C - \mathbf{v} \cdot \frac{\partial f}{\partial \mathbf{r}} - \frac{\mathbf{F}}{m} \cdot \frac{\partial f}{\partial \mathbf{v}} \right). \tag{2.6.6}$$

Assuming that the distribution vanishes at $|\mathbf{v}| \to \infty$, this expression can be recast in the following form:

$$\frac{\partial H_q}{\partial t} + \nabla \cdot \mathbf{j}_q = G_q, \tag{2.6.7}$$

where the current, $\mathbf{j}_q$, and the source, G_q, are given by

$$\mathbf{j}_q(\mathbf{r}, t) = \int d^3\mathbf{v}\mathbf{v}(f)^q \ln_q(f), \tag{2.6.8a}$$

$$\begin{aligned} G_q(\mathbf{r},\ t) = \int d\omega d^3\mathbf{v} d^3\mathbf{v}_1 \sigma V_r [1 + q(f)^{q-1} \ln_q(f)] \\ R_q(f, f_1; f', f_1'), \quad (2.6.8b) \end{aligned}$$

respectively. To evaluate G_q, we employ a symmetry consideration. First of all, G_q should be invariant under the interchange, $\mathbf{v} \leftrightarrow \mathbf{v}_1$, as the cross section is. In addition, $d^3\mathbf{v} d^3\mathbf{v}_1 = d^3\mathbf{v}' d^3\mathbf{v}_1'$ holds. Therefore, substituting Eq. (2.6.4) into Eq. (2.6.8b) and utilizing these symmetries, we find G_q to be

$$G_q(\mathbf{r}, t) = -\frac{q}{4} \int d\omega d^3\mathbf{v} d^3\mathbf{v}_1 \sigma V_r [\ln_{q*}(f') + \ln_{q*}(f_1') - \ln_{q*}(f) - \ln_{q*}(f_1)]$$

$$\{e_q[\ln_{q*}(f') + \ln_{q*}(f'_1)] - e_q[\ln_{q*}(f) + \ln_{q*}(f_1)]\}, \quad (2.6.9)$$

where we have set $(x)^{q-1}\ln_q(x) = \ln_{q*}(x)$ with $q* = 2 - q$. The factors, $\ln_{q*}(f') + \ln_{q*}(f'_1) - \ln_{q*}(f) - \ln_{q*}(f_1)$ and $e_q[\ln_{q*}(f') + \ln_{q*}(f'_1)] - e_q[\ln_{q*}(f) + \ln_{q*}(f_1)]$, have the same sign since the q-exponential function is monotonic. So, G_q in Eq. (2.6.9) is never positive. Thus we have

$$\frac{\partial H_q}{\partial t} + \nabla \cdot \mathbf{j}_q = G_q \leq 0, \quad (2.6.10)$$

which generalizes the ordinary H-theorem.

Finally, let us look at the stationary state. In this case, $G_q = 0$, which implies that

$$\ln_{q*}(f') + \ln_{q*}(f'_1) = \ln_{q*}(f) + \ln_{q*}(f_1), \quad (2.6.11)$$

which can be regarded as a generalization of the detailed balance condition. Eq. (2.6.11) describes the additive invariance of the $q*$-logarithmic quantities. Kinematically, the additive invariants are the total mass, energy and momentum. This fact allows us to write that $\ln_{q*}(f) = a_0 + \mathbf{a}_1 \cdot \mathbf{v} + a_2\mathbf{v}^2$, or, equivalently, $\ln_{q*}(f) = b_0 - b_1(\mathbf{v} - \mathbf{v}_0)^2$, where a_0, a_2, b_0, and b_1 are constants, and $\mathbf{a}_1$ and $\mathbf{v}_0$ are constant vectors. Consequently, we obtain the generalized Maxwellian distribution

$$f = Ae_{q*}[-\beta^*(\mathbf{v} - \mathbf{v}_0)^2], \quad (2.6.12)$$

where $A = [1 + (1 - q*)b_0]_+^{1/(1-q*)}$ and $\beta^* = b_1/[1 + (1 - q*)b_1]$.

Thus, we see that the stationary solution to the generalized Boltzmann equation is given by the distribution in nonextensive statistical mechanics in the homogeneous limit.

2.7 Concluding remarks

In this particle, we have surveyed the foundations of nonextensive statistical mechanics. In particular, we have reviewed both statistical and kinetic ones, which reveal remarkable parallelisms between Boltzmann-Gibbs statistical mechanics and nonextensive statistical mechanics. However, one should always notice that the latter is essentially concerned with complex systems with broken ergodicity (e.g., at the edge of chaos of nonlinear dynamical systems) and, thus, with nontrivial phase space structures (e.g., multifractals), but in their long-persisting nonequilibrium stationary states.

It is our opinion that Tsallis' form of a generalized entropy and associated nonextensive theory may perhaps be one among many for which we were able to provide generalizations of the foundations and yet preserving all the tenets of statistical mechanics. It seems that Nature has diverse complexities, which certainly need modifications of Boltzmann-Gibbs statistical mechanics for simple systems. Therefore, an important problem to be addressed is to classify the complexities and identify universality classes. We confidently believe that further investigations along this line will significantly extend the horizon of statistical mechanics.

Acknowledgment

Significant parts of the works reviewed here have been done in collaboration with A. K. Rajagopal, whom the present author would like to thank for a lot of enlightening discussions.

References

[1] C. Anteneodo and C. Tsallis, Phys. Rev. Lett. **80**, 5313 (1988).
[2] C. Anteneodo and R. O. Vallejos, Phys. Rev. E **65**, 016210 (2001).
[3] S. Abe and Y. Okamoto eds., *Nonextensive Statistical Mechanics and Its Applications* (Springer-Verlag, Heidelberg, 2001).
[4] V. Latora, A. Rapisarda, and C. Tsallis, Phys. Rev. E **64**, 056134 (2001).
[5] V. Latora, A. Rapisarda, and C. Tsallis, Physica A **305**, 129 (2002).
[6] A. Pluchino, V. Latora, and A. Rapisarda, Physica A **338**, 60 (2004).
[7] F. Baldovin, L. G. Moyano, A. P. Majtey, A. Robledo, C. Tsallis, Physica A **340**, 205 (2004).
[8] A. Pluchino, G. Andronico, and A. Rapisarda, Physica A **349**, 143 (2005).
[9] Special issue of Physica A **305** (2002).
[10] M. Gell-Mann and C. Tsallis eds., *Nonextensive Entropy: Interdisciplinary Applications* (Oxford University Press, Oxford, 2004).
[11] Special issue of Physica A **340** (2004).
[12] Topical issue of Continuum Mech. Thermodyn. **16** (2004).
[13] C. Tsallis, J. Stat. Phys. **52**, 479 (1988)
[14] C. Beck and F. Schlögl, *Thermodynamics of Chaotic Systems: An Introduction* (Cambridge University Press, Cambridge, 1993).
[15] S. Abe and G. B. Bagci, Phys. Rev. E **71**, 016139 (2005).
[16] C. Tsallis, R. S. Mendes, and A. R. Plastino, Physica A **261**, 534 (1998).
[17] S. Abe and A. K. Rajagopal, Phys. Rev. Lett. **91**, 120601 (2003).
[18] L. G. Moyano, F. Baldovin, and C. Tsallis, e-print cond-mat/0305091.
[19] S. Abe, e-print cond-mat/0504036, to appear in Physica A.
[20] R. Balian and N. L. Balazs, Ann. Phys. (NY) **179**, 97 (1987).
[21] R. Kubo, H. Ichimura, T. Usui, and N. Hashitsume, *Statistical Mechanics* (North-Holland, Amsterdam, 1988).
[22] S. Abe and A. K. Rajagopal, J. Phys. A **33**, 8733 (2000).
[23] S. Abe and A. K. Rajagopal, Phys. Lett. A **272**, 341 (2000).
[24] S. Abe and A. K. Rajagopal, Europhys. Lett. **55**, 6 (2001).
[25] B. V. Gnedenko and A. N. Kolmogorov, *Limit Distributions for Sums of Independent Random Variables* (Addison-Wesley, Massachusetts, 1954).
[26] J. A. S. Lima, R. Silva, and A. R. Plastino, Phys. Rev. Lett. **86**, 2938 (2001).

3

Critical Attractors and the Physical Realm of q-statistics

A. Robledo

Instituto de Física, Universidad Nacional Autónoma de México, Apartado Postal 20-364, México 01000 D.F., Mexico.
E-mail: robledo@fisica.unam.mx

Summary. Here we give an account of recent understanding on the dynamics at critical attractors of simple one-dimensional nonlinear maps. This dynamics is relevant to a discussion about the applicability of the Tsallis generalization of canonical statistical mechanics. The critical attractors considered are those at the familiar pitchfork and tangent bifurcations and the period-doubling onset of chaos in unimodal maps of general nonlinearity $\zeta > 1$. The nonexponential sensitivity to initial conditions ξ_t and the related spectra of q-generalized Lyapunov coefficients λ_q have been determined with the use of known properties of the fixed-point maps under Feigenbaum's renormalization group (RG) transformation. We have found an equality between the λ_q and the corresponding rates of entropy production K_q holds at the critical attractors provided the rate K_q is obtained from the q-entropy S_q. We identify the Mori singularities in the Lyapunov coarse-grained function $\lambda(\mathsf{q})$ at the onset of chaos with the appearance of special values for the entropic index q. The physical area of the q-statistics is further probed by considering the dynamics of critical fluctuations and of glass formation in thermal systems. In both cases a close connection is made with critical attractors in unimodal maps.

3.1 Outline

The study of the singular dynamical properties at critical attractors is important because it provides insights into the limits of validity of the canonical or Boltzmann-Gibbs (BG) statistical mechanics and helps inspect the form of its possible generalization when phase space mixing and ergodicity break down. At one-dimensional critical attractors of nonlinear maps the ordinary Lyapunov coefficient λ_1 vanishes and the sensitivity to initial conditions ξ_t for large iteration time t ceases to obey exponential behavior, exhibiting instead power-law or faster than exponential behavior. As it is generally understood, the standard exponential divergence of trajectories in chaotic attractors provides a mechanism to justify the assumption of irreversibility in the BG statistical mechanics [1]. In contrast, the onset of chaos in (necessarily dissipative) unimodal maps, the prototypical critical attractor, imprints memory

A. Robledo: *Critical Attractors and the Physical Realm of q-statistics*, StudFuzz **206**, 72–113 (2006)
www.springerlink.com

preserving, nonmixing, phase space properties to its trajectories, and we consider them here with a view to assess a recent generalization of the usual BG statistics.

This generalization, that we refer to as q-statistics, is the statistical-mechanical framework based on the Tsallis entropy S_q [2], [3]. We review the rigorous evidence that has accumulated [4]-[9] for its suitability in describing the dynamical properties of critical attractors in unimodal maps of general nonlinearity $\zeta > 1$, such as the pitchfork and tangent bifurcations and the onset of chaos associated to these. We also describe connections between the properties of these attractors and those of systems with many degrees of freedom at extremal or transitional states. Two specific suggestions have been recently developed. In one case the dynamics at the tangent bifurcation has been shown to be related to that of intermittent clusters at thermal critical states [10]. In the second case the dynamics at the noise-perturbed period-doubling onset of chaos has been demonstrated to be closely analogous to the glassy dynamics observed in supercooled molecular liquids [11].

At critical attractors qualitative changes in the behavior of a dynamical system take place with variation in a control parameter [12] - [14]. Perhaps the most interesting types of critical attractors are strange nonchaotic attractors (SNAs) [15]. These are geometrically involved (multifractal) sets with trajectories on them that typically exhibit a nonexponential ξ_t. Important examples are the onset of chaos via period doubling, intermittence and quasi-periodicity, the three universal routes to chaos exhibited by the prototypical logistic and circle maps [12] - [14]. Other critical attractors occur at transitions between periodic orbits such as the so-called pitchfork bifurcations in the period-doubling cascades [12] - [14]. On the other hand, crises, qualitative changes in chaotic attractors that suddenly expand or disappear [16], are not critical attractors as these display positive (leading) Lyapunov coefficients and do not appear to entail absence of phase-space mixing properties.

The focal point of the q-statistical description for the dynamics of critical attractors is a sensitivity to initial conditions ξ_t associated to the q-exponential functional form, i.e. the 'q-deformed' exponential function $\exp_q(x) \equiv [1-(q-1)x]^{-1/(q-1)}$. From such ξ_t one or several spectra of q-generalized Lyapunov coefficients λ_q can be determined. The λ_q are dependent on the initial position x_0 and each spectrum can be examined by varying this position. The λ_q satisfy a q-generalized identity $\lambda_q = K_q$ of the Pesin type [14] where K_q is an entropy production rate based on S_q, defined in terms of the q-logarithmic function $\ln_q y \equiv (y^{1-q} - 1)/(1-q)$, the inverse of $\exp_q(x)$.

The allowed values of the entropic index q are obtained from the universality class parameters to which the attractor belongs. For the simpler pitchfork and tangent bifurcations there is a single well-defined value for the index q for each type of attractor [4], [5]. For SNAs the situation is more complicated and there appear to be a multiplicity of indexes q but with precise values given by the attractor scaling functions. They appear in pairs and are related to the occurrence of couples of conjugate dynamical 'q-phase' transitions [9], [17],

[18] and these are identified as the source of the special values for the entropic index q. The q-phase transitions connect qualitatively different regions of the attractor. For the case of the Feigenbaum attractor at the period-doubling onset of chaos an infinite family of such transitions take place but of rapidly decreasing strength.

In all cases any small change in the control parameter μ of the map leads to a crossover from q-statistics to BG statistics. An infinitesimal shift in the value of this parameter makes the attractor periodic or chaotic, its sensitivity becomes a decreasing or increasing exponential, and the value of the entropic indexes q all become unity. As it is known, for chaotic attractors the ordinary identity $\lambda_1 = K_1 > 0$ holds, where the rate K_1 is based on the canonical BG entropy expression S_1. However, for small shifts of μ and sufficiently short times t the dynamics is still given by the q-statistical description. Part I of this review presents details of these developments.

The manifestation of q-statistics in systems with many degrees of freedom can be explored via connections that have been established between the dynamics of critical attractors and the dynamics taking place in, for example, thermal systems under conditions when mixing and ergodic properties are not easily fulfilled. A remarkable relationship between intermittency and critical phenomena has been recently suggested [19] [20]. This development brings together fields of research in nonlinear dynamics and condensed matter physics, specifically, the dynamics in the proximity of a critical attractor appears associated to the dynamics of fluctuations of an equilibrium state with well-known scaling properties [21]. We examine this connection in some detail with special attention to several unorthodox properties, such as, the extensivity of the Tsallis entropy S_q of fractal clusters of order parameter, and the anomalous - faster than exponential - sensitivity to initial conditions, together with aging scaling features, in their time evolution.

As a second example we describe our finding [11] that the dynamics at the noise-perturbed onset of chaos in unimodal maps is analogous to that observed in supercooled liquids close to vitrification. We demonstrate that four major features of glassy dynamics in structural glass formers are displayed by orbits with vanishing Lyapunov coefficient. These are: two-step relaxation, a relationship between relaxation time and configurational entropy, aging scaling properties, and evolution from diffusive to subdiffusive behavior and finally arrest. The known properties in control-parameter space of the noise-induced bifurcation gap in the period-doubling cascade [12] play a central role in determining the characteristics of dynamical relaxation at the chaos threshold. These two applications of the dynamical properties of critical attractors are presented in Part II.

3.2 Part I. Anomalous dynamics at onset of chaos and other critical attractors in unimodal maps

3.2.1 Two routes to chaos in unimodal maps

A unimodal map (a one-dimensional map with one extremum) contains infinite families of critical attractors at which the ergodic and mixing properties breakdown [22]. These are the tangent (or saddle-node) bifurcations that give rise to windows of periodic trajectories within chaotic bands and the accumulation point(s) of the pitchfork bifurcations, the so-called period-doubling onset of chaos [12] - [14] at which these periodic windows come to an end. There are other attractors for which the Lyapunov coefficient λ_1 diverges to minus infinity, where there is faster than exponential convergence of orbits. These are the superstable attractors located between successive pitchfork bifurcations. They are present at the initial period doubling cascade and at all the other cascades within periodic windows, whose accumulation points are replicas of the Feigenbaum attractor. See Figs. 3.1 and 3.2.

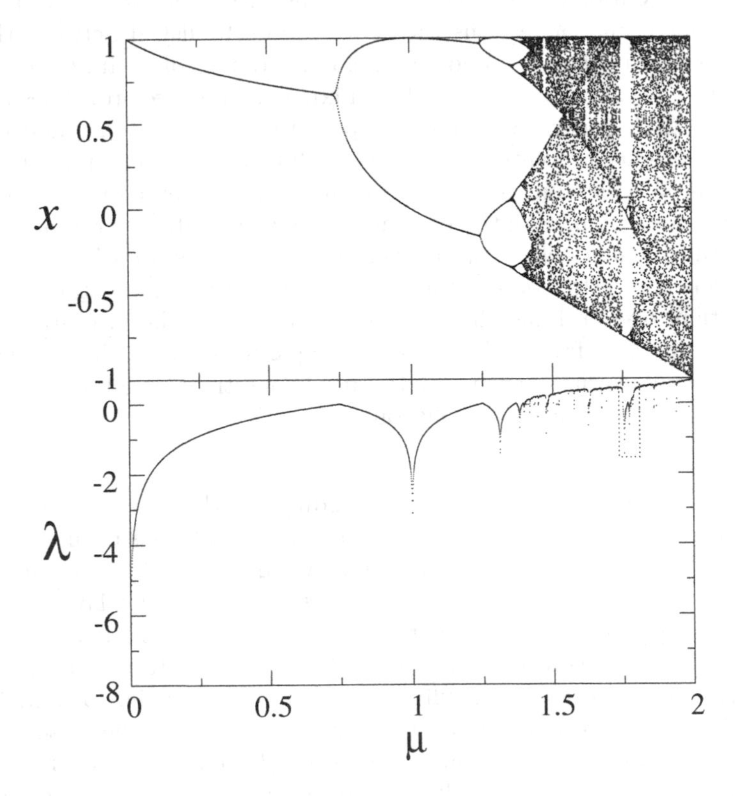

Fig. 3.1. Logistic map attractor and its Lyapunov coefficient as function of control parameter μ.

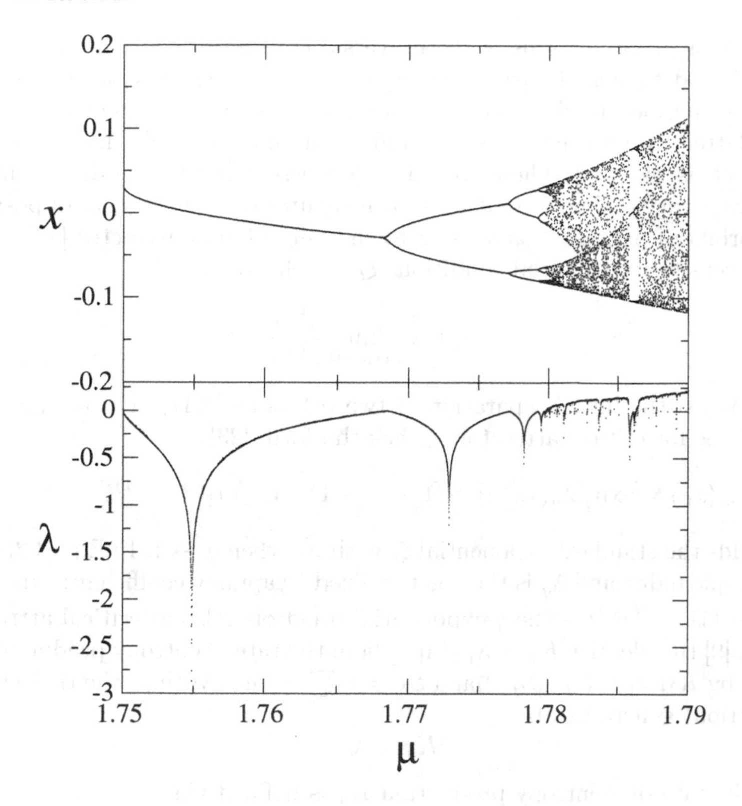

Fig. 3.2. Enlargement of the boxes shown in Fig.3.1 within the window of periodicity three.

The properties of the critical attractors are universal in the renormalization group (RG) sense, that is, all maps $f(x)$ that lead to the same fixed-point map $f^*(x)$ under a repeated functional composition and rescaling transformation share the same scaling properties. For unimodal maps this transformation takes the form $Rf(x) \equiv \alpha f(f(x/\alpha))$, where α assumes a fixed value (positive or negative real number) for each universality class and $f^*(x)$ is given by

$$f^*(x) \equiv \lim_{n\to\infty} R^{(n)} f(x) = \lim_{n\to\infty} \alpha^n f^{(2^n)}(x/\alpha^n), \qquad (3.2.1)$$

and satisfies

$$f^*(x) = \alpha f^*(f^*(x/\alpha)). \qquad (3.2.2)$$

The universality of the static or geometrical properties of critical attractors is understood since long ago [12] - [14]. This is represented, for example, by the generalized dimensions D_q or the spectrum $f(\widetilde{\alpha})$ that characterize the multifractal attractor at the period-doubling onset of chaos [13], [14]. The dynamical properties of critical attractors also display universality, as we see below, the entropic index q in the sensitivity ξ_t and the Lyapunov spectra λ_q is given in terms of the universal constant α. For the cases of the pitchfork and tangent bifurcations the results are relatively straightforward but for the period-doubling accumulation point the situation is more complex. In the latter case an infinite set of universal constants, of which α is most prominent, is required. These constants are associated to the discontinuities in the trajectory scaling function σ that measures the convergence of positions in the orbits of period 2^n as $n \to \infty$ to the Feigenbaum attractor [12].

The sensitivity to initial conditions ξ_t is defined as

$$\xi_t(x_0) \equiv \lim_{\Delta x_0 \to 0} \frac{\Delta x_t}{\Delta x_0} \tag{3.2.3}$$

where Δx_0 is the initial separation of two orbits and Δx_t that at time t. As we shall see for critical attractors ξ_t has the form [23]

$$\xi_t(x_0) = \exp_q[\lambda_q(x_0)\, t] \equiv [1 - (q-1)\lambda_q(x_0)\, t]^{-1/(q-1)}, \tag{3.2.4}$$

that yields the standard exponential ξ_t with λ_1 when $q \to 1$. In Eq. (3.2.4) q is the entropic index and λ_q is the q-generalized Lyapunov coefficient; $\exp_q(x) \equiv [1-(q-1)x]^{-1/(q-1)}$ is the q-exponential function. Also at critical attractors [23], [7], [9] the identity $K_1 = \lambda_1$ [14] (where the rate of entropy production K_1 is given by $K_1 t = S_1(t) - S_1(0)$and $S_1 = -\sum_i p_i \ln p_i$ with p_i the trajectories' distribution) generalizes to

$$K_q = \lambda_q, \tag{3.2.5}$$

where the rate of q-entropy production K_q is defined via

$$K_q t = S_q(t) - S_q(0), \tag{3.2.6}$$

and where

$$S_q \equiv \sum_i p_i \ln_q \left(\frac{1}{p_i}\right) = \frac{1 - \sum_i^W p_i^q}{q-1} \tag{3.2.7}$$

is the Tsallis entropy. (Recall that $\ln_q y \equiv (y^{1-q} - 1)/(1-q)$ is the inverse of $\exp_q(y)$).

We take as a starting point and framework for the study of fixed-point map properties the prototypical logistic map, or its generalization to non-linearity of order $\zeta > 1$,

$$f_\mu(x) = 1 - \mu\,|x|^\zeta, \qquad -1 \le x \le 1, \quad 0 \le \mu \le 2, \tag{3.2.8}$$

where x is the phase space variable, μ the control parameter, and $\zeta = 2$ corresponds to the familiar logistic map. Our results relate to the anomalous ξ_t and its associated spectrum λ_q for the above-mentioned critical attractors that are involved in the two routes to chaos exhibited by unimodal maps, the intermittency and the period doubling routes. For the Feigenbaum attractor we describe the relationship of ξ_t and λ_q with Mori's q-phase transitions [17], one of which was originally observed numerically [17], [18].

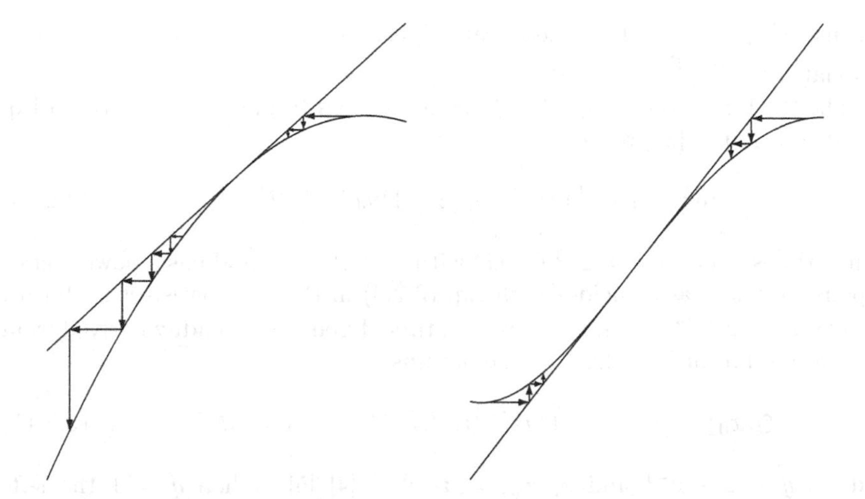

Fig. 3.3. Schematic form of $f^{(n)}$ at the tangent bifurcation(left) and pitchfork bifurcation (right).

3.2.2 Tangent and pitchfork bifurcations

The exact geometric or *static* solution of the RG Eq. (3.2.2) for the tangent bifurcations, known since long ago [12] - [14], has been shown to describe as well the *dynamics* of iterates at this attractor [4], [5]. Also, a straightforward extension of this approach was shown to apply to the pitchfork bifurcations [4], [5]. In Fig. 3.3 we show a sketch of the neighborhoods of maps at these bifurcations. We recall that the period-doubling and intermittency transitions are based on the pitchfork and the tangent bifurcations, respectively, and that at these critical attractors the ordinary Lyapunov coefficient $\lambda_1 = 0$. The sensitivity ξ_t can be determined analytically and its relation with the rate of entropy production examined [4]. The fixed-point expressions have the specific form that corresponds to the temporal evolution suggested by the q-statistics. Refs. [4], [5] contain the derivation of the q-Lyapunov coefficients λ_q and the description of the different possible types of sensitivity ξ_t. The pitchfork and the left-hand side of the tangent bifurcations display weak insensitivity to initial conditions, while the right-hand side of the tangent bifurcations presents a 'super-strong' (faster than exponential) sensitivity to initial conditions [5].

For the transition to periodicity of order n in the ζ-logistic map the composition $f_\mu^{(n)}$ is first considered. In the neighborhood of one of the n points tangent to the line with unit slope one obtains

$$f^{(n)}(x) = x + u\,|x|^z + o(|x|^z), \tag{3.2.9}$$

where u is the expansion coefficient. At the tangent bifurcations one has $z = 2$ and $u > 0$, whereas for the pitchfork bifurcations one has instead $z = 3$, because $d^2 f_\mu^{(2^k)}/dx^2 = 0$ at these transitions, and $u < 0$ is now the coefficient associated to $d^3 f_\mu^{(2^k)}/dx^3 < 0$.

The RG fixed-point map $x' = f^*(x)$ associated to maps of the form in Eq. (3.2.9) was found [24] to be

$$x' = x \exp_z(ux^{z-1}) = x[1-(z-1)ux^{z-1}]^{-1/(z-1)}, \tag{3.2.10}$$

as it satisfies $f^*(f^*(x)) = \alpha^{-1} f^*(\alpha x)$ with $\alpha = 2^{1/(z-1)}$ and has a power-series expansion in x that coincides with Eq. (3.2.9) in the two lowest-order terms. (Above $x^{z-1} \equiv |x|^{z-1}\,\mathrm{sgn}(x)$). The long time dynamics is readily derived from the static solution Eq. (3.2.10), one obtains

$$\xi_t(x_0) = [1-(z-1)ax_0^{z-1}t]^{-z/(z-1)}, \qquad u = at, \tag{3.2.11}$$

and so, $q = 2 - z^{-1}$ and $\lambda_q(x_0) = zax_0^{z-1}$ [4] [5]. When $q > 1$ the left-hand side ($x < 0$) of the tangent bifurcation map, Eq. (3.2.9), exhibits a weak insensitivity to initial conditions, i.e. power-law convergence of orbits. However at the right-hand side ($x > 0$) of the bifurcation the argument of the q-exponential becomes positive and this results in a 'super-strong' sensitivity to initial conditions, i.e. a sensitivity that grows faster than exponential [5]. For the tangent bifurcation one has $z = 2$ in $q = 2 - z^{-1}$ and so $q = 3/2$. For the pitchfork bifurcation one has $z = 3$ in $q = 2 - z^{-1}$ and one obtains $q = 5/3$. Notably, these specific results for the index q are valid for all $\zeta > 1$ and therefore define the existence of only two universality classes for unimodal maps, one for the tangent and the other one for the pitchfork bifurcations [5]. See Figs. 3.4 and 3.5.

There is an interesting scaling property displayed by ξ_t in Eq. (3.2.11) similar to the scaling property known as *aging* in systems close to glass formation. This property is observed in two-time functions (e.g. time correlations) for which there is no time translation invariance but scaling is observed in terms of a time ratio variable t/t_w where t_w is a 'waiting time' assigned to the time interval for preparation or hold of the system before time evolution is observed through time t. This property can be seen immediately in ξ_t if one assigns a waiting time t_w to the initial position x_0 as $t_w = x_0^{1-z}$. Eq. (3.2.11) reads now

$$\xi_{t,t_w} = [1-(z-1)at/t_w]^{-z/(z-1)}. \tag{3.2.12}$$

The sensitivity for this critical attractor is dependent on the initial position x_0 or, equivalently, on its waiting time t_w, the closer x_0 is to the point of tangency the longer t_w but the sensitivity of all trajectories fall on the same q-exponential curve when plotted against t/t_w. Aging has also been observed for the properties of the map in Eq.(9) but in a different context [27].

Notice that our treatment of the tangent bifurcation differs from other studies of intermittency transitions [28] in that there is no feed back mechanism of iterates into the origin of $f^{(n)}(x)$ or of its associated fixed-point map

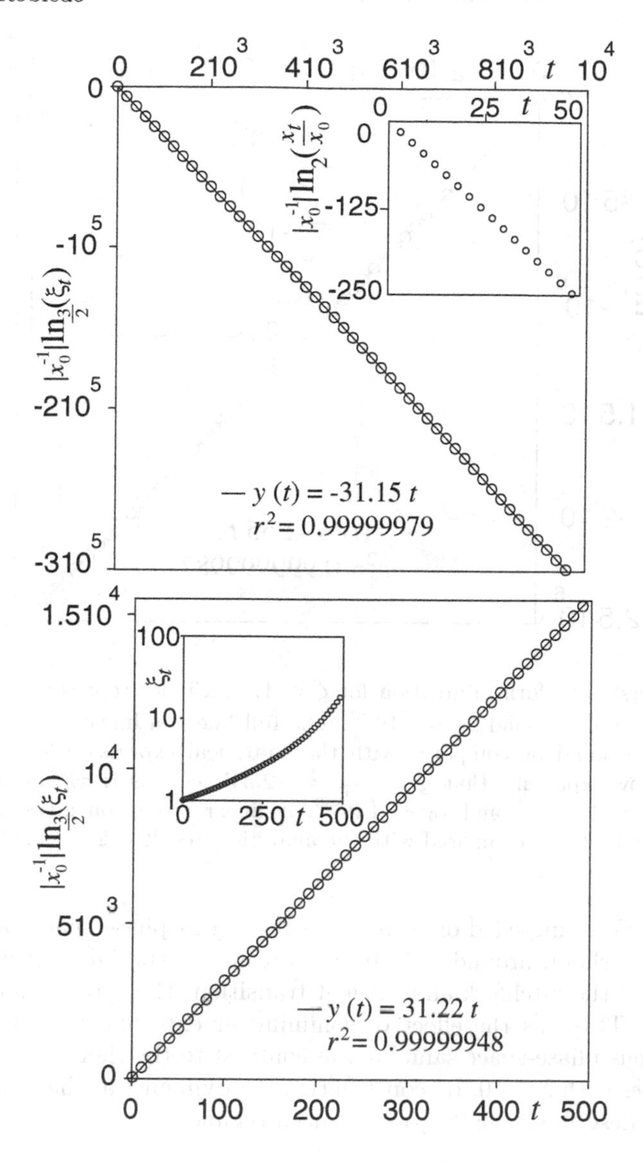

Fig. 3.4. Above is left side of the first tangent bifurcation for $\zeta = 2$. Circles represent $|x_0^{-1}| \ln_{3/2}(\xi_t)$ for the iterates of $f^{(3)}$ and $x_0 \sim -10^{-4}$. The full line is a linear regression, which slope, -31.15, should be compared with the exact expression for the generalized Lyapunov exponent, that gives $\lambda_q = 31.216 \cdots$. Inset: $|x_0^{-1}| \ln_2(x_t/x_0)$, for $t = 3m$, $(m = 1, 2, ...)$ and $x_0 \sim -10^{-6}$. A linear regression gives in this case a slope of -5.16, to be compared with the exact result $u/3 = 5.203 \cdots$. Below is right side of the first tangent bifurcation for $\zeta = 2$. Circles represent $|x_0^{-1}| \ln_{3/2}(\xi_t)$ for the iterates of $f^{(3)}$ and $x_0 \sim +10^{-4}$. The full line is a linear regression with slope equal to 31.22. Inset: Log-linear plot of ξ_t evidencing a super-exponential growth.

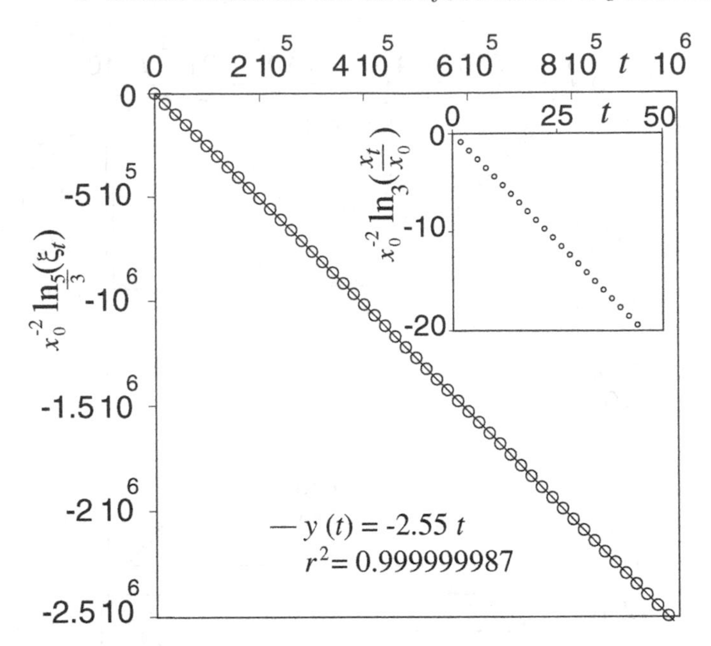

Fig. 3.5. First pitchfork bifurcation for $\zeta = 1.75$. Circles represent $x_0^{-2}\ln_{5/3}(\xi_t)$ for the iterates of $f^{(2)}$ and $x_0 \sim +10^{-3}$. The full line is a linear regression, which slope, -2.55, should be compared with the analytical expression for the generalized Lyapunov exponent, that gives $\lambda_q = -2.547\cdots$. Inset: $x_0^{-2}\ln_3(x_t/x_0)$, for $t = 2m$, $(m = 1, 2, \cdots)$ and $x_0 \sim +10^{-3}$. A linear regression gives in this case a slope of -0.44, to be compared with the analytical result $u/2 = -0.424\cdots$.

$f^*(x)$. Therefore, impeded or incomplete mixing in phase space (a small interval neighborhood around $x = 0$) arises from the special 'tangency' shape of the map at the pitchfork and tangent transitions that produces monotonic trajectories. This has the effect of confining or expelling trajectories causing anomalous phase-space sampling, in contrast to the thorough coverage in generic states with $\lambda_1 > 0$. By construction the dynamics at the intermittency transitions, describe a purely q-exponential regime.

3.2.3 Period-doubling accumulation point

The dynamics at the Feigenbaum attractor has been analyzed recently [6], [7]. For the ζ-logistic map this attractor is located at μ_c, the accumulation point of the control parameter values for the pitchfork bifurcations μ_n, $n = 1, 2, ...$, that is also that for the superstable orbits $\overline{\mu}_n$, $n = 1, 2, ...$ The same attractor reappears in multiples together with the precursor cascade of period-doubling bifurcations in the infinite number of windows of periodic trajectories that interpose the chaotic attractors beyond μ_c. The number of cascades within each window is equal to the period of the orbit that emerges at the tangent bifurcation at its opening. See Fig. 1. By taking as initial condition $x_0 = 0$ at μ_c, or equivalently $x_1 = 1$, it is found that the resulting orbit, a superstable orbit of period 2^∞, consists of trajectories made of intertwined power laws that asymptotically reproduce the entire period-doubling cascade that occurs for $\mu < \mu_c$. This orbit captures the properties of the superstable orbits that precedes it. Here again the Lyapunov coefficient λ_1 vanishes (although the attractor is also the limit of a sequence of supercycles with $\lambda_1 \to -\infty$) and in its place there appears a spectrum of q-Lyapunov coefficients $\lambda_q^{(k)}$. This spectrum was originally studied in Refs. [29], [17] and our interest has been to examine its properties in relation with the expressions of the Tsallis statistics. We found that the sensitivity to initial conditions has precisely the form of a set of interlaced q-exponentials, of which we determine the q-indexes and the associated $\lambda_q^{(k)}$. As mentioned, the appearance of a specific value for the q index (and actually also that for its conjugate value $Q = 2 - q$) turns out to be due to the occurrence of Mori's 'q-phase transitions' [17] between 'local attractor structures' at μ_c. Furthermore, it has also been shown [7], [9] that the dynamical and entropic properties at μ_c are naturally linked through the q-exponential and q-logarithmic expressions, respectively, for the sensitivity to initial conditions ξ_t and for the entropy S_q in the rate of entropy production $K_q^{(k)}$. We have corroborated analytically the equality $\lambda_q^{(k)} = K_q^{(k)}$. Our results support the validity of the q-generalized Pesin identity for critical attractors in low-dimensional maps.

More specifically, the absolute values for the positions x_τ of the trajectory with $x_{t=0} = 0$ at time-shifted $\tau = t + 1$ have a structure consisting of subsequences with a common power-law decay of the form $\tau^{-1/1-q}$ with

$$q = 1 - \frac{\ln 2}{(\zeta - 1)\ln \alpha(\zeta)}, \tag{3.2.13}$$

where $\alpha(\zeta)$ is the Feigenbaum universal constant for nonlinearity $\zeta > 1$ that measures the period-doubling amplification of iterate positions [6]. That is, the Feigenbaum attractor can be decomposed into position subsequences generated by the time subsequences $\tau = (2k + 1)2^n$, each obtained by proceeding through $n = 0, 1, 2, ...$ for a fixed value of $k = 0, 1, 2, ...$ See Fig. 3.6. The $k = 0$ subsequence can be written as $x_t = \exp_{2-q}(-\lambda_q^{(0)} t)$ with $\lambda_q^{(0)} = (\zeta - 1)\ln \alpha(\zeta)/\ln 2$. These properties follow from the use of $x_0 = 0$ in the scaling relation [6]

$$x_\tau \equiv \left| g^{(\tau)}(x_0) \right| = \tau^{-1/1-q} \left| g(\tau^{1/1-q} x_0) \right|, \tag{3.2.14}$$

where $g(x)$ is the Feigenbaum fixed-point map [12]-[14].

The sensitivity associated to trajectories with other starting points $x_0 \neq 0$ within the multifractal attractor (but located within either its most sparse or

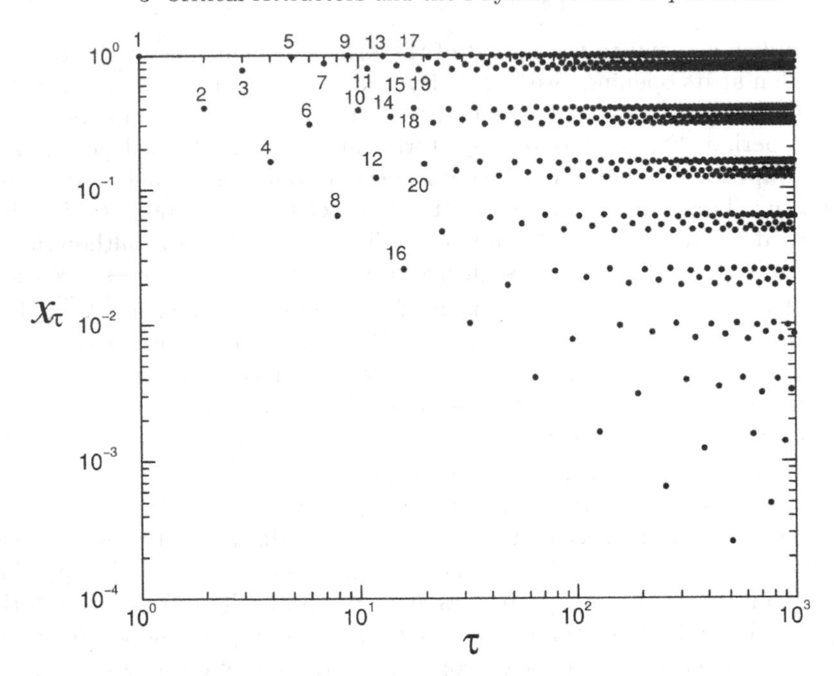

Fig. 3.6. Absolute values of positions in logarithmic scales of the first 1000 iterations τ for a trajectory of the logistic map at the onset of chaos $\mu_c(0)$ with initial condition $x_0 = 0$. The numbers correspond to iteration times. The power-law decay of the time subsequences described in the text can be clearly appreciated.

most crowded regions) can be determined similarly with the use of the time subsequences $\tau = (2k+1)2^n$. One obtains

$$\lambda_q^{(k)} = \frac{(\zeta - 1)\ln \alpha(\zeta)}{(2k+1)\ln 2} > 0, \qquad k = 0, 1, 2, ..., \tag{3.2.15}$$

for the positive branch of the Lyapunov spectrum, when the trajectories start at the most crowded ($x_{\tau=0} = 1$) and finish at the most sparse ($x_{\tau=2^n} = 0$) region of the attractor. By inverting the situation we obtain

$$\lambda_Q^{(k)} = -\frac{2(\zeta - 1)\ln \alpha(\zeta)}{(2k+1)\ln 2} < 0, \qquad k = 0, 1, 2, ..., \tag{3.2.16}$$

for the negative branch of $\lambda_q^{(k)}$, i.e. starting at the most sparse ($x_{\tau=0} = 0$) and finishing at the most crowded ($x_{\tau=2^n+1} = 1$) region of the attractor. Notice that $Q = 2 - q$ as $\exp_Q(y) = 1/\exp_q(-y)$. For the case $\zeta = 2$ see Refs. [6] and [7], for general $\zeta > 1$ see Refs. [8] and [9] where also different and more direct derivations are presented. So, when considering these two dominant families of orbits all the q-Lyapunov coefficients appear associated to only two specific values of the Tsallis index, q and $2-q$ with q given by Eq.

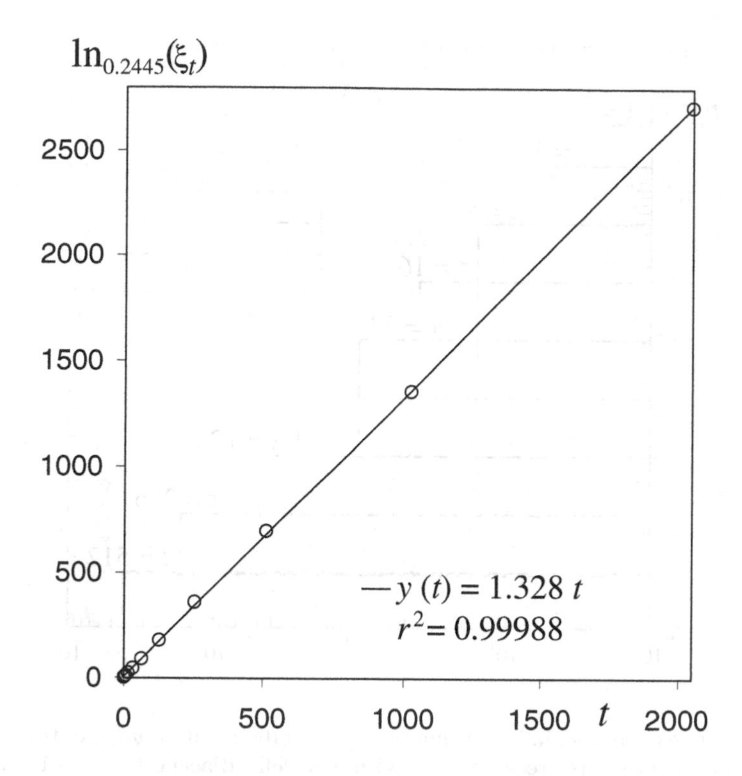

Fig. 3.7. The q-logarithm of sensitivity to initial conditions ξ_t vs t, with $q = 1 - \ln 2/\ln\alpha = 0.2445...$, and initial conditions $x_0 = 0$ and $x_0 = \delta \simeq 10^{-8}$ (circles). The full line is the linear regression $y(t)$. As required, the numerical results reproduce a straight line with a slope very close to $\lambda_q = \ln\alpha/\ln 2 = 1.3236...$.

(3.2.13). In Fig. 3.7 we show the q-logarithm of $\xi_t(x_0 = 1)$ vs t for the $k = 0$ time subsequence $\tau = 2^n$ when $\zeta = 2$ and $q = 1 - \ln 2/\ln\alpha(2) = 0.2445...$ and $\lambda_q^{(0)} = \ln\alpha(2)/\ln 2 = 1.3236...$

Ensembles of trajectories with starting points close to $x_{\tau=0} = 1$ expand in such a way that a uniform distribution of initial conditions remains uniform for all later times $t \leq T$ where T marks the crossover to an asymptotic regime. As a consequence of this the identity of the rate of entropy production $K_q^{(k)}$ with $\lambda_q^{(k)}$ was established [7]. See Figs. 3.8 and 3.9. A similar reasoning can be generalized to other starting positions [9].

Notably, the appearance of a specific value for the q index (and actually also that for its conjugate value $Q = 2 - q$) works out [9] to be due to the occurrence of Mori's 'q-phase transitions' [17] between 'local attractor structures' at μ_c. To see this in more detail we observe that the sensitivity $\xi_t(x_0)$ can be obtained [9] from

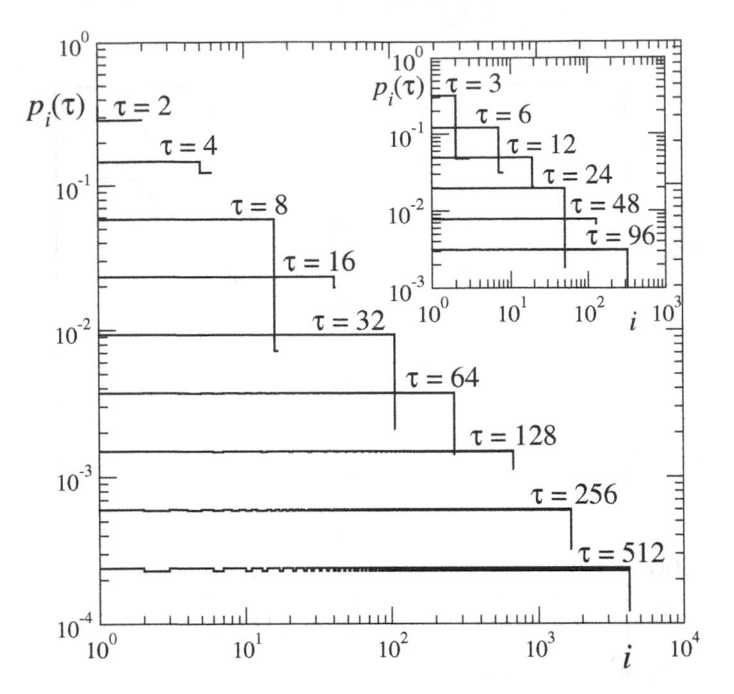

Fig. 3.8. Time evolution, in logarithmic scales, of a distribution $p_i(\tau)$ of trajectories at μ_∞. Initial positions are contained within a cell adjacent to $x = 1$ and i is the relative number of cells. Iteration time is shown for the first two subsequences ($k = 0, 1$).

$$\xi_t(m) \simeq \left| \frac{\sigma_n(m-1)}{\sigma_n(m)} \right|^n, \qquad t = 2^n - 1,\ n \tag{3.2.17}$$

where $\sigma_n(m) = d_{n+1,m}/d_{n,m}$ and where $d_{n,m}$ are the diameters that measure adjacent position distances that form the period-doubling cascade sequence [12]. Above, the choices $\Delta x_0 = d_{n,m}$ and $\Delta x_t = d_{n,m+t}$, $t = 2^n - 1$, have been made for the initial and the final separation of the trajectories, respectively. In the large n limit $\sigma_n(m)$ develops discontinuities at each rational of the form $m/2^{n+1}$ [12], and according to the expression above for $\xi_t(m)$ the sensitivity is determined by these discontinuities. For each discontinuity of $\sigma_n(m)$ the sensitivity can be written in the forms [9]

$$\xi_t = \exp_q[\lambda_q t], \qquad \lambda_q > 0 \tag{3.2.18}$$

and

$$\xi_t = \exp_{2-q}[\lambda_{2-q} t], \qquad \lambda_{2-q} < 0, \tag{3.2.19}$$

where q and the spectra λ_q and λ_{2-q} depend on the parameters that describe the discontinuity [9]. This result reflects the multi-region nature of the multifractal attractor and the memory retention of these regions in the dynamics.

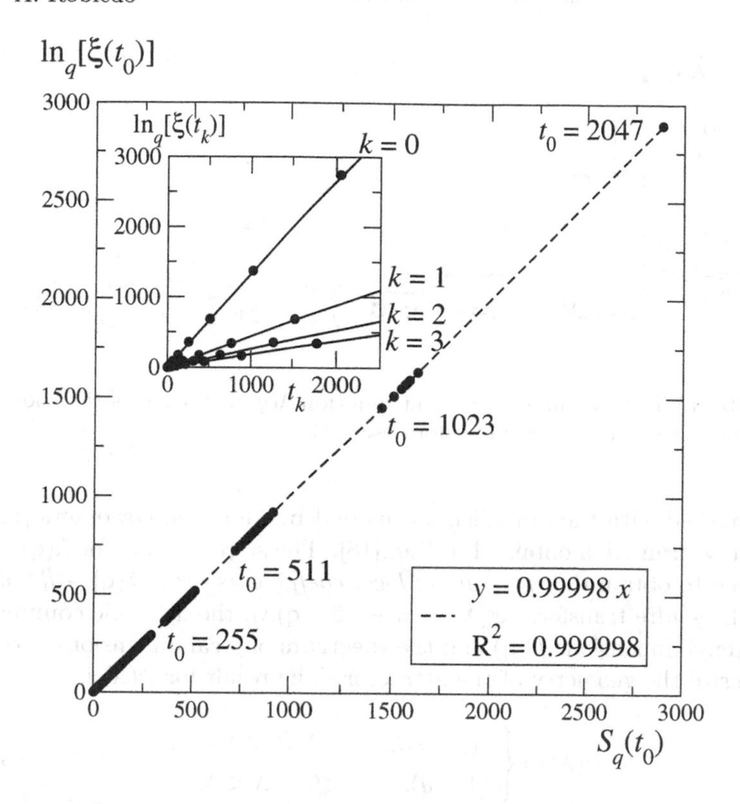

Fig. 3.9. Numerical corroboration (full circles) of the generalized Pesin identity $K_q^{(k)} = \lambda_q^{(k)}$ at μ_∞. On the vertical axis we plot the q-logarithm of ξ_{t_k} (equal to $\lambda_q^{(k)}t$) and in the horizontal axis S_q (equal to $K_q^{(k)}t$). In both cases $q = 1 - \ln 2/\ln\alpha = 0.2445....$ The dashed line is a linear fit. In the inset the full lines are from analytical results.

The pair of q-exponentials correspond to a departing position in one region and arrival at a different region and vice versa, the trajectories expand in one sense and contract in the other. The largest discontinuity of $\sigma_n(m)$ at $m = 0$ is associated to trajectories that start and finish at the most crowded ($x \simeq 1$) and the most sparse ($x \simeq 0$) regions of the attractor. In this case one obtains again Eq. (3.2.15), the positive branch of the Lyapunov spectrum, when the trajectories start at $x \simeq 1$ and finish at $x \simeq 0$. By inverting the situation one obtains Eq. (3.2.16), the negative branch of the Lyapunov spectrum. So, when considering these two dominant families of orbits all the q-Lyapunov coefficients appear associated to only two specific values of the Tsallis index, q and $Q = 2 - q$ with q given by Eq. (3.2.13).

As a function of the running variable $-\infty < \mathsf{q} < \infty$ the q-Lyapunov coefficients become a function $\lambda(\mathsf{q})$ with two steps located at $\mathsf{q} = q = 1 - \ln 2/(\zeta - 1)\ln\alpha(\zeta)$ and $\mathsf{q} = Q = 2 - q$. See Fig. 3.10. In this manner contact can

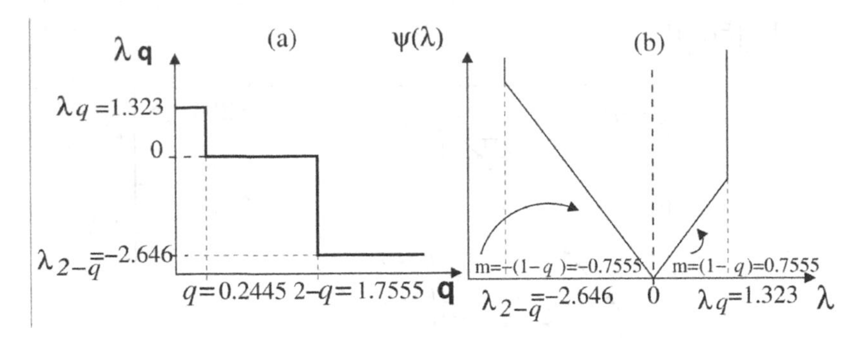

Fig. 3.10. a) The Lyapunov coefficient function $\lambda(q)$ at the chaos threshold at μ_c and b) the spectrum $\psi(\lambda)$. See text for description.

be established with the formalism developed by Mori and coworkers [17] and the q-phase transition obtained in Ref. [18]. The step function for $\lambda(\mathsf{q})$ can be integrated to obtain the *spectrum of local coefficients* $\phi(\mathsf{q})$ $(\lambda(\mathsf{q}) \equiv d\phi/d\lambda(\mathsf{q}))$ and its Legendre transform $\psi(\lambda)$ $(\equiv \phi - (1-\mathsf{q})\lambda)$, the dynamic counterparts of the Renyi dimensions $D(\mathsf{q})$ and the spectrum of local dimensions $f(\widetilde{\alpha})$ that characterize the geometry of the attractor. The result for $\psi(\lambda)$ is

$$\psi(\lambda) = \begin{cases} (1-Q)\lambda, & \lambda_Q^{(0)} < \lambda < 0, \\ (1-q)\lambda, & 0 < \lambda < \lambda_q^{(0)}. \end{cases} \tag{3.2.20}$$

As with ordinary thermal 1st order phase transitions, a q-phase transition is indicated by a section of linear slope $1-q$ in the spectrum (free energy) $\psi(\lambda)$, a discontinuity at q in the Lyapunov function (order parameter) $\lambda(\mathsf{q})$, and a divergence at q in the variance (susceptibility) $v(\mathsf{q})$. For the onset of chaos at $\mu_c(\zeta = 2)$ a q-phase transition was determined numerically [17], [18]. According to $\psi(\lambda)$ above we obtain a conjugate pair of q-phase transitions that correspond to trajectories linking two regions of the attractor, the most crowded and most sparse. See Fig. 3.10. Details appear in Ref. [9].

3.2.4 Discussion of Part I

The search and evaluation of the applicability of the q-statistics involves an examination of the domain of validity of the BG canonical formalism. The suggested physical circumstances for which BG statistics fails to be applicable are thought to be associated to situations that lack the full degree of chaotic irregular dynamics that probes phase space thoroughly, a requisite for true equilibrium. In nonlinear one-dimensional maps such anomalous circumstances are signaled by the vanishing of the Lyapunov coefficient and exhibit non-ergodicity or unusual power-law mixing. At the period-doubling onset of chaos the (long time) trajectories are confined to a multifractal subset of phase space with fractal dimension $d_f < 1$, and the trajectories are nonmixing. The dynamics at the intermittency transitions and at the Feigenbaum attractor describe a purely q-statistical regime, since the maps studied here do not consider access of trajectories to an adjacent or neighboring chaotic region, as in the setting of Refs. [28] or as in trajectories in conservative maps with weakly developed chaotic regions [12] - [14]. Hence there is no reappearance of trajectories from chaotic regions that would cause the relaxation from the q-statistical regime with vanishing ordinary Lyapunov exponent to a BG regime with a positive one at some crossover iteration time τ.

We have determined the dynamical behavior at the pitchfork and tangent bifurcations of unimodal maps of arbitrary nonlinearity $\zeta > 1$. This was accomplished via the consideration of the solution to the Feigenbaum RG recursion relation for these types of critical attractors. Our studies have made use of the specific form of the ζ-logistic map but the results have a universal validity as conveyed by the RG approach. The RG solutions are exact and have the analytical form of q-exponentials, we have shown that they are the time (iteration number) counterpart of the static fixed-point map expression found by Hu and Rudnick for the tangent bifurcations and that is applicable also to the pitchfork bifurcations [4]. The q-exponential form of the time evolution implies an analytical validation of the expression for ξ_t suggested by the q-statistical mechanics. It also provides straightforward predictions for q and λ_q in terms of the fixed-point map properties [4]. We found that the index q is independent of ζ and takes one of two possible values according to whether the transition is of the pitchfork or the tangent type. The generalized Lyapunov exponent λ_q is simply identified with the leading expansion coefficient u, together with the starting position x_0.

We have shown that for incipient chaotic states the identity between the Lyapunov coefficient and the rate of entropy change holds rigorously, although in a q-generalized form. Because the entropic index q (as is the case of λ_q and K_q) is obtainable in terms of Feigenbaum's α we are able to address the much-asked question regarding the manner in which the index q and related quantities are determined in a concrete application. The generic chaotic attractor is that associated to $\lambda_1 > 0$, but it is evident that the critical attractor with $\lambda_1 = 0$ carries with it different properties. The analysis was specifically carried out for the Feigenbaum attractor of the logistic map but our findings clearly have a universal validity for the entire class of unimodal maps and its generalization to other degrees of nonlinearity. In a more general context our results indicate a limit of validity to the BG theory based on S_1 and the appropriateness of the Tsallis S_q for this kind of critical dynamics.

Our most striking finding is that the dynamics at the period-doubling accumulation point is constituted by an infinite family of Mori's q-phase transitions, each associated to orbits that have common starting and finishing positions located at specific regions of the attractor. Each of these transitions is related to a discontinuity in the trajectory scaling function σ, or 'diameters ratio' function, and this in turn implies a q-exponential ξ_t and a spectrum of

q-Lyapunov coefficients for each set of orbits. The transitions come in pairs with specific conjugate indexes q and $Q = 2 - q$, as these correspond to switching starting and finishing orbital positions. Since the amplitude of the discontinuities in σ diminishes rapidly, in practical terms there is only need of evaluation for the first few of them. The dominant discontinuity is associated to the most crowded and sparse regions of the attractor and this alone provides a very reasonable description, as found in earlier studies [23], [30], [6], [7]. Thus, the special values for the Tsallis entropic index q in ξ_t are equal to the special values of the variable q in the formalism of Mori and colleagues at which the q-phase transitions take place.

The ergodic hypothesis lies at the foundation of statistical mechanics implying that trajectories in phase space cover thoroughly the entire pertinent regions. But is this hypothesis always correct? Already many years ago the answer to this question has been probed by the study of simple dynamical systems with only a few degrees of freedom [12] - [14]. Besides their uncomplicated definition these systems display extremely convoluted motion in phase space that is neither regular nor simply ergodic, and the mechanism by which ergodicity emerges in these and more complex deterministic systems has been effectively explored by studying the sensitivity to initial conditions and the associated Lyapunov coefficients [12] - [14]. The distinction between periodic and chaotic motion is signaled, respectively, by the long-time exponential approach or departure of trajectories with close initial positions. Here we have reviewed some properties of the borderline critical attractors in one-dimensional nonlinear maps at which the exponential sensitivity law stops working and recalled that the universal dynamical behavior under these circumstances actually follow the predictions of the *q-statistics*.

We have examined the dynamical properties at the Feigenbaum attractor of unimodal maps and obtained further understanding about their nature. We exhibited links between original developments, such as Feigenbaum's trajectory scaling function σ and Mori's dynamical q-phase transitions, with more recent advances, such as q-exponential sensitivity to initial conditions [6] and q-generalized Pesin identity [7].

3.3 Part II. Critical attractor dynamics in thermal systems at phase transitions and glass formation

3.3.1 Two manifestations of incipient chaos in thermal systems

It is of interest to know if the anomalous dynamics found for critical attractors in low-dimensional maps bears some correlation with the anomalous dynamical behavior at extremal or transitional states in systems with many degrees of freedom. Two specific examples have been recently developed. In one case the dynamics at the period-doubling onset of chaos has been demonstrated to be closely analogous to the glassy dynamics observed in supercooled molecular liquids [11]. In the second case the dynamics at the tangent bifurcation has been shown to be related to that at of fluctuations at thermal critical states [10].

We examine first the intermittent properties of clusters of order parameter at critical thermal states [10]. Our interest is on the local fluctuations of a system undergoing a second order phase transition, for example, in the Ising model as the magnetization fluctuates and generates magnetic domains on all size scales at its critical point. In particular, the object of study is a single cluster of order parameter ϕ at criticality. This is described by a coarse-grained free energy or effective action, like in the Landau-Ginzburg-Wilson (LGW) continuous spin model portrayal of the equilibrium configurations of Ising spins at the critical temperature and zero external field. At criticality the LGW free energy takes the form

$$\Psi_c[\phi] = a \int dr^d \left[\frac{1}{2} (\nabla \phi)^2 + b \left| \phi \right|^{\delta+1} \right], \tag{3.3.1}$$

Where a and b are constants, δ is the critical isotherm exponent and d is the spatial dimension ($\delta = 5$ for the $d = 3$ Ising model with short range interactions). As we describe below, a cluster of radius R is an unstable configuration whose amplitude in ϕ grows in time and eventually collapses when an instability is reached. This process has been shown [19], [20] to be reproduced by a nonlinear map with tangency and feedback characteristics, such that the time evolution of the cluster is given in the nonlinear system as a laminar event of intermittent dynamics.

The method employed to determine the cluster's order parameter profile $\phi(\mathbf{r})$ adopts the saddle-point approximation of the coarse-grained partition function Z, so that $\phi(\mathbf{r})$ is its dominant configuration and is determined by solving the corresponding Euler-Lagrange equation. The procedure is equivalent to the density functional approach for stationary states in equilibrium nonuniform fluids. Interesting properties have been derived from the thermal average of the solution found for $\phi(\mathbf{r})$, evaluated by integrating over its amplitude ϕ_0, the remaining degree of freedom after its size R has been fixed. These are the fractal dimension of the cluster [31] [32] and the intermittent behavior in its time evolution [19], [20]. Both types of properties are given in terms of the critical isotherm exponent δ.

As we describe below, the dominance of $\phi(\mathbf{r})$ in Z depends on a condition that can be expressed as an inequality between two lengths in space. This is $r_0 \gg R$, where r_0 is the location of a divergence in the expression for $\phi(\mathbf{r})$ that decreases as an inverse power of the cluster amplitude ϕ_0. When $r_0 \gg R$ the profile is almost horizontal but for $r_0 \gtrapprox R$ the profile increases from its center faster than an exponential. It is this feature that gives the cluster the properties that we present and discuss here. These properties relate to the dependence of the number of cluster configurations on size R, and the sensitivity to initial conditions ξ_t of order-parameter evolution on time t.

The above-mentioned properties appear to be at variance with the usual BG statistics but compatible [10] with the q-statistics [2], [3]. A condition for these properties to arise is criticality but also is the situation that phase space has only been partially represented by selecting only dominant configurations. Hence, the motivation to examine this problem rests on explaining the physical and methodological basis under which proposed generalizations of the BG statistics may apply.

A second example of a link between critical attractor and thermal system properties is the realization [11] that the dynamics at the noise-perturbed onset of chaos in unimodal maps is analogous to that observed in supercooled liquids close to vitrification. Four major features of glassy dynamics in structural glass formers, two-step relaxation, aging, a relationship between relaxation time and configurational entropy, and evolution from diffusive to subdiffusive behavior and finally arrest, are shown to be displayed by the properties of orbits with vanishing Lyapunov coefficient. The previously known properties in control-parameter space of the noise-induced bifurcation gap [12], [33] play a central role in determining the characteristics of dynamical relaxation at the chaos threshold.

In spite of the vast knowledge and understanding gathered together on the dynamics of glass formation in supercooled liquids this condensed matter phenomenon continues to attract interest [34]. This is so because there remain basic unanswered questions that are both intriguing and difficult [34]. A very pronounced slowing down of relaxation processes is the principal expression of the approach to the glass transition [34], [35], and this is generally interpreted as a progressively more imperfect realization of phase space mixing. Because of this extreme condition an important question is to find out whether under conditions of ergodicity and mixing breakdown the BG statistical mechanics remains capable of describing stationary states in the immediate vicinity of glass formation.

The basic ingredient of ergodicity failure is obtained for orbits in the limit towards vanishing noise amplitude. Our study supports the idea of a degree of universality underlying the phenomenon of vitrification, and points out that it is present in different classes of systems, including some with no explicit consideration of their molecular structure. The map has only one degree of freedom but the consideration of external noise could be taken to be the effect of many other systems coupled to it, like in the so-called coupled map lattices [36]. Our interest has been to study a system that is gradually forced into a nonergodic state by reducing its capacity to sample regions of its phase space that are space filling, up to a point at which it is only possible to move within a multifractal subset of this space. The logistic map with additive external noise reads,

$$x_{t+1} = f_\mu(x_t) = 1 - \mu x_t^2 + \sigma\chi_t, \qquad -1 \le x_t \le 1, \quad 0 \le \mu \le 2, \qquad (3.3.2)$$

where χ_t is a Gaussian-distributed random variable with average $\langle \chi_t \chi_{t'} \rangle = \delta_{t.t'}$, and σ measures the noise intensity [12], [33].

3.3.2 Critical clusters

The method considers a one dimensional system with unspecified range of interactions. The analysis can be carried out for higher dimensions with no further significant assumptions and with comparable results [19]-[32]. The LGW free energy reads now

$$\Psi_c[\phi] = a \int_0^R dx \left[\frac{1}{2} \left(\frac{d\phi}{dx} \right)^2 + b \left| \phi \right|^{\delta+1} \right], \tag{3.3.3}$$

and we assume the saddle-point approximation - valid for $a \gg 1$ - to get around the nontrivial task of carrying out the path integration in Z. The saddle-point configurations are obtained from the Euler-Lagrange equation

$$\frac{d^2\phi}{dx^2} = -\frac{dV}{d\phi}, \tag{3.3.4}$$

where $V = -b \left| \phi \right|^{\delta+1}$. Integration of Eq. (3.3.4) yields

$$U = \frac{1}{2} \left(\frac{d\phi}{dx} \right)^2 - b \left| \phi \right|^{\delta+1}. \tag{3.3.5}$$

Subsequent integration of Eq. (3.3.5) with $U = 0$ leads to profiles for critical clusters of the form [19]-[32]

$$\phi(x) = A \left| x - x_0 \right|^{-2/(\delta-1)}, \tag{3.3.6}$$

where

$$A = \left[\sqrt{b/2}\, (\delta - 1) \right]^{-2/(\delta-1)}$$

and

$$x_0 = \left[\sqrt{b/2}\, (\delta - 1) \right]^{-1} \phi_0^{-(\delta-1)/2}, \tag{3.3.7}$$

where x_0 is a system-dependent reference position and $\phi_0 = \phi(0)$. The value of ϕ at the edge of the cluster is $\phi_R = \phi(R)$ and the cluster free energy is

$$\Psi_c[\phi] = 2ab \int_0^R dx\ \phi(x)^{\delta+1}. \tag{3.3.8}$$

This family of solutions give the largest contributions to Z. Similar solutions are obtained for small $U \approx 0$ [31], [32] where now the position of the singularity x_0 depends also on U. These solutions enter Z with a weight $\exp(-\alpha R \left| U \right|)$ and therefore their relevance diminishes as $\left| U \right|$ increases.

Extensivity of critical cluster entropy

The profile $\phi(x)$ in Eq. (3.3.6), as well as those solutions for $U \simeq 0$, can be rewritten in the q-exponential form

$$\phi(x) = \phi_0 \exp_q(kx), \tag{3.3.9}$$

with $q = (1+\delta)/2$ and $k = \sqrt{2b}\phi_0^{(\delta-1)/2}$. Because $\delta > 1$ one has $q > 1$ and $\phi(x)$ grows faster than an exponential as $x \to x_0$ and diverges at x_0. See Fig. 3.11. It is important to notice that only configurations with $R \ll x_0$ have a nonvanishing contribution to the path integration in Z [31], [32] and that these configurations vanish for the infinite cluster size system. There are some characteristics of nonuniform convergence in relation to the limits $R \to \infty$ and $x_0 \to \infty$, a feature that is significant for our connection with q-statistics. By taking $\delta = 1$ the system is set out of criticality, then $q = 1$ and the profile $\phi(x)$ becomes the exponential $\phi(x) = \phi_0 \exp(k_0\, x)$, $k_0 = \sqrt{a_0 t}$.

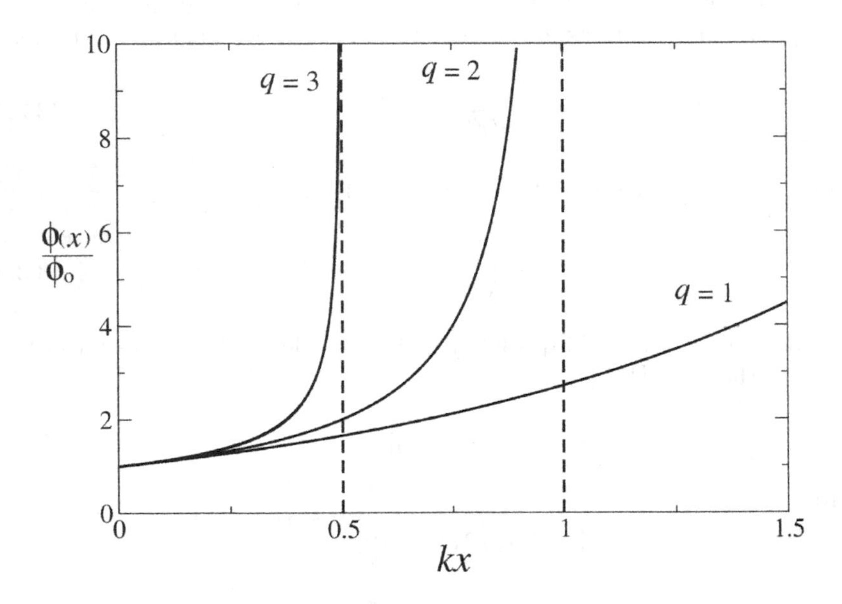

Fig. 3.11. Cluster order-parameter profile $\phi(x)$ according to Eq. (3.3.9) for $q = 1$, 2 and 3 that correspond, respectively, to $\delta = 1$, 3 and 5.

The quantity

$$\Phi(R) = \int_0^R dx \phi(x), \tag{3.3.10}$$

or total 'magnetization' of the cluster, is given by

$$\Phi(R) = \Phi_0 \left\{ \left[\exp_q(kR)\right]^{2-q} - 1 \right\}, \qquad R < x_0, \tag{3.3.11}$$

where $\Phi_0 = \mathrm{sgn}(3-\delta)[2\phi_0/(\delta-3)k]$ with $q = (1+\delta)/2$. See Fig. 3.12. We have not elaborated the special case $\delta = 3$. For $\delta = 1$, one has $\Phi(R) = \phi_0 k_0^{-1}[\exp(k_0 R) - 1]$. The rate at which $\Phi(R)$ grows with R, $d\Phi(R)/dR$, is necessarily equal to ϕ_R, the value of $\phi(x)$ at the edge of the cluster, therefore

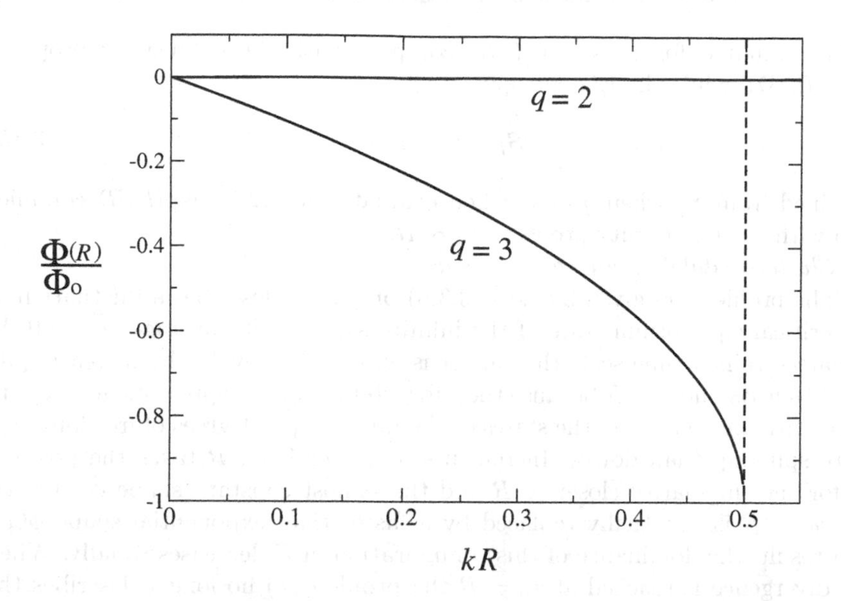

Fig. 3.12. Total 'magnetization' of the cluster $\Phi(R)$ as function of cluster size R according to Eq. (3.3.11) for $q = 2$ and 3, that correspond, respectively, to $\delta = 3$ and 5.

$$\frac{d\Phi(R)}{dR} = \phi_0 \exp_q(kR), \qquad R < x_0, \tag{3.3.12}$$

while for $\delta = 1$ it is $d\Phi(R)/dR = \phi_0 \exp(k_0 R)$.

The expressions above may be used to estimate the dependence on cluster size R of the number of microscopic configurations $\Omega[\phi]$ that make up the partial partition function Z_ϕ [10] for the dominant coarse-grained $\phi(x)$. This dependence may be obtained in a way analogous to that of how the dependence with time of the number of configurations Ω for an ensemble of trajectories in a one-dimensional dynamical system is determined. Here 'trajectory positions' are given by the values of ϕ in microscopic configurations and 'time' is given by the cluster size R. Initially adjacent positions stay adjacent and Ω is almost constant but at later times they spread and Ω increases rapidly. For chaotic orbits the increment is exponential [12] - [14] but for marginally chaotic orbits at the tangent bifurcation Ω increases as a q-exponential with $q > 1$ [4], [5]. The ensemble of trajectories is initially contained in the interval $[0, \phi_0]$ and at time R they occupy the interval $[0, \phi_R]$, therefore we assume [10]

$$\Omega(R) \sim \phi_0^{-1} d\Phi(R)/dR = \phi_0^{-1}\phi_R. \tag{3.3.13}$$

Then, it is significant to note that the Tsallis entropy [2],

$$S_q = \ln_q \Omega \equiv \frac{\Omega^{1-q} - 1}{1-q}, \tag{3.3.14}$$

when evaluated for $\Omega \sim \exp_q(kR)$ complies with the extensivity property $S_q \sim R$ [37], while the BG entropy

$$S_{BG}(t) = \ln \Omega, \tag{3.3.15}$$

obtained from S_q when $q = 1$, when evaluated for $\Omega \sim \exp(k_0 R)$ complies also with the extensivity property $S_1 \sim R$.

Cluster instability and intermittency

The profile $\phi(x)$ given by Eqs. (3.3.6) or (3.3.9) describes a fluctuation of the critical equilibrium state of the infinite system with average $\langle\overline{\phi}\rangle = 0$. In a coarse-grained time scale the cluster is expected to evolve by increasing its amplitude ϕ_0 and size R because the subsystem studied represents an environment with unevenness in the states of the microscopic degrees of freedom (e.g. more spins up than down). Increments in ϕ_0 for fixed R takes the position x_0 for the singularity closer to R and the almost constant shape $\phi(x) \simeq \phi_0$ for $x_0 \gg R$ is eventually replaced by a faster than exponential shape $\phi(x)$, as a result, the dominance of this configuration in Z decreases rapidly. When the divergence is reached at $x_0 = R$ the profile $\phi(x)$ no longer describes the spatial region where the subsystem is located. But a later fluctuation would again be represented by a cluster $\phi(x)$ of the same type. From this renewal process we obtain a picture of intermittency. A similar situation would occur if R is increased for fixed ϕ_0.

A connection was revealed [19], [20] between the fluctuation properties of a critical cluster described by Eqs. (3.3.6) and (3.3.9) and the dynamics of marginally chaotic intermittent maps. This relationship was demonstrated in various ways in Refs. [19], [20] by considering the properties of the thermal average

$$\langle \Phi(R) \rangle = Z^{-1} \int D[\phi]\ \Phi(R)\ \exp(-\Psi_c[\phi]), \tag{3.3.16}$$

for fixed R. When $x_0 \gg R$ the profile is basically flat $\phi(x) \simeq \phi_0$, the LGW free energy is $\Psi_c \simeq 2abR\phi_0^{\delta+1}$, and the path integral in Eq. (3.3.16) becomes an ordinary integral over $0 \leq \phi \leq \phi_0$. One obtains

$$\langle \Phi(R) \rangle \simeq \frac{\phi_0 R}{2} \exp\left(-uR\phi_0^{\delta+1}\right), \tag{3.3.17}$$

where $u = 2ab(\delta + 1)/(\delta + 2)(\delta + 3)$.

The procedure that resembles the intermittency picture mentioned above is to consider the value of $\langle \Phi \rangle$ at successive times $t = 0, 1, ...$, and assume that this quantity changes by a fixed amount η per unit time, that is

$$\langle \Phi_{t+1} \rangle = \langle \Phi_t \rangle + \eta. \tag{3.3.18}$$

Making use of Eq. (3.3.17) one obtains [19], [20] for small values of ϕ_t the map

$$\phi_{t+1} = \epsilon + \phi_t + \omega \phi_t^{\delta+1}, \tag{3.3.19}$$

where the shift parameter is $\epsilon \sim R^{-1}$ and the amplitude of the nonlinear term is $\omega = u\eta$.

Eq. (3.3.19) can be recognized as that describing the intermittency route to chaos in the vicinity of a tangent bifurcation [12] - [14]. The complete form of the map [19], [20] displays a 'superexponentially' decreasing region that takes back the iterate close to the origin in approximately one step. Thus the parameters of the thermal system determine the dynamics of the map. The mean number of iterations in the laminar region was seen to be related to the mean magnetization within a critical cluster of radius R. There is a corresponding power law dependence of the duration of the laminar region on the shift parameter ϵ of the map [19]. For $\epsilon > 0$ the (small) Lyapunov exponent is simply related to the critical exponent δ [20].

3.3.3 Vitrification

We quote briefly the main dynamical properties displayed by supercooled liquids on approach to glass formation. One is the growth of a plateau in two-time correlations e.g. the intermediate scattering function F_k [34], [35], and consequently a two-step process of relaxation. This consists of a primary power-law decay in time difference $t = t_2 - t_1$ (so-called β relaxation) that leads into the plateau, the duration t_x of which diverges also as a power law of the difference $T - T_g$ as the temperature T decreases to a critical value T_g. After t_x there is a secondary power law decay (so-called α relaxation) that leads to a conventional equilibrium state [34], [35]. This behavior is shown by molecular dynamics simulations [38] and it is successfully reproduced by mode coupling (MC) theory [39].

A second important dynamic property of glasses is the loss of time translation invariance, a characteristic known as aging [25] [26], that is due to the fact that properties of glasses depend on the procedure by which they are obtained. The time reduction of correlations display a scaling dependence on the ratio t/t_w where t_w is a waiting time. A third notable property is that the experimentally observed relaxation behavior of supercooled liquids is effectively described, via reasonable heat capacity assumptions [34], by the so-called Adam-Gibbs equation [40],

$$t_x = A \exp\left(\frac{B}{TS_c}\right), \tag{3.3.20}$$

where the relaxation time t_x can be identified with the viscosity or the inverse of the diffusivity, and the configurational entropy S_c is related to the number of minima of the fluid's potential energy surface (and A and B are constants) [34]. Eq. (3.3.20) implies that the reason for viscous slow-down in supercooled liquids is a progressive reduction in the number of configurations that the system is capable of sampling as $T - T_g \to 0$. A first principles derivation of this equation has not been developed at present.

Finally, the sharp slow down of dynamics in supercooled liquids on approach to vitrification is illustrated by the progression from normal diffusiveness to subdiffusive behavior and at last to a stop in the growth of the molecular mean square displacement, all this within a small range of temperatures or densities [41], [42]. This deceleration of the dynamics is caused by the confinement of any given molecule by a 'cage' formed by its neighbors; and it is the breakup and rearrangement of the cages which drives structural relaxation, letting molecules diffuse throughout the system. Evidence indicates that lifetime of the cages increases as conditions move towards the glass transition, probably because cage rearrangements involve a larger number of molecules as the glass transition is approached [41], [42].

The erratic motion of a Brownian particle is usually described by the Langevin theory [43]. As it is well known, this method finds a way to avoid the detailed consideration of many degrees of freedom by representing with a noise source the effect of collisions with molecules in the fluid in which the particle moves. The approach to thermal equilibrium is produced by random forces, and these are sufficient to determine dynamical correlations, diffusion, and a basic form for the fluctuation-dissipation theorem [43]. In the same spirit, attractors of nonlinear low-dimensional maps under the effect of external noise can be used to model states in systems with many degrees of freedom. Notice that the general map formula

$$x_{t+1} = x_t + h_\mu(x_t) + \sigma\chi_t, \tag{3.3.21}$$

is a discrete form for a Langevin equation with nonlinear 'friction force' term h_μ and χ_t is the same Gaussian white noise random variable as in Eq. (3.3.2) and σ the noise intensity. With the choice $h_\mu(x) = 1 - x - \mu x^2$ we recover Eq. (3.3.2).

Noise-perturbed onset of chaos

When $\sigma > 0$ the noise fluctuations smear the fine structure of the periodic attractors as the iterate visits positions within a set of bands or segments like those in the chaotic attractors (see Fig. 3.13), however there is still a distinct transition to chaos at $\mu_c(\sigma)$ where the Lyapunov exponent λ_1 changes sign. The period doubling of bands ends at a finite maximum period $2^{N(\sigma)}$ as $\mu \to \mu_c(\sigma)$ and then decreases at the other side of the transition. This effect displays scaling features and is referred to as the bifurcation gap [12], [33]. When σ is small the trajectories visit sequentially the set of $2^{N(\sigma)}$ disjoint bands leading to a cycle, but the behavior inside each band is chaotic. The trajectories represent ergodic states as the accessible positions have a fractal dimension equal to the dimension of phase space. When $\sigma = 0$ the trajectories correspond to a nonergodic state, since as $t \to \infty$ the positions form only a Cantor set of fractal dimension $d_f = 0.5338....$ Thus the removal of the noise $\sigma \to 0$ leads to an ergodic to nonergodic transition in the map.

As shown in Ref. [11] when $\mu_c(\sigma > 0)$ there is a 'crossover' or 'relaxation' time $t_x = \sigma^{r-1}$, $r \simeq 0.6332$, between two different time evolution regimes. This crossover occurs when the noise fluctuations begin erasing the fine structure

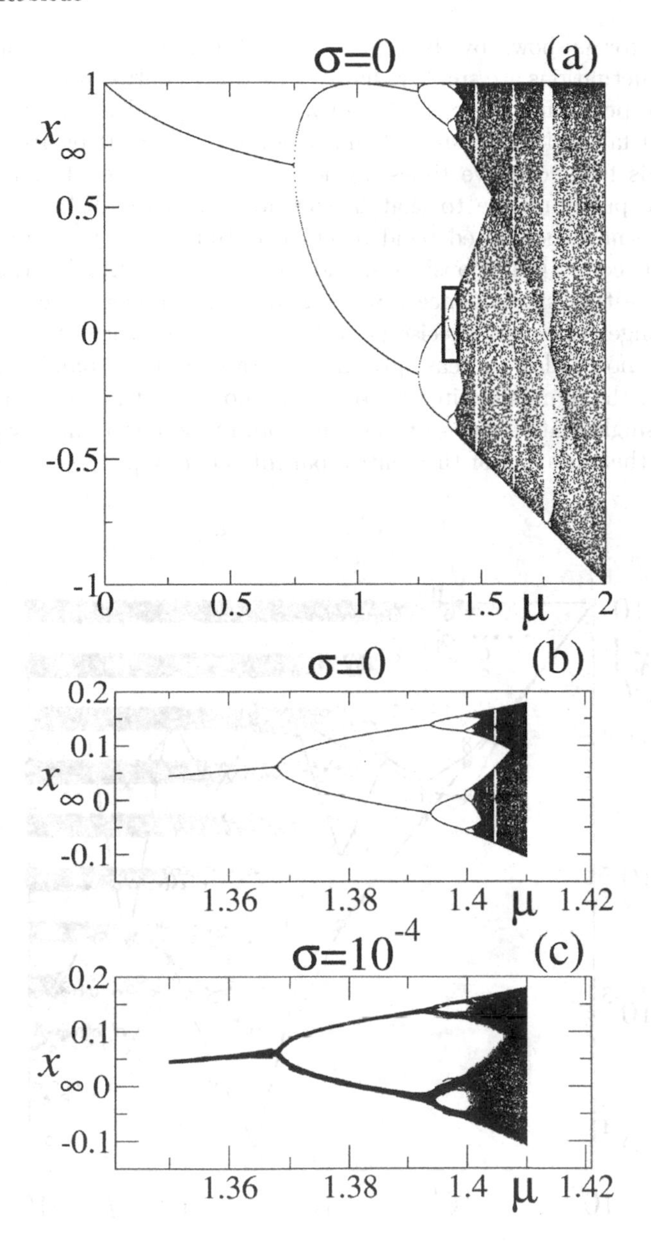

Fig. 3.13. (a) Logistic map attractor. (b) Magnification of the box in (a). (c) Noise-induced bifurcation gap in the magnified box.

of the attractor as shown by the superstable orbit with $x_0 = 0$ in Fig. 3.6. For $t < t_x$ the fluctuations are smaller than the distances between the neighboring subsequence positions of the $x_0 = 0$ orbit at $\mu_c(0)$, and the iterate position with $\sigma > 0$ falls within a small band around the $\sigma = 0$ position for that t. The bands for successive times do not overlap. Time evolution follows a subsequence pattern close to that in the noiseless case. When $t \sim t_x$ the width of the noise-generated band reached at time $t_x = 2^{N(\sigma)}$ matches the distance between adjacent positions, and this implies a cutoff in the progress along the position subsequences. See Fig. 3.14. At longer times $t > t_x$ the orbits no longer trace the precise period-doubling structure of the attractor. The iterates now follow increasingly chaotic trajectories as bands merge with time. This is the dynamical image - observed along the time evolution for the orbits of a single state $\mu_c(\sigma)$ - of the static bifurcation gap initially described in terms of the variation of the control parameter μ [33].

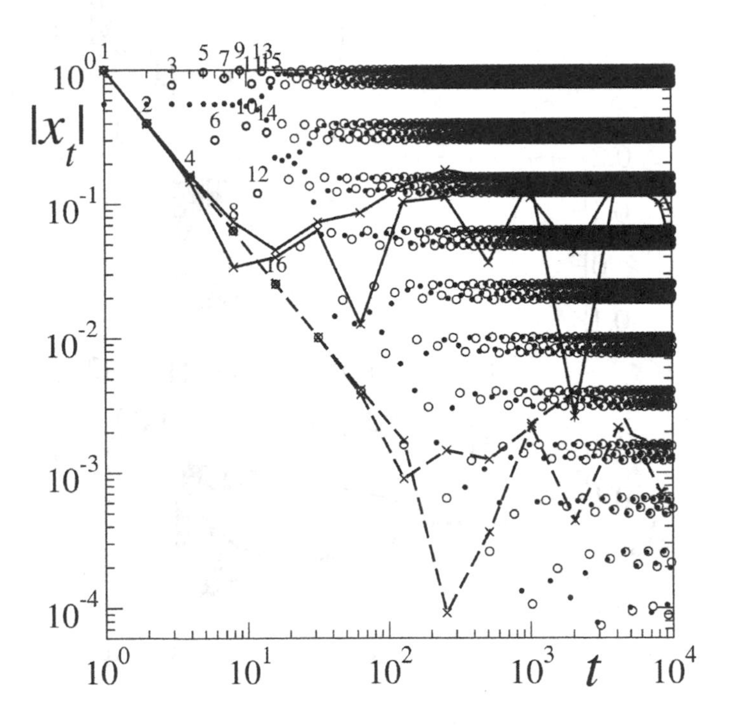

Fig. 3.14. Absolute values of positions in logarithmic scales of iterations t for various trajectories with and without additive noise. Setting $\mu = \mu_c$ and $\sigma = 0$, empty circles correspond to the absolute values of the attractor positions obtained by iterating $x_0 = 0$ (the numbers label time $t = 1, \cdots, 16$) , while small dots correspond to $x_0 = 0.56023$, close to a repeller, the unstable solution of $x = 1 - \mu_c x^2$. Full (and dashed) lines represent two trajectories for $\mu = \mu_c$, plotted at times $t = 2^n$, $n = 0, 1, \ldots$, for different values of noise amplitude $\sigma = 10^{-3}$ (and $\sigma = 10^{-6}$).

Two-step relaxation

In Ref. [11] it was suggested that in the map at $\mu_c(\sigma)$ the analog of the β relaxation would be obtained by considering initial conditions x_0 outside the critical attractor since the resultant orbits display a power-law transient as the positions approach asymptotically those of the attractor. After this transient an intermediate plateau may be observed and would correspond to the regime $t < t_x$, described before, when the iterates are confined to nonintersecting bands before they reach the bifurcation gap, its duration t_x grows as an inverse power law of σ. The analog of the α relaxation was proposed to be the band merging crossover process that takes place for $t > t_x$.

This relaxation processes were studied [11] by evaluation of the two-time correlation function

$$c_e(t_2 - t_1) = \frac{\langle x_{t_2} x_{t_1} \rangle - \langle x_{t_2} \rangle \langle x_{t_1} \rangle}{\sigma_{t_1} \sigma_{t_2}}, \tag{3.3.22}$$

for different values of the noise amplitude σ. In Eq. (3.3.22) $\langle ... \rangle$ represents an average over an ensemble of trajectories starting with initial conditions randomly distributed across a small interval around x_0 and $\sigma_{t_i} = \sqrt{\langle x_{t_i}^2 \rangle - \langle x_{t_i} \rangle^2}$. In Fig. 3.15a we show the behavior of $c_e(t)$ for $x_0 = 0.56023$, the initial relaxation, that is, the transient fall into the attractor, is captured by the choice of initial condition that corresponds to a repeller or unstable periodic orbit. As shown in the figure this initial decay of $c_e(t)$ disappears as $\sigma \to 0$. In Fig.3.15b we show the behavior of $c_e(t)$ for $x_0 = 0$, the choice of initial position at the attractor eliminates the 'β relaxation' and shows the development of the plateau followed by the encounter with the bifurcation gap. This secondary relaxation process can be clearly appreciated. In Fig. 15b is also shown the location of the crossover time t_x at which the bifurcation gap is reached. (We note that $t_x = \sigma^{r-1}$ is approximately obtained by retaining only the first order term in a perturbation [11]).

Adam-Gibbs relation

As the counterpart to the Adam-Gibbs formula, the noise-perturbed map model for glassy dynamics exhibits a relationship between the plateau duration t_x, and the entropy S_c for the attractor with the largest number of bands allowed by the bifurcation gap - the noise-induced cutoff in the period-doubling cascade [12]. This entropy is obtained from the probability of band occupancy at position x, i.e. the distribution of the iterate positions within the 2^N bands, and has the form $S_c = 2^N \sigma s$, where s is the entropy associated to a single band. Use of $2^N = t_x + 1$ and $t_x = \sigma^{r-1}$, $r - 1 \simeq -0.3668$ [11], leads to

$$t_x = \left(\frac{s}{S_c} \right)^{(1-r)/r}. \tag{3.3.23}$$

Since $(1 - r)/r \simeq 0.5792$ then $t_x \to \infty$ and $S_c \to 0$ as $\sigma \to 0$. See [11] for details on the derivation. Clearly, at variance with the exponential Adam-

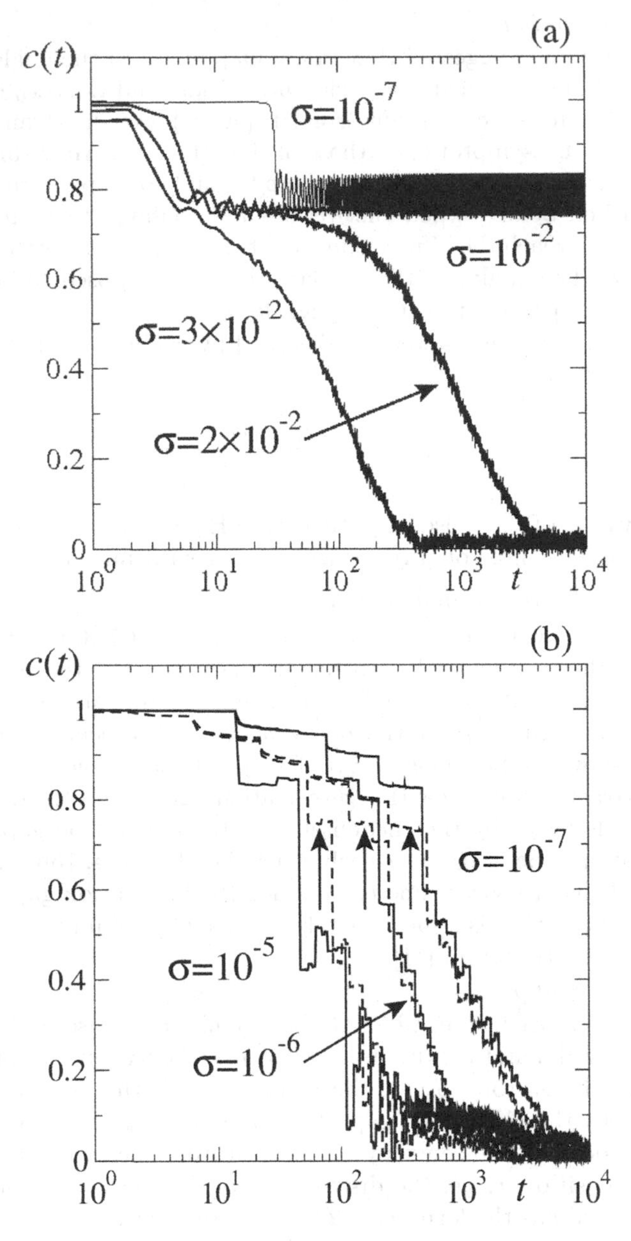

Fig. 3.15. Two step relaxation. We plot the two-time correlation function as defined in Eq. (3.3.22) for an ensemble of 5000 trajectories initially distributed at random around $x_0 = 0.56023$ (a) and $x_0 = 0$ (b) (in both cases $\mu = \mu_c$). The width of the initial interval is of order 10^{-7}. Curves are labeled with their corresponding values of noise amplitude σ. In (a) we used a gray line to make visible the curve obtained for $\sigma = 10^{-2}$. Full (dashed) lines in (b) refer to a 'waiting time' $t_1 = 50$ ($t_1 = 10$). Vertical arrows, from left to right, indicate the crossover time $t_x(\sigma)$ for $\sigma = 10^{-5}, 10^{-6}, 10^{-7}$, respectively.

Gibbs equation in structural glass formers [34], this expression turned out to have a power law form.

Aging

As indicated in Ref. [11] the interlaced power-law position subsequences that constitute the superstable orbit of period 2^∞ within the noiseless attractor at $\mu_c(0)$ imply a built-in aging scaling property for the single-time function x_t. These subsequences are relevant for the description of trajectories that are at first 'held' at a given attractor position for a waiting period of time t_w and then 'released' to the normal iterative procedure. The holding positions were chosen to be any of those along the top band shown in Fig. 3.6 with $t_w = 2k+1$, $k = 0, 1, ...$ One obtains [11]

$$x_{t+t_w} \simeq \exp_q(-\lambda_q t/t_w), \tag{3.3.24}$$

where $\lambda_q = \ln\alpha/\ln 2$, with $\alpha = 2.50290....$ This property is gradually removed when noise is turned on. The presence of a bifurcation gap limits its range of validity to total times $t_w + t < t_x(\sigma)$ and so progressively disappears as σ is increased [11].

Aging and its related scaling property at the onset of chaos for $\sigma = 0$ was studied [11] via the evaluation of two-time correlation functions obtained through a time average. The form

$$c(t+t_w, t_w) = (1/N)\sum_{n=1}^{N} \phi^{(n)}(t+t_w)\phi^{(n)}(t) \tag{3.3.25}$$

was chosen, where $\phi(\tau) = f_{\mu_c}^{(\tau)}(0)$ and $f_\mu(x) = 1 - \mu x^2$. In Fig. 3.16a we show $c(t+t_w, t_w)$ for different values of $t_w = 2k+1$, $k = 0, 1, ...$, and in Fig. 3.16b the same data where the rescaled variable $t/t_w = 2^n - 1$ has been used. The characteristic scaling of aging behavior is especially clear.

Subdiffusion and arrest.

To investigate subdiffusion and arrest close to glass formation in the map at $\mu_c(\sigma)$, we constructed a periodic map with repeated cells. This setting has being used to study deterministic diffusion in nonlinear maps, in which the trajectories migrate into neighboring cells due to chaotic motion. For fully chaotic maps diffusion is normal [12] but for marginally chaotic maps it is anomalous [27]. In our case we design the map in such a way that diffusion is due only to the random noise term, otherwise motion is confined to a single cell. So, we have the periodic map $x_{t+1} = F(x_t)$, $F(l+x) = l + F(x)$, $l = ... -1, 0, 1, ...$, where

$$F(x) = \begin{cases} -\left|1 - \mu_c x^2\right| + \sigma\xi, & -1 \le x < 0, \\ \left|1 - \mu_c x^2\right| + \sigma\xi, & 0 \le x < 1. \end{cases} \tag{3.3.26}$$

Fig. 3.17a shows the periodic map Eq. (3.3.26) together with a portion of one of its trajectories. As it can be observed, the escape from the central cell into any of its neighbors occurs when $|F(x)| > 1$ and this can only happen

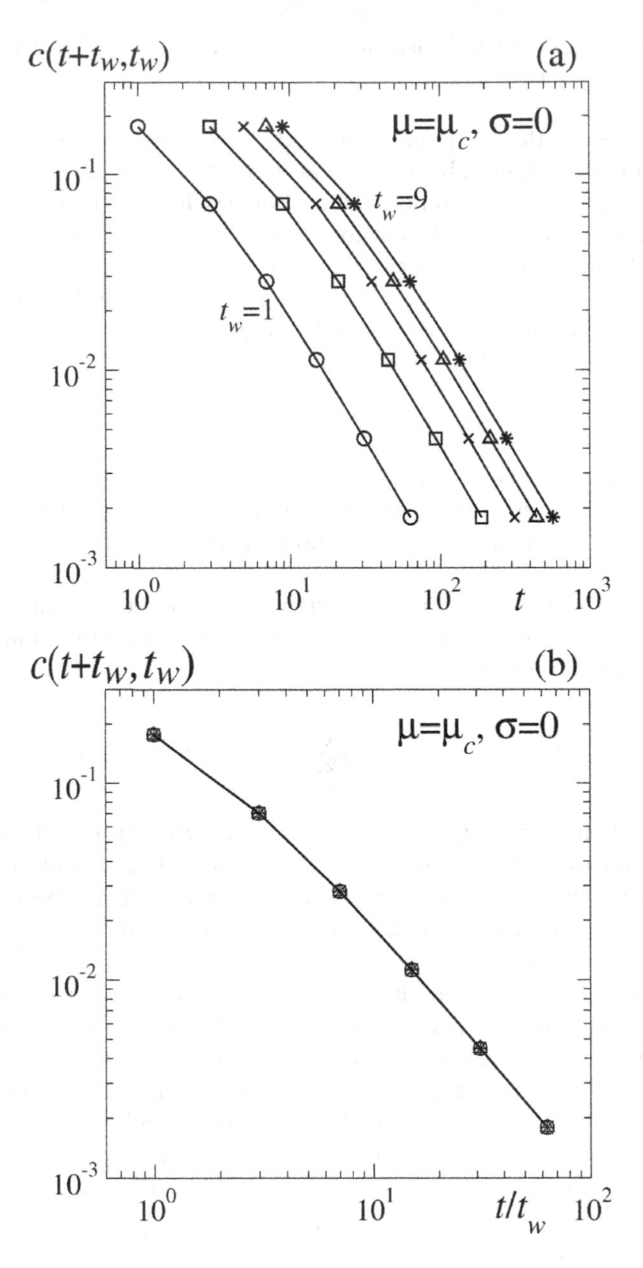

Fig. 3.16. Aging of correlations $c(t + t_w, t_w)$ according to Eq. (3.3.25) for the Feigenbaum attractor ($\mu = \mu_c$, $\sigma = 0$). Total observation time is $n = 1000$. In (a) is shown the explicit dependence on the waiting time of two-time correlations (from left to right $t_w = 1, 3, 5, 7, 9$). In (b) all curves collapse upon rescaling t/t_w.

when $\sigma > 0$. As $\sigma \to 0$ the escape positions are confined to values of x increasingly closer to $x = 0$, and for $\sigma = 0$ the iterate position is trapped within the cell. Likewise for any other cell. Fig. 3.17b shows the mean square displacement $\langle x_t^2 \rangle$ as obtained from an ensemble of trajectories initially distributed throughout the interval $[-1, 1]$ for several values of the noise amplitude. The progression from normal diffusion to subdiffusion and to final arrest can be plainly observed as $\sigma \to 0$. For small σ ($\leq 10^{-2}$) $\langle x_t^2 \rangle$ shows a down turn and later an upturn similar to those observed in colloidal glass experiments [41] and attributed to cage rearrangements. In the map this feature reflects cell crossings.

Nonuniform convergence

As we have recounted, the dynamics at the chaos threshold in the presence of noise displays the characteristic elements of glassy dynamics observed in molecular glass formers. The limit of vanishing noise amplitude $\sigma \to 0$ (the counterpart of the limit $T - T_g \to 0$ in the supercooled liquid) leads to loss of ergodicity. This nonergodic state with $\lambda_1 = 0$ corresponds to the limiting state, $\sigma \to 0$, $t_x \to \infty$, of a family of small σ states with glassy properties. It is of interest to note that at $\mu_c(\sigma)$ the trajectories and its resultant sensitivity to initial conditions are expressed for $t < t_x$ via the q-exponentials of the Tsallis formalism [11]. For $\sigma = 0$ this analytical forms are exact [5], [9] and a Pesin identity, linking accordingly generalized Lyapunov coefficients and rates of entropy production, holds [6], [9]. There is nonuniform convergence related to the limits $\sigma \to 0$ and $t \to \infty$. If $\sigma \to 0$ is taken before $t \to \infty$ orbits originating within the attractor remain there and exhibit fully-developed aging properties, whereas if $t \to \infty$ is taken before $\sigma \to 0$ a chaotic orbit with exponential sensitivity to initial conditions would be observed.

3.3.4 Discussion of Part II

We have examined the study of clusters at criticality in thermal systems by means of the saddle-point approximation in the LGW free energy model [19], [20], [31], [32]. The retention of only one coarse-grained configuration leads to cluster properties that are physically reasonable but also appear to fall outside the limits of validity of the BG theory. The fractal geometry and the intermittent behavior of critical clusters obtained from this method [19], [20] [31], [32] are both consistent with equivalent properties found for clusters at the critical points of the $d = 2$ Ising and Potts models [44], [45]. On the other hand, it was found that the entropy expression that provides the property of extensivity for the estimate of the number of cluster configurations is not the usual BG expression but that for the q-statistics. Likewise, the nonlinear map and its corresponding sensitivity to initial conditions linked to the intermittency of clusters do not follow the fully-chaotic trajectories of BG statistics but display the features of q-statistics.

With regards to the extensivity of entropy, what our assumptions and results mean basically is that extensivity of entropy (BG or q-generalized) of

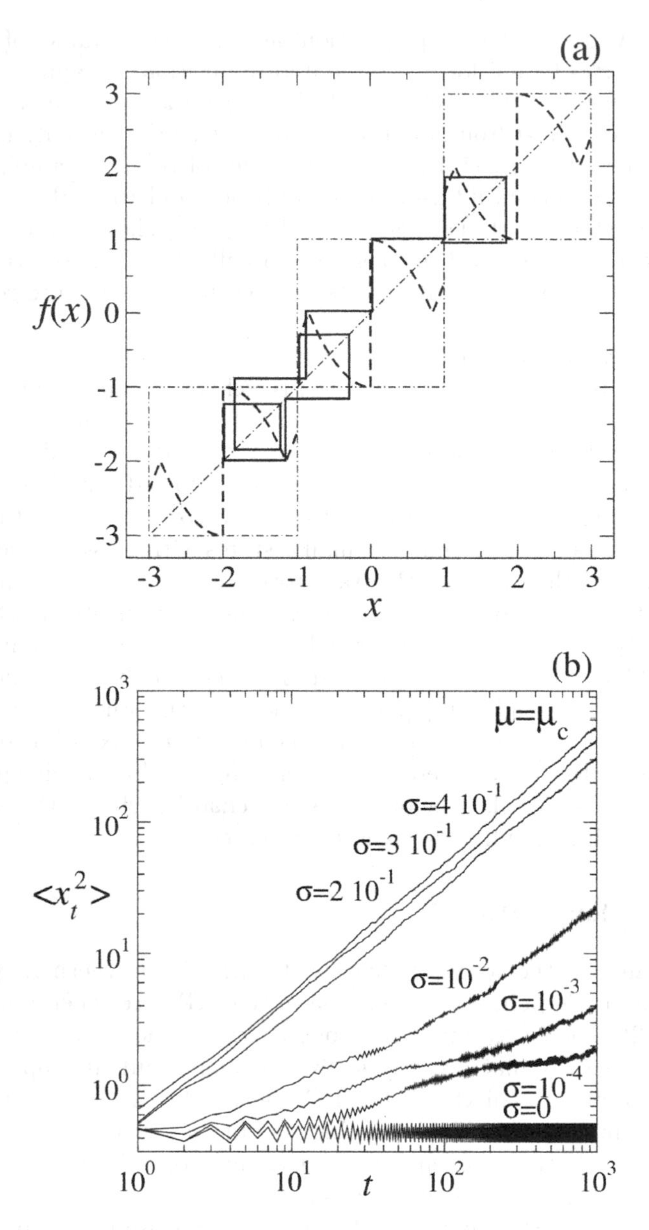

Fig. 3.17. Glassy diffusion in the noise-perturbed onset of chaos. (a)Repeated-cell map (thick dashed line) and trajectory (full line). (b)Time evolution of the mean square displacement $\langle x_t^2 \rangle$ for an ensemble of 1000 trajectories with initial conditions randomly distributed inside $[-1, 1]$. Curves are labeled by the value of the the noise amplitude.

a cluster and the linear growth with time of entropy (BG or q-generalized) of trajectories for the attractor of a nonlinear map are equivalent. The crossover from the expression for S_q to that for S_1 is obtained when the system is taken out of criticality, because $\delta \rightarrow 1$ makes $q \rightarrow 1$. When at criticality the crossover between q-statistics and BG statistics in the dynamical behavior is related to the subsystem's size as follows. We observe that the map shift parameter depends on the domain size as $\epsilon \sim R^{-1}$. So, the time evolution of ϕ displays laminar episodes of duration $< t > \sim \epsilon^{-\delta/(\delta+1)}$ and the Lyapunov coefficient in this regime is $\lambda_1 \sim \epsilon$ [20]. Within the first laminar episode the dynamical evolution of $\phi(x)$ obeys q-statistics, but for very large times the occurrence of many different laminar episodes leads to an increasingly chaotic orbit consistent with the small $\lambda_1 > 0$ and BG statistics would be recovered. As R increases ($R \ll x_0$ always) the time duration of the q-statistical regime increases and in the limit $R \rightarrow \infty$ there is only one infinitely long q-statistical laminar episode with $\lambda_1 = 0$ with no crossover to BG statistics. On the other hand when $R > x_0$ the clusters $\phi(x)$ are no longer dominant, for the infinite subsystem $R \rightarrow \infty$ their contribution to Z vanishes and no departure from BG statistics is expected to occur [4], [10].

In view of the results presented here the departure from BG statistics and the applicability of q-statistics is due in part to the presence of the long-ranged correlations in space and in time that take place at criticality. These correlations give the integrand in the LGW Ψ_c a power-law dependence of the form $|\phi|^{\delta+1}$ with $\delta > 1$ (commonly $\delta \geq 3$ as $\delta = 3$ gives the Gaussian critical point) and this in turn determines the q-exponential expression for $\phi(x)$ and the properties derived from it. On the other hand, the neglect of all coarse-grained configurations other than the most dominant implies that phase space has not been properly sampled, and that the ergodic and mixing properties characteristic of equilibrium BG statistics are not guaranteed. It should be clear that the properties studied are those of a single order-parameter cluster. The configurations of the total system at criticality obey ordinary BG statistics.

With regards to the dynamics of noise-perturbed logistic maps at the chaos threshold we have seen that it exhibits the most prominent features of glassy dynamics in supercooled liquids. Specifically our results are: 1) The two-step relaxation occurring in two-time correlations when $\sigma \rightarrow 0$ was determined in terms of the map bifurcation gap properties. 2) The map equivalent of the Adam-Gibbs law was obtained as a power-law relation between $t_x(\sigma)$ and the entropy $S_c(\sigma)$ associated to the noise broadening of chaotic bands. 3) The trajectories and their two-time correlations at $\mu_c(\sigma \rightarrow 0)$ were shown to obey an aging scaling property. 4) The progression from normal diffusiveness to subdiffusive behavior and finally to a stop in the growth of $\langle x_t^2 \rangle$ was demonstrated with the use of a repeated-cell map. These properties were determined from the trajectories of iterates at $\mu_c(\sigma)$, and use was made of the fixed-point map solution $g(x)$ of the RG doubling transformation consisting of functional composition and rescaling, $\mathbf{R}f(x) \equiv \alpha f(f(x/\alpha))$. Positions for time subsequences within these trajectories are expressed analytically in terms of the q-exponential function.

The existence of this analogy cannot be considered accidental since the limit of vanishing noise amplitude $\sigma \to 0$ involves loss of ergodicity. The occurrence of these properties in this simple dynamical system with degrees of freedom represented via a random noise term, and no reference to molecular interactions, suggests a universal mechanism lying beneath the dynamics of glass formation. As already proved [5], the dynamics of deterministic unimodal maps at the onset of chaos is a genuine example of the pertinence of the q-statistics in describing states with $\lambda_1 = 0$. As we have seen this nonergodic state corresponds to the limiting state, $\sigma \to 0$, $t_x \to \infty$, for a family of small σ noisy states with glassy properties, that are described for $t < t_x$ via the q-exponentials of the generalized formalism [5]. The fact that these features transform into the usual BG exponential behavior for $t > t_x$ provides an opportunity for investigating the crossover from the ordinary BG to the q-statistics in the physical circumstance of loss of mixing and ergodic properties.

It has been suggested on several occasions [3] that the setting in which q-statistics appears to emerge is linked to the incidence of nonuniform convergence, such as that involving the thermodynamic $N \to \infty$ and infinite time $t \to \infty$ limits. Here it is clear that a similar situation takes place, that is, if $\sigma \to 0$ is taken before $t \to \infty$ a nonergodic orbit confined to the Feigenbaum attractor and with fully developed glassy features is obtained, whereas if $t \to \infty$ is taken before $\sigma \to 0$ a typical $q = 1$ chaotic orbit is observed.

The point of view that our studies suggest is that the observed slow dynamics in a given system can be seen to be composed of the ideal glassy features arising from ergodicity breakdown with other superimposed system-dependent features. The differences to be found between supercooled-liquid dynamics (from experimental or from fluid model calculations) and the ideal map dynamics would then be credited to the presence of molecular structure and other specific effects. Finally, it is worth mentioning that while the properties displayed by the map capture in a qualitative, heuristic, way the phenomenological issues of vitrification, they are obtained in a quantitative and rigorous manner as the map is concerned.

3.4 Clarifying remarks

To demarcate the scope of the results presented here a few explicatory comments might be useful. We stated that the dynamical properties at critical attractors in unimodal maps bear a significant relation to a theoretical scheme based on the Tsallis entropy S_q. With regards to nonlinear dynamics the aim of this scheme, here called q-statistics, is to describe the departure from exponential sensitivity and the fate of the identity between the Lyapunov coefficient and the rate of entropy production characteristic of chaotic attractors. Chaotic attractors have ergodic and mixing properties and display exponential sensitivity ξ_t with Lyapunov coefficients λ_1 independent from initial conditions x_0. These coefficients satisfy a Pesin identity $\lambda_1 = K_1$ (qualified below) where the rate of entropy production K_1 is based on the BG entropy expression S_1. In its basic - early - proposition [23] q-statistics consisted of simple suggestions: (i) A q-exponential expression for ξ_t from which a q-Lyapunov coefficient λ_q could be extracted as the coefficient of iteration time t (just as λ_1 can be read from the exponential ξ_t). And (ii) a generalized identity $\lambda_q = K_q$ where the rate K_q would be obtained from the entropy expression S_q. The value for the entropic index q would be presumably determined in terms of the main parameters of the specific problem in hand. After the examination of the dynamics at the critical attractors in unimodal maps we can make an assessment of this scheme.

The first important – and expected – characteristic of the coefficients λ_q that we determined is their dependence on the initial condition x_0. The $\lambda_q(x_0)$ describe an spectrum of expansion (or contraction) rates of orbits. Interestingly, the spectrum $\lambda_q(x_0)$ can be reduced to a single value with the introduction of a waiting time t_w related to x_0 (see Eqs. (3.2.12) and (3.3.24)) and this reveals the incidence of a scaling property known as aging that we exploited in our study of glassy dynamics. The nonergodic source for the dependence on x_0 can be linked to a breakdown in time translation invariance.

A second relevant corroborated feature is that all the coefficients in each spectrum $\lambda_q(x_0)$ appear associated to the same well-defined value of q and this in turn is easily identified with a universal parameter for the critical attractor. The cases of the pitchfork and tangent bifurcations are essentially straightforward but the period-doubling onset of chaos is much more involved and requires a careful description that considers the multifractal properties of the attractor. For the pitchfork and tangent bifurcations one obtains $q = 2 - z^{-1}$ where z is the leading nonlinearity in Eq. (3.2.9), and as we saw there are only two possible values, $q = 5/3$ or $q = 3/2$, according to whether the transition is of one type or the other. In our discussion of critical clusters there is a simple link between q and the critical isotherm exponent δ which defines the thermal transition universality class.

We found that the sensitivity for the Feigenbaum attractor does not have the form of a single q-exponential but of infinitely many interlaced q-exponentials. If time evolution is followed starting from an arbitrary position x_0 within the attractor and recorded at every time t what is observed [18] are strongly fluctuating quantities that persist in time and with a tangled pattern structure that displays memory retention. On the other hand, if specific initial positions with known location within the multifractal are chosen, and subsequent positions are observed only at pre-selected times, when the trajectories visit another chosen region, a well-defined q-exponential form for ξ_t emerges. The specific value of q and the associated Lyapunov spectrum λ_q can be determined, and for each case the value of q is given by that of a discontinuity in the trajectory scaling function σ of Feigenbaum. The corresponding spectrum λ_q reflects all starting positions in the multifractal region where the trajectories originate.

Trajectories that connect two regions with different local structure display a dynamical phase transition with features analogous to those in thermal first-order phase transitions. An statistical-mechanical formalism was developed a decade and a half ago [17] capable of describing these phenomena and was applied to the Feigenbaum attractor [18]. This formalism describes transitions between different 'q-phases' that are signalled by jump discontinuities in a Lyapunov function $\lambda(\mathsf{q})$ (the order parameter) at a special value $\mathsf{q} = q$, or equivalently, sections of constant slope $1 - q$ in a 'free energy' function $\psi(\lambda)$; q is the 'ordering field'. We have shown that the value q of q at the dynamical phase transition is the same as the value of the Tsallis q index in ξ_t and that the jump in amplitude of Mori's coarse-grained Lyapunov function $\lambda(\mathsf{q})$ coincides with the Lyapunov spectrum λ_q obtained from ξ_t. Therefore, we have identified the cause or source of the special values for the entropic index q in the Feigenbaum attractor.

Because the dynamics within the Feigenbaum attractor retains memory we can distinguish the two senses, outgoing or incoming, of trajectories connecting two regions of the attractor. As a consequence of this the q-phase transitions occur in conjugate pairs, with q-indexes q and $2-q$, with spectra, λ_q and λ_{2-q}, and the function $\psi(\lambda)$ displays two sections of constant slopes with opposite signs, $1 - q$ and $q - 1$. The dynamics at the tangent bifurcation can also be couched within the same two-region framework.

But the multiregion nature of the Feigenbaum attractor leads to a multiplicity of Mori's q-phase transitions and a hierarchy of q-indexes. As we have seen these can be put into a one to one correspondence with the jump discontinuities of the trajectory scaling function σ originally calculated by Feigenbaum for nonlinearity $\zeta = 2$. Each discontinuity in σ quantifies the difference in proximity of elements between two regions of the attractor that is relevant to the dynamics of trajectories that connect them, information that is difficult to extract, or partially missing, from the multifractal spectrum of dimensions $f(\widetilde{\alpha})$. It is central to our discussion to observe that each discontinuity in σ leads to explicit expressions for a conjugate pair of q-indexes, of q-exponential sensitivities, of q-Lyapunov spectra and of q-phase transitions. Since the amplitude of the discontinuities in σ decreases rapidly, one of the q-index pairs dominates, the values of which we found to be associated to the most crowded and most sparse regions of the multifractal.

The fact that the expression for ξ_t can be put in the form of one or more q-exponentials is evidently useful as it is a clear cut method for extracting the spectra for the λ_q, the counterparts of λ_1 for chaotic attractors. The λ_q in turn can be used to determine the Lyapunov function $\lambda(\mathsf{q})$ in a manner independent from the coarse-grained formalism proposed by Mori [17] [18], and as we have seen this permitted us to identify a hierarchy of q-phase transitions when the original numerical calculations had uncovered only one [18].

There is another significant consequence of the q-exponential basis for the expressions for ξ_t. This relates to the rate of entropy production of ensembles of trajectories under the action of the attractor. We have shown that for initial positions within a small interval around an attractor point x_0 the equality $\lambda_q = K_q$ holds up to a time T dependent on the interval size, where K_q is the difference in Tsallis entropy between time t and $t = 0$. For chaotic attractors a single equality $\lambda_1 = K_1$ holds, whereas for the Feigenbaum attractor one has infinitely many identities $\lambda_q^{(k)} = K_q^{(k)}$, but remarkably, for each of them the entropy S_q is evaluated with the same value of the q-index and the same pre-selected times t_k for the trajectories that link with starting and finishing positions the multifractal regions under consideration.

It is important to clarify the circumstances under which the equalities $\lambda_q^{(k)} = K_q^{(k)}$ are obtained as these could be interpreted as shortcomings of the formalism. First of all, the rate K_q does not generalize the trajectory-based Kolmogorov-Sinai (KS) entropy $\mathcal{K}_1$ that is involved in the well-known Pesin identity $\lambda_1 = \mathcal{K}_1$ [12] - [14]. The q-generalized KS entropy $\mathcal{K}_q$ would be defined in the same manner as $\mathcal{K}_1$ but with the use of S_q. The rate K_q is determined from values of S_q at two different times [14]. The relationship between $\mathcal{K}_1$ and K_1 has been investigated for several chaotic maps [46] and it has been established that the equality $\mathcal{K}_1 = K_1$ occurs during an intermediate stage in the evolution of the entropy $S_1(t)$, after an initial transient dependent on the initial distribution and before an asymptotic approach to a constant saturation value. Here we have looked into the analogous intermediate regime in which one would expect $\mathcal{K}_q = K_q$, as we argue below.

We have only considered initial conditions within small distances outside the positions of the Feigenbaum attractor and have not focused on the initial transient behavior referred to in the above paragraph. A noteworthy exception has been that for the noise-perturbed onset of chaos employed to study glassy dynamics, where this transient behavior was identified with the primary relaxation process known as β relaxation. As for the final asymptotic regime mentioned above it should be kept in mind that the distance between trajectories, which defines λ_q, always saturates because of the finiteness of the available phase space ($[-1, 1]$ or the multifractal subset that is the Feigenbaum attractor). Separation of incipiently chaotic trajectories, just as separation of chaotic ones, undergo two different processes, stretching which leads to the q-exponential regime in ξ_t and folding which keeps the orbits bounded. Therefore for t sufficiently large Eq. (3.2.4) would be no longer valid, just like the exponential ξ_t of chaotic attractors. This is the reason there is a saturation time T in our determination of $\lambda_q^{(k)}$ and consequently justifies our use of the rates $K_q^{(k)}$ in $\lambda_q^{(k)} = K_q^{(k)}$. It is widely known that special care needs to be taken to avoid saturation due to folding in determining λ_1 and similar limitations occur for $\lambda_q^{(k)}$.

As expected all the properties determined here in relation to q-statistics stem from universal quantities, such as the trajectory scaling function σ originally derived with the use of the RG functional composition transformation. Had this not been the case, the source of the q-indexes in *bona fide* applications of this generalized BG scheme could not be understood. On the other hand, to steer clear of the terminology of the q-statistics would unnecessarily obscure the details of the dynamics here considered. The dynamics at the Feigenbaum attractor, with ordinary $\lambda_1 = 0$, has long been seen as a hardly manageable collection of strongly fluctuating local expansion rates $\lambda(x_t, x_0) = \ln |dx_t/dx_0|$ which do not converge to any constant and only diverge logarithmically for $t \to \infty$ [29], [18]. By evaluating these rates for specific classes of initial positions with necessarily coordinated observation times, and with use of quantities connected with the q-statistics, corroborated to hold, we have resolved in great detail the temporal structure of these coefficients.

Acknowledgments

I thank Fulvio Baldovin and Estela Mayoral for their essential participation in the studies here described. I have also very much benefitted from discussions with Constantino Tsallis, Sumiyoshi Abe and Hugo Hernández-Saldaña, and received valuable help from Dan Silva-López. Work partially supported by DGAPA-UNAM and CONACyT (Mexican Agencies).

References

[1] It should be noted that for the chaotic dynamics of a hard sphere gas there is no dissipation nor strange attractor. It is of course this dynamics that is relevant to Boltzmann's assumption of molecular chaos.
[2] C. Tsallis, J. Stat. Phys. **52**, 479 (1988).
[3] For recent reviews see, *Nonextensive Entropy – Interdisciplinary Applications*, M. Gell-Mann and C. Tsallis, eds., (Oxford University Press, New York, 2004). See http://tsallis.cat.cbpf.br/biblio.htm for full bibliography.
[4] A. Robledo, Physica **A 314**, 437 (2002); Physica **D 193**, 153 (2004).
[5] F. Baldovin and A. Robledo, Europhys. Lett. **60**, 518 (2002).
[6] F. Baldovin and A. Robledo, Phys. Rev. **E 66**, 045104 (2002).
[7] F. Baldovin and A. Robledo, Phys. Rev. **E 69**, 045202 (2004).
[8] E. Mayoral and A. Robledo, Physica **A 340**, 219 (2004).
[9] E. Mayoral and A. Robledo, Phys. Rev. **E 72**, 026209 (2005).
[10] A. Robledo, Physica **A 344**, 631 (2004); Molec. Phys. **103**, 3025 (2005).
[11] A. Robledo, Phys. Lett. **A 328**, 467 (2004); Physica **A 342**, 104 (2004); F. Baldovin and A, Robledo, Phys. Rev. **E 72**, 066213 (2005).
[12] H.G. Schuster, *Deterministic Chaos. An Introduction*, 2nd Revised Edition (VCH Publishers, Weinheim, 1988).

[13] C. Beck and F. Schlogl, *Thermodynamics of Chaotic Systems* (Cambridge University Press, UK, 1993).
[14] R.C. Hilborn, *Chaos and nonlinear dynamics*, 2nd Revised Edition (Oxford University Press, New York, 2000),
[15] C. Grebogi, E. Ott, S. Pelikan, and J.A. Yorke, Physica **D 13**, 261 (1984).
[16] C. Grebogi, E. Ott and J.A. Yorke, Phys. Rev. Lett. **48**, 1507 (1982); C. Grebogi and E. Ott, Physica **7D**, 181 (1983).
[17] H. Mori, H. Hata, T. Horita and T. Kobayashi, Prog. Theor. Phys. Suppl. **99**, 1 (1989).
[18] H. Hata, T. Horita and H. Mori, Progr. Theor. Phys. **82**, 897 (1989); G. Anania and A. Politi, Europhys. Lett. **7** (1988).
[19] Y.F. Contoyiannis and F.K. Diakonos, Phys. Lett. **A268**, 286 (2000).
[20] Y.F. Contoyiannis, F.K. Diakonos, and A. Malakis, Phys. Rev. Lett. **89**, 035701 (2002).
[21] See, for example, H.E. Stanley, *Introduction to Phase Transitions and Critical Phenomena* (Oxford University Press, New York, 1987).
[22] Ergodicity is referred here in relation to the total phase space on which the map is defined.
[23] C. Tsallis, A.R. Plastino and W.-M. Zheng, Chaos, Solitons and Fractals **8**, 885 (1997).
[24] B. Hu and J. Rudnick, Phys. Rev. Lett. **48**, 1645 (1982).
[25] See, for example, J.P. Bouchaud, L.F. Cugliandolo, J. Kurchan and M. Mezard, in *Spin Glasses and Random Fields*, A.P. Young, editor (World Scientific, Singapore, 1998).
[26] See also, C.A. Angell, K.L. Ngai, G.B. McKenna, P.F. McMillan and S.W. Martin, J. Appl. Phys. **88** (2000) 3113.
[27] E. Barkai, Phys. Rev. Lett. **90**, 104101 (2003).
[28] See, for example, P. Gaspard and X.-J. Wang, Proc. Natl. Acad. Sci. USA, **85**, 4591 (1988).
[29] G. Anania and A. Politi, Europhys. Lett. **7** (1988).
[30] M.L. Lyra and C. Tsallis, Phys. Rev. Lett. **80**, 53 (1998).
[31] N.G. Antoniou, Y.F. Contoyiannis, F.K. Diakonos, and C.G. Papadoupoulos, Phys. Rev. Lett. **81**, 4289 (1998).
[32] N.G. Antoniou, Y.F. Contoyiannis, and F.K. Diakonos, Phys. Rev. E **62**, 3125 (2000).
[33] J.P. Crutchfield, J.D. Farmer and B.A. Huberman, Phys. Rep. **92**, 45 (1982).
[34] For a review see, P.G. De Benedetti and F.H. Stillinger, Nature **410**, 267 (2001).
[35] P.G. De Benedetti, *Metastable Liquids*. Concepts and Principles (Princeton Univ. Press, Princeton, 1996).
[36] K. Kaneko, Chaos **2**, 279 (1992).
[37] C. Tsallis, M. Gell-Mann and Y. Sato, Proc. Nat. Acad. Sci. **102**, 15377 (2005).
[38] W. Kob and H.C. Andersen, Phys. Rev. E **51**, 4626 (1995).

[39] W. Götze and L. Sjögren, Rep. Prog. Phys. **55**, 241 (1992).
[40] G. Adam and J.H. Gibbs, J. Chem. Phys. **43**, 139 (1965).
[41] E. R. Weeks and D.A. Weitz, Chem. Phys. **284**, 361 (2002); Phys. Rev. Lett. **89**, 095704 (2002).
[42] A. Lawlor, D. Reagan, G.D. McCullagh, P. De Gregorio, P. Tartaglia, and K.A. Dawson, Phys. Rev. Lett. **89**, 245503 (2002).
[43] See, for example, P.M. Chaikin and T.C. Lubensky, *Principles of condensed matter physics*, Cambridge University Press, Cambridge, UK, 1995.
[44] A. L. Stella and C. Vanderzande, Phys. Rev. Lett. **62**, 1067 (1989); C Vanderzande and A. L. Stella, J. Phys. A: Math. Gen. **22**, L445 (1989); A. Coniglio, Phys. Rev. Lett. **62**, 3054 (1989).
[45] S. Gupta, P. Lacock and H. Satz, Nucl. Phys. **B 362**, 583 (1991); Z. Burda, J. Wosiek and K. Zalewski, Phys. Lett. **B 266**, 439 (1991).
[46] V. Latora and M. Baranger, Phys. Rev. Lett. **82**, 520 (1999).

4

Non-Boltzmannian Entropies for Complex Classical Systems, Quantum Coherent States and Black Holes

A. G. Bashkirov

Institute Dynamics of Geospheres, Russian Academy of Sciences, 119334, Moscow, RUSSIA
E-Mail: abas@idg.chph.ras.ru

Summary. The Renyi entropy is derived as a cumulant average of the Boltzmann entropy in the same way as the Helmholtz free energy can be obtained by cumulant averaging of a Hamiltonian. Such a form of the information entropy and the principle of entropy maximum (MEP) for it are justified by the Shore–Johnson theorem. The application of MEP to the Renyi entropy gives rise to the Renyi distribution. Thermodynamic entropy in the Renyi thermostatistics increases with system complexity (gain of an order parameter $\eta = 1 - q$) and reaches its maximal value at $q_{\min}$. The Renyi distribution for such q becomes a pure power–law distribution. Because a power–law distribution is characteristic for self-organizing systems the Renyi entropy can be considered as a potential that drives the system to a self-organized state. The derivative of difference of entropies in the Renyi and Gibbs thermostatistics in respect to η exhibits a jump at $\eta = 0$. This permits us to consider the transfer to the Renyi thermostatistics as a peculiar kind of a phase transition into a more organized state.

The last section is devoted to development of thermodynamics of coherent states of quantum systems and black holes. The entropy of the quantum field is found to be proportional to the surface area of the static source. The Bekenstein–Hawking entropy of a black hole can also be interpreted as the thermodynamic entropy of coherent states of a physical vacuum in a vicinity of a horizon surface.

4.1 Introduction

By the early 20th century basic principles of statistical description of thermodynamic systems were well established by efforts of Boltzmann, Gibbs, Einstein and many others. In particular, a statistical definition of an entropy was found to be very important in the theory of non-equilibrium processes and fluctuations. As for an equilibrium statistical thermodynamics, the entropy was nothing more than one of characteristic thermodynamic functions. This situation changed drastically when methods of an information theory penetrated into statistical mechanics.

A.G. Bashkirov: *Non-Boltzmannian Entropies for Complex Classical Systems, Quantum Coherent States and Black Holes*, StudFuzz **206**, 114–162 (2006)
www.springerlink.com

In 1957 Jaynes [1] proposed a rule to assign numerical values to probabilities in circumstances where certain partial information is available. Jaynes showed in particular how his rule, when applied to statistical mechanics, leads to the usual Gibbs' canonical distribution. Today this rule, known as the Maximum Entropy Principle (MEP), is used in many fields, ranging from physics and chemistry to stock market analysis. Applicability in physics is, however, much wider. Aside of statistical thermodynamics, MEP has now become a powerful tool in non–equilibrium statistical physics [2, 3]. For the latest developments in classical MEP the interested reader may consult ref. [4] and citations therein.

Nevertheless the MEP has always remained controversial. The controversy is related to the very goal of Janes' approach, namely, to remove statistical mechanics from the field of physics and reconstruct it as a branch of logic or epistemology. This aspect of the MEP met the resistance from the physics community. The works of Jaynes [5], Penrose [6], Lavis and Milligan [7], Buck and Macaulay [8], Balian [9], Denbigh and Denbigh [10], Dougherty [11] and other cited there (see also [12]) provide insight in the pros and cons of the MEP in relation to statistical physics. It may be supposed that this discussion played a part in a general revision of relationship between physics and information theory noticed by Wheeler [13]: *"I, like other searchers, attempt formulation after formulation of the central issues...taking for a working hypothesis the most effective one that has survived this winnowing: It from bit. Otherwise put, every it – every particle, every force, even the space–time continuum itself – derives its function, its meaning, its very existence entirely-even if in some contexts indirectly-from the apparatus-elicited answers to yes-or-no questions, binary choices, bits. It from bit symbolizes the idea that the physical world has at bottom - at a very deep bottom, in most instances - an immaterial source and explanation; that which we call reality arises in the last analysis from the posing of yes-no questions and the registering of equipment-evoked responses; in short, that all things physical are information-theoretic in origin and this is a participatory universe."*

For our purposes it suffices to note that the term MEP is not a physical principle in the proper sense. Nevertheless the MEP may be used as a very useful heuristic method for construction of probability distributions. According to the MEP for the Boltzmann-Gibbs statistics, the distribution of probabilities $p = \{p_i\}$ is determined by the requirement of maximum of the Gibbs–Shannon entropy

$$S^{(G)} = -\sum_i p_i \ln p_i$$

under additional constraints of fixed value

$$U = \langle H \rangle_p \equiv \sum_i H_i p_i, \tag{4.1.1}$$

and normalization of p. Then, the distribution $\{p_i\}$ is determined from the extremum of the functional

$$L_G(p) = S^{(G)} - \beta_0 \sum_i^W H_i p_i - \alpha_0 \sum_i^W p_i; \tag{4.1.2}$$

where β_0 and α_0 are Lagrange multipliers . Its maximum is ensured by the Gibbs canonical distribution. Indeed, equating its functional derivative with respect to p_i to zero we find

$$\ln p_i + \beta_0 H_i + \alpha_0 + 1 = 0$$

whence

$$p_i = e^{-\alpha_0 - 1 - \beta_0 H_i}$$

With account of normalization condition $\sum_i p_i = 1$ we get $e^{\alpha_0+1} = Z_G$ and

$$p_i = p_i^{GZ} \equiv Z_G^{-1} e^{-\beta_0 H_i}, \qquad Z_G \equiv \sum_i e^{-\beta_0 H_i} \tag{4.1.3}$$

On the other side, multiplying Eq. (1.4) by p_i and summing up over i, with account of normalization condition $\sum_i p_i = 1$ we get $\alpha_0 + 1 = S^{(G)} - \beta_0 U$ and

$$p_i = p_i^{GS} \equiv e^{-S^{(G)} - \beta_0 \Delta H_i}, \qquad \Delta H_i = H_i - U \tag{4.1.4}$$

The expressions (4.1.3) and (4.1.4) may be called a Z-form and S-form of the Gibbs' distribution respectively in which β_0 is determined by condition of correspondence between Gibbs thermostatistics and classical thermodynamics as $\beta_0 = 1/k_B T_0$ where T_0 is the thermodynamic temperature. They are equivalent due to equivalence of two definitions of the Helmholtz free energy, $F = U - T_0 S$ and $F = -k_B T_0 \ln Z_G$.

However, when investigating complex physical systems (for example, fractal and self-organizing structures, turbulence) and a variety of social and biological systems, it appears that the Gibbs distribution does not correspond to observable phenomena. In particular, it is not compatible with a power-law distribution that is typical [14] for such systems. Introducing into Eq. (4.1.2) of additional constraints to a sought distribution in the form of conditions of true average values $\langle X^{(m)} \rangle_p$ of some physical parameters of the system $X^{(m)}$ gives rise to generalized Gibbs distributions with additional terms in the exponent (4.1.3) but does not change its exponential form.

Montroll and Shlesinger [15] investigated this problem and found that MEP for the Gibbs–Shannon entropy could give rise to the power–law distribution under only very special constraint that "has not been considered as a natural one for use in auxiliary conditions."

These problems will be the subject of next Secs. 2–10. The last Sec. 11 is devoted to development of thermodynamics of a quantum mechanical system in a coherent state. This approach provides also fresh insight (in Subsec. 11.3) into the problem of the black hole entropy.

4.2 Helmholtz free energy and Renyi entropy

On the other hand, there is no doubts about the MEP as itself, because of it is a kind of "a maximum honesty principle" according to which we demand a maximal uncertainty from the distribution apart from true description of prescribed average values. In the opposite case we risk to introduce a false information into the description of the system. Thus, it only remains for us to throw doubt on the information entropy form.

The well–known Boltzmann formula defines a statistical entropy, as a logarithm of a number of states W attainable for the system

$$S_W^{(B)} = \ln W \tag{4.2.1}$$

Here and below the entropy is written as dimensionless value without the Boltzmann constant k_B. Besides, we will use a natural logarithm instead of binary logarithm accepted in information theory.

This definition is valid not only for physical systems but for much more wide class of social, biological, communication and other systems described with the use of statistical approach. The only but decisive restriction on the validity of this equation is the condition that all W states of the system have equal probabilities (such systems are described in statistical physics by a microcanonical ensemble). It means that probabilities $p_i = p_W \equiv 1/W$ (for all $i = 1, 2, ..., W$) that permits to rewrite the Boltzmann formula (4.2.1) as

$$S_W^{(B)} = -\ln p_W.$$

When the probabilities p_i are not equal we can introduce an ensemble of microcanonical subsystems in such a manner that all W_i states of the i-th subsystem have equal probabilities p_i and its Boltzmann entropy is $S_i^{(B)} = -\ln p_i$. The simple averaging of the Boltzmann entropy $S_i^{(B)}$ leads to the Gibbs–Shannon entropy

$$S^{(G)} = \langle S_i^{(B)} \rangle_p \equiv -\sum_i p_i \ln p_i. \tag{4.2.2}$$

Just such derivation of $S^{(G)}$ is used in some textbooks (see, e. g. [16, 17]). This entropy is generally accepted in statistical thermodynamics and communication theory but needs in modification for complex systems. To seek out a direction of modification of the Gibbs–Shannon entropy we consider first extremal properties of an equilibrium state in thermodynamics.

A direct calculation of the average energy of a system gives the internal energy (4.1.1), its extremum is characteristic of an equilibrium state of rest for a mechanical system, other than a thermodynamic system that can change heat with a heat bath. An equilibrium state of the latter system is characterized by extremum of the Helmholtz free energy F. To derive it statistically from the Hamiltonian without use of thermodynamics we, following Balescu [5], introduce generating function for the random value H_i

$$\Phi_H(\lambda) = \sum_i e^{\lambda H_i},$$

where λ is the arbitrary constant, and construct a cumulant generating function

$$\Psi_H(\lambda) = \ln \Phi_H(\lambda)$$

that becomes the Helmholtz free energy F when devided by λ that is chosen as $\lambda = -1/k_B T_0$. Such a choice of the pre-factor $1/\lambda$ ensures a limiting passing of the Helmholtz free energy F into the internal energy U that is the simple equilibrium average of the Hamiltonian. Really, when $\lambda \to \infty$ ($T_0 \to 0$) with the use of the l'Hopital rule we get

$$F(T_0)|_{T_0 \to 0} = \lim_{T_0 \to 0} \left[-k_B T_0 \ln \sum_i e^{-H_i/k_B T_0} \right] = \sum_i p_i^{(G)} H_i = U.$$

It means that an equilibrium state of the thermodynamic system is defined by minimum of the internal energy as well as of the free energy at very low temperature of a heat bath T_0.

Now we return to the problem of a generalized entropy for open complex systems. Exchange by both energy and entropy is characteristic for them. As an illustration, there is a description by Kadomtsev [19] of self-organized structure in a plasma sphere: "The entropy is being born continuously within the sphere and flowing out into surroundings. If the entropy flow had been blocked, the plasma would 'die'. It is necessary to remove continuously 'slag' of newly produced entropy".

That is the reason why the Gibbs–Shannon entropy, derived by the simple averaging of the Boltzmann entropy can not be à function of which extremum characterizes a steady state of a complex system which exchange entropy with surroundings, just as the minimum of the internal energy does not characterize an equilibrium state of the thermodynamic system being in heat contact with a heat bath.

It is pertinent to introduce the noun of an *entropy bath* (or *information bath*). Coupling with the entropy bath is a necessary condition for self-organization of a complex system. As a result of such coupling the system under consideration can not reach a state of thermodynamic equilibrium that is characterized by minimum of the Helmholtz free energy. It is necessary to look for any other function to characterize its steady state resulted from the coupling with the entropy bath.

An effort may be made to find a "free entropy" of a sort by the same way that was used above for derivation of the Helmholtz free energy for a system coupled with a heat bath. The generating function is introduced as

$$\Phi_S(\lambda) = \sum_i e^{\lambda S_i^{(B)}}$$

Then the cumulant generating function is

$$\Psi_S(\lambda) = \ln \Phi_S(\lambda) = \ln \sum_i p_i^{-\lambda}.$$

To obtain the desired generalization of the entropy we are to find a λ-dependent numerical pre-factor which ensures a limiting pass of the new entropy into the Gibbs–Shannon entropy. Such the coefficient is $1/(\lambda+1)$. Indeed, the new λ-family of entropies

$$S(\lambda) = \frac{1}{1+\lambda} \ln \sum_i p_i^{-\lambda}.$$

includes the Gibbs–Shannon entropy as a particular case when $\lambda \to -1$.

Thus, it has appeared that the desired "free entropy" coincides with the known Renyi entropy [20, 21]. It is conventional to write it with the parameter $q = -\lambda$ in the form

$$S_q^{(R)}(p) = \frac{1}{1-q} \ln \sum_i p_i^q \tag{4.2.3}$$

Renyi introduced his entropy on a base of strictly formal motivation. Renyi wanted to find the most general class of entropies which preserved the additivity of statistically independent systems and were compatible with Kolmogorov–Nagumo [22, 23] generalized averages of the form

$$\langle x \rangle_\phi = \phi^{-1}\left(\sum_i p_i \phi(x_i)\right) \tag{4.2.4}$$

where $\phi(x)$ is an arbitrary continuous and strictly monotonic function, $\phi^{-1}(x)$ is the inverse function and $x_i = S_i^{(B)}$. Renyi then proved that when the postulate of entropy additivity for independent events is applied to Eq.(4.2.4) it dramatically restricts the class of possible $\phi(x)$'s. In fact, only two types are possible; $\phi(x) = c\,x + d$ which implies the Gibbs-Shannon entropy (4.2.2) and $\phi(x) = c\,e^{(1-q)x}$ with $q > 0$ (c and d are arbitrary constants) which implies the Renyi entropy (4.2.3). Note that for linear $\phi(x)$'s the Kolmogorov–Nagumo generalized average turns out to be the ordinary linear mean and hence Gibbs-Shannon entropy is the averaged entropy in the usual sense as it was shown above.

On the other hand, the exponential Kolmogorov–Nagumo function $\phi(x)$ leads us to the same expression that was derived above as a "free entropy". Physically, such a choice of $\phi(x)$ on its own appears accidental here until it is not pointed to the fact that the same exponential function of the Hamiltonian provides derivation of the free energy which is extremal at an equilibrium

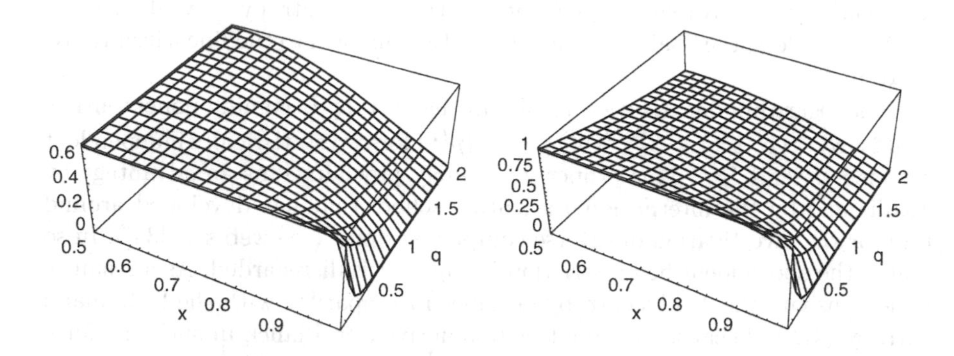

Fig. 4.1. The entropies $S_q^{(R)}(p)$ (left) and $S_q^{(Ts)}(p)$ (right) for the particular case $W = 2$, $p_1 = x$, $p_2 = 1 - x$ and $0.5 < x < 1$, $0 < q < 2$.

state of a thermodynamic system exchanging heat with a heat bath. This fact permits to suppose that the Renyi entropy derived in the same manner should be extremal at a steady state of a complex system which exchange entropy with its surroundings very actively.

When the postulate of entropy additivity for independent events is rejected, the class of possible $\phi(x)$ is enlarged sufficiently. Nevertheless, for any acceptable $\phi(x)$, an entropy S_ϕ does pass into the Boltzmann entropy in the case of equally probable distribution, and in my opinion such a behavior is to be considered as a criterion of self-consistency of any generalized form of entropy.

Different properties of the Renyi entropy are discussed in particular in Refs. [20, 24, 25]. It is positive ($S_q^{(R)} \geq 0$), convex at $q \leq 1$, passes into the Gibbs–Shannon entropy $\lim_{q\to 1} S_q^{(R)} = S^{(G)}$ and into Boltzmann entropy (4.2.1) for any q in the case of equally probable distribution p. An important general property of the Renyi entropy is that it is a monotonically decreasing function of q:

$$S_q^{(R)}(p) \geq S_{q'}^{(R)}(p) \qquad \text{for } q < q'.$$

Renyi's entropy hass proved to be important in variety of practical applications. Coding theory [26], statistical inference [27], quantum mechanics [28], chaotic dynamical systems [29, 30, 31, 32] and multifractals provide examples.

In the case of $|1 - \sum_i p_i^q| \ll 1$ (which, in view of normalization of the distribution $\{p_i\}$, corresponds to the condition $|1 - q| \ll 1$), one can restrict oneself to the linear term of logarithm expansion in the expression for $S_q^{(R)}(p)$ over this difference, and $S_q^{(R)}(p)$ changes to the Tsallis entropy [33]

$$S_q^{(Ts)}(p) = \frac{1}{1-q}\left(\sum_i^W p_i^q - 1\right).$$

This entropy is derivable independently of the Renyi entropy, as well (see Sec. 4). Nevertheless, the Tsallis entropy is identical in form to the linearized Renyi entropy.

The logarithm linearization results in the entropy becoming nonextensive, that is, $S_q^{(Ts)}(p^I \cdot p^{II}) \neq S_q^{(Ts)}(p^I) + S_q^{(Ts)}(p^{II})$ for two statistically independent systems Σ^I and Σ^{II}. This property is widely proclaimed as an advantage by Tsallis and by the international scientific school that has developed around him for the investigation of diverse complex systems (see web site [34]). In so doing, the above-identified restriction $|1-q| \ll 1$ is disregarded. As a result of nonextensivity the Tsallis entropy becomes incompatible with the Boltzmann entropy (4.2.1) because of the latter was derived by Planck in such the form just from the extensivity condition $S(p^I \cdot p^{II}) = S(p^I) + S(p^{II})$.

Both entropies are illustrated in Fig. 1 for the particular case $W = 2$ and $p_1 = x$, $p_2 = 1 - x$ at the range $0.5 < x < 1$ and $0 < q < 2$. It is seen that $S_q^{(R)}(p) < S_q^{(Ts)}(p)$ as would be expected for the logarithm function and its linearization. Another special feature of the Renyi entropy is that for $p_1 = p_2 = 1/2$ $(x = 0.5)$ it becomes independent on q and equal to $S^{B)}(W = 2) = \ln 2 \simeq 0.69$ in contrast to the Tsallis entropy that does not pass into the Boltzmann entropy for such p and remains to be q-dependent.

4.3 On stability of Renyi and Tsallis entropies

Currently a lot of papers by Tsallis and his coworkers have appeared (see e.g. [35, 36, 37, 38, 39]) where statements are repeated that the Renyi entropy is unstable (discontinuous) under arbitrary small deformations of any given probability distribution in contrast to stability (continuity) of the Tsallis entropy. These works are based on the papers by Abe [40] and Leshe [41].

Two counterexamples (for $q < 1$ and $q > 1$) of instability of the Renyi entropy were discussed in Ref. [40] following Leshe [41] and it was shown that the Tsallis entropy is stable for these counterexamples. From the time of its publication this work is often referred as a mortal verdict for the Renyi entropy. On the other hand, the Renyi entropy is widely used now. Because of this, the main points of Ref. [40] are to be revised carefully. This question became the subject of our discussion with Abe [42, 43]. Abe [40] calculated responses $\left|\Delta S_q^{(R)}\right|$ and $\left|\Delta S_q^{(Ts)}\right|$ to small variation of initial model distributions over W states of a system and then passed to the limit $W \to \infty$ treating an amplitude δ of the variation as a finite constant. As a result, he found a loss of continuity of a response of the Renyi entropy to small perturbations. We feel that such a conclusion is inadmissible on two counts. First, the normalized values

$$\left|\Delta\left(\frac{S_q^{(R)}}{S_{\max}^{(R)}}\right)\right| \quad \text{and} \quad \left|\Delta\left(\frac{S_q^{(Ts)}}{S_{\max}^{(Ts)}}\right)\right|$$

rather than $|\Delta S_q^{(R)}|$ and $|\Delta S_q^{(Ts)}|$ per se, were considered with different W-dependent normalization factors, $S_{max}^{(R)} = \ln W$ and $S_{max}^{(Ts)} = (W^{1-q} - 1)/(1-q)$, respectively. Such different normalizations influenced their limiting properties and, consequently, conclusions about their stabilities. Indeed, their ratio at large W is

$$\frac{S_{\max}^{(Ts)}}{S_{\max}^{(R)}} = \frac{1}{1-q} \frac{W^{1-q}-1}{\ln W}\bigg|_{W\to\infty} \longrightarrow W^{1-q}.$$

It is just this additional multiplier which ensures continuity of the gain of the normalized Tsallis entropy $S_q^{(Ts)}/S_{\max}^{(Ts)}$ at $q < 1$ in contrast to $S_q^{(R)}/S_{\max}^{(R)}$. It would be more reasonable to normalize both entropies by the common denominator, say, $\sup\{S_{\max}^{(R)}, S_{\max}^{(Ts)}\}$ or number of states W. But the most essential point is that we have to deal only with unbounded entropies because of their bounded versions $S_q^{(R)}/S_{\max}^{(R)}$ and $S_q^{(Ts)}/S_{\max}^{(Ts)}$ are irrelevant both to thermostatistics and information theory. Second, continuity is to be checked with the use of the opposite iterated limiting process: first, $\delta \to 0$ and then $W \to \infty$. Such an order corresponds to a traditional approach in statistical physics where all properties are calculated first for finite systems and the thermodynamic limit is performed after all calculations (see, e. g. [44]).

In what follows, results of Ref. [40] are reconsidered with due regard to both these observations. For the sake of brevity, the first of Abe's counterexamples alone will be discussed here. It is especially important, as it refers to the range $0 < q < 1$ of most if not all of the applications [45] of the Renyi entropy. The second counterexample may be discussed in the same manner.

The examined small ($\delta \ll 1$) deformation of distribution $\{p\}$ over W states ($W \gg 1$) for $0 < q < 1$ is

$$p_i = \delta_{i1}, \qquad p_i' = \left(1 - \frac{\delta}{2}\frac{W}{W-1}\right) p_i + \frac{\delta}{2}\frac{1}{W-1}. \tag{4.3.1}$$

Using the well-known definitions of the Tsallis and Renyi entropies, we get

$$\left|\Delta S_q^{(R)}\right| = \frac{1}{1-q} \ln\left[\left(1-\frac{\delta}{2}\right)^q + \left(\frac{\delta}{2}\right)^q (W-1)^{1-q}\right] \tag{4.3.2}$$

$$\left|\Delta S_q^{(Ts)}\right| = \frac{1}{1-q} \left[\left(1-\frac{\delta}{2}\right)^q + \left(\frac{\delta}{2}\right)^q (W-1)^{1-q} - 1\right] \tag{4.3.3}$$

It is seen from here that the gain of the Tsallis entropy is no more than the linearization of the logarithm form of the gain of the Renyi entropy. As a consequence of concavity of the logarithm function we have

$$\left|\Delta S_q^{(R)}\right| < \left|\Delta S_q^{(Ts)}\right| \qquad \text{for all } \delta \text{ and } W.$$

Thus, stability of the Renyi entropy for the counterexample (4.3.1) is at least not lower stability of the Tsallis entropy.

Because he considered their normalized values with different W-dependent normalization factors, $S_{\max}^{(R)}$, $S_{\max}^{(Ts)}$, Abe was not concerned with this evident inequality. This difference in normalization influenced their limiting properties and consequently, their stabilities. In particular, Abe found that

$$\left|\frac{\Delta S_q^{(R)}}{S_{\max}^{(R)}}\right| \to 1 \qquad \text{and} \qquad \left|\frac{\Delta S_q^{(Ts)}}{S_{\max}^{(Ts)}}\right| \to \left(\frac{\delta}{2}\right)^q$$

for all q. Then, we can take the case $q \simeq 1$ when $S_q^{(R)} \simeq S_q^{(Ts)}$. According to [40], one of this entropies, $S_q^{(R)}$ is unstable, while other one, $S_q^{(Ts)}$ is stable due to different name only.

As for stability of the Renyi and Tsallis entropies, there is no double limit ($\delta \to 0$, $W \to \infty$) of $\left|\Delta S_q^{(R)}\right|$ and $\left|\Delta S_q^{(Ts)}\right|$ as functions of δ and W but there is a repeated limit ($\delta \to 0$ and then $W \to \infty$) and it is equal to zero. Indeed, for any finite W and $\delta/2 \ll (W-1)^{-|1-q|/q}$ these functions, Eqs. (4.3.2) and (4.3.3), become infinitesimal

$$\left|\Delta S_q^{(R)}(p)\right| \simeq \frac{1}{1-q}\left(\frac{\delta}{2}\right)^q (W-1)^{1-q}, \qquad 0<q<1,$$

$$\left|\Delta S_q^{(Ts)}(p)\right| \simeq \frac{1}{1-q}\left(\frac{\delta}{2}\right)^q (W-1)^{1-q}, \qquad 0<q<1,$$

In terms of $\varepsilon - \delta$ the continuity condition is formulated in the next form: For every given $\varepsilon > 0$ we are to find such δ that both ΔS become less ε if $\sum_i |p_i - p_i'| \leq \delta$. Here it means that $\delta < 2[(1-q)W^{1-q}\varepsilon]^{1/q}$. Dependence of this condition on W is a result of the fact that both entropies are not bounded functionals.

Another attack on the Renyi entropy was recently launched by Lesche [46]. The most important points of his paper related to the subject are outlined below. He considers first an initial macrostate that is characterized by eigenvalues $a_I, b_I, ..., h_I$ of macroscopic commeasurable observables $A, B, ..., H$. The special characteristics of this kind of state are the following: (1) The number of occupied microstates $W_I = W_{a_I,b_I,...,h_I}$ is of the order M^N and therefore any individual probability $(W_I)^{-1}$ is extremely small (of the order of M^{-N}). (2) The number of empty microstates is larger than the number of occupied microstates by a huge factor, which is also of the order of $\tilde{M}^N$. Essentially these two characteristics make this sort of state a problem case.

This initial state is of the form

$$\rho = \sum_j^{W_I} |a_I, b_I, ..., h_I, j\rangle \frac{1}{W_I} \langle a_I, b_I, ..., h_I, j|$$

where $|a_I, b_I, ..., h_I, j\rangle$ is a basis of common eigenstates (Dirac's ket-vectors) and j is an index of degeneracy of the initial macrostate.

The Renyi entropy of this state becomes the Boltzmann entropy of the macrostate

$$S_q^{(R)} = S^{(B)} = \ln W_I \tag{4.3.4}$$

Then Leshe proposes to imagine that a friend of ours enters the laboratory and criticizes our experiment. He claims our preparation of state may in some cases result in the macrostate $[\bar{a}, \bar{b}, ..., \bar{h}]$ with the number of occupied microstates $\overline{W} = W_{\bar{a},\bar{b},...,\bar{h}}$. Now, if the probability of such cases, say, $\delta = 10^{-100}$, we will not be able to convince our friend that our probability assignment is better than his by showing experimental results. His density operator would be

$$\tilde{\rho} = \sum_j^{W_I} |a_I, b_I, ..., h_I, j\rangle \frac{1-\delta}{W_I} \langle a_I, b_I, ..., h_I, j| + \sum_j^{\overline{W}} |\bar{a}, \bar{b}, ..., \bar{h}, j\rangle \frac{\delta}{\overline{W}} \langle \bar{a}, \bar{b}, ..., \bar{h}, j|$$

The Renyi entropy of the friend's probability assignment is

$$\begin{aligned} \tilde{S}_q^{(R)}(\delta) &= \frac{1}{1-q} \ln \left\{ (1-\delta)^q \, W_I^{1-q} + \delta^q \, \overline{W}^{1-q} \right\} \\ &= \frac{1}{1-q} \ln \left\{ \delta^q \, \overline{W}^{1-q} \left(1 + \frac{(1-\delta)^q}{\delta^q} \frac{W_I^{1-q}}{\overline{W}^{1-q}} \right) \right\} \\ &= \ln \left(\delta^{q/1-q} \, \overline{W} \right) + \frac{1}{1-q} \ln \left(1 + \frac{(1-\delta)^q}{\delta^q} \frac{W_I^{1-q}}{\overline{W}^{1-q}} \right) \end{aligned}$$

The first term is of order N. To estimate the second term we now distinguish the following two cases: (1) If $q > 1$, we shall assume that our friend thought of a state $[\bar{a}, \bar{b}, ..., \bar{h}]$ with smaller entropy than the main state $[a_I, b_I, ..., h_I]$, that is,

$$\overline{W} \ll \delta^{q/q-1} W_I \qquad \text{for } q > 1 \tag{4.3.5}$$

If $S^{(B)}(a_I, b_I, ..., h_I) - S^{(B)}(\bar{a}, \bar{b}, ..., \bar{h})$ is macroscopic (of the order N), the third term is clearly also negligible as compared to the first one. (2) If $q < 1$, we assume that the friend thought of a state $[\bar{a}, \bar{b}, ..., \bar{h}]$ whose entropy is macroscopically larger than $S^{(B)}(a_I, b_I, ..., h_I)$, that is

$$\overline{W} \gg \delta^{-q/1-q} W_I \qquad \text{for } q < 1 \tag{4.3.6}$$

Again the second term will be negligible.

$$\tilde{S}_q^{(R)}(\delta) = \ln \left(\delta^{q/1-q} \overline{W} \right) \tag{4.3.7}$$

or

$$\tilde{S}_q^{(R)}(\delta) = \ln \overline{W} + \frac{q}{1-q} \ln \delta = \ln \left(\delta^{-1} \overline{W} \right) + \frac{1}{1-q} \ln \delta \tag{4.3.8}$$

Then, according to Lesche, the second terms of these equations are negligible as compared to the first ones (for instance, with $N \approx 10^{24}$ and $\delta = 10^{-100}$ his argument is restriction to q values with $|1-q| \gg 10^{-22}$). So, in either case, the Renyi entropy of the friend's probability assignment would essentially be the entropy of the irrelevant state $(\bar{a}, \bar{b}, ..., \bar{h})$

$$\tilde{S}_q^{(R)} \simeq \ln \overline{W} \simeq \ln\left(\delta^{-1}\overline{W}\right), \tag{4.3.9}$$

which is far away from our initial value, Eq. (4.3.4).

I begin my comments with the remark that Lesche did not touch the same problem for the Tsallis entropy in spite of that he thanks C. Tsallis and S. Abe for bringing his attention to the subject that he abandoned many years ago.

I have made an effort [47] to make up this deficiency. Following are corresponding speculations for the Tsallis entropy. The Tsallis entropy of the initial state ρ is

$$S_q^{(Ts)} = \frac{1}{1-q}\left(W_I^{1-q} - 1\right) \tag{4.3.10}$$

For the friend's probability assignment $\tilde{\rho}$ it becomes

$$\begin{aligned}\tilde{S}_q^{(Ts)}(\delta) &= \frac{1}{1-q}\{(1-\delta)^q\, W_I^{1-q} + \delta^q\, \overline{W}^{1-q} - 1\} \\ &= \frac{1}{1-q}\left(\delta^{q/1-q}\,\overline{W}\right)^{1-q}\left\{1 + \frac{W_I^{1-q} - 1}{(\delta^{q/1-q}\,\overline{W})^{1-q}}\right\}\end{aligned}$$

With the use of the same speculations that have led Lesche to Eqs. (4.3.7)-(4.3.9) the Tsallis entropy for both cases $q > 1$ and $q < 1$ takes the form

$$\tilde{S}_q^{(Ts)}(\delta) = \frac{1}{1-q}\left((\delta^{q/1-q}\,\overline{W})^{1-q} - 1\right)$$

So, we see that the Tsallis entropy is far away from its initial value, Eq. (4.3.10), as well as the Renyi entropy.

Under restrictions (4.3.5) and (4.3.6) they both correspond to an equally probable distribution $p_j = (\delta^{q/1-q}\overline{W})^{-1}$ (for all j) related to the friend's state $[\bar{a}, \bar{b}, ..., \bar{h}]$ without regard for the initial state $[a_I, b_I, ..., h_I]$. Just this result should be expected. Indeed, states with greatest (least) probabilities contribute significantly to both q-entropies at $q > 1$ ($q < 1$). The restrictions (4.3.5) and (4.3.6) may be rewritten for the probabilities as

$$\begin{aligned}\frac{\delta}{\overline{W}} &\gg 10^{100/(q-1)}\frac{1-\delta}{W_I} \qquad \text{for } q > 1; \\ &\ll 10^{100/(q-1)}\frac{1-\delta}{W_I} \qquad \text{for } q < 1.\end{aligned}$$

Then, the resulting equations for $\tilde{S}_q^{(R)}(\delta)$ and $\tilde{S}_q^{(Ts)}(\delta)$ are to be accepted as quite natural, but not as evidences of their instabilities.

It is interesting to discuss an alternative form of the Renyi and Tsallis entropies expansions. The more natural form of the Renyi entropy expansion should start with the Boltzmann entropy, Eq. (4.3.4), as a leading term of the expansion. Then

$$\begin{aligned} S_q^{(R)}(\delta) &= \frac{1}{1-q}\ln\left\{(1-\delta)^q W_I^{1-q}\left(1+\frac{\delta^q}{(1-\delta)^q}\frac{\overline{W}^{1-q}}{W_I^{1-q}}\right)\right\} \\ &= \ln W_I + \frac{q}{1-q}\ln(1-\delta) + \frac{1}{1-q}\ln\left(1+\frac{\delta^q}{(1-\delta)^q}\frac{\overline{W}^{1-q}}{W_I^{1-q}}\right) \end{aligned}$$

The second term of this expansion is negligible when $|1-q| \gg \delta$. The third term becomes negligible when the macrostate proposed by the friend fulfills the inequalities

$$\frac{\overline{W}^{1-q}}{W_I^{1-q}} \ll \delta^{-q}$$

or, taking into consideration that $\delta = 10^{-100}$,

$$\begin{aligned} \frac{\delta}{\overline{\overline{W}}} &\ll 10^{100/(q-1)}\frac{1-\delta}{W_I} \qquad \text{for } q>1; \\ &\gg 10^{100/(q-1)}\frac{1-\delta}{W_I} \qquad \text{for } q<1 \end{aligned}$$

We see that δ-pre-factors $10^{100/(q-1)}$ in both last inequalities alleviate restrictions imposed on $\overline{W}$, in contrast to the role of such δ-pre-factors in the inequalities (3.20), (3.21) of the Lesche's expansion. It means that the third term of our expansion can be neglected in much more numerous cases than the one of the Lesche expansion. In particular, cases of $\overline{W} \simeq W_I$ are included here.

It may be supposed that the macrostate $[\bar{a}, \bar{b}, ..., \bar{h}]$ chosen stochastically would fulfill inequalities (3.24) or (3.25), but not (3.20) or (3.21) with overwhelming probability, in contrast to the choice by the Lesche's friend who takes the part of Maxwell's demon of a sort. So, the Renyi entropy of the more probable assignment would be essentially the same Boltzmann entropy of our initial state

$$S_q^{(R)}(\delta) \simeq \ln W_I.$$

$$\begin{aligned} \tilde{S}_q^{(Ts)}(\delta) &= \frac{1}{1-q}\{(1-\delta)^q W_I^{1-q} + \delta^q \overline{W}^{1-q} - 1\} \\ &= \frac{1}{1-q}(1-\delta)^q W_I^{1-q}\left\{1+\frac{\delta^q \overline{W}^{1-q} - 1}{(1-\delta)^q W_I^{1-q}}\right\} \\ &\simeq \frac{1}{1-q}\left\{(1-\delta)^q W_I^{1-q} - 1\right\} \end{aligned}$$

$$\simeq \frac{1}{1-q}\left\{W_I^{1-q}-1\right\},$$

Similar alternative expansion of the Tsallis entropy under the same restrictions gives rise to which is the Tsallis entropy of the initial state. In general, it should be proposed to Lesche's experimenters to perform a coarse-graining with a corresponding weight function over all possible initial states if they are not sure in their initial choice in such a degree that they are ready to accept the friend's version. Discussion of such an approach would be too far from the subject of the Lesche original paper [46], so I have confined myself to indication of the alternative initial states.

The above speculations count in favor of stability of both Renyi and Tsallis entropies relative to stochastic perturbations of an initial macrostate with the uniform distribution of probabilities of microstates.

4.4 Axiomatics of an information entropy

The above physical foundation may appear as yet imperfectly conclusive to revise the traditional Gibbs-Shannon inforamtion entropy.

To find the most rigorous foundation of the new information entropy it is desirable to axiomatize it. Shannon [48] was the first who investigated axiomatic foundation of the Gibbs-Shannon entropy. These axioms were advanced by Khinchin [49] and are as follows, see also Ref. [24]):

1. $S(p)$ is a function of the probabilities p_i only and has to take its maximum value for the uniform distribution of probabilities $p_i = 1/W$: $S(1/W, ..., 1/W) \geq S(p')$, where p' is any other distribution.
2. The second axiom refers to a composition Σ of a master subsystem Σ^I and subordinate subsystem Σ^{II} for which probability of a composed state is
$$p_{ij} = Q(j|i)p_i^I \tag{4.4.1}$$
where $Q(j|i)$ is the conditional probability to find the subsystem Σ^{II} in the state j if the master subsystem Σ^I is in the state i. Then the axiom requires that
$$S(p) = S(p^I) + S(p^I|p^{II}) \tag{4.4.2}$$
where
$$S(p^I|p^{II}) = \sum_i p_i^I S(Q|i) \tag{4.4.3}$$
is the conditional entropy and $S(Q|i)$ is the partial conditional entropy of the subsystem Σ^{II} when the subsystem Σ^I is in the i-th state.
3. $S(p)$ remains unchanged if the sample set is enlarged by a new, impossible event with zero probability: $S(p_1, ..., p_W) = S(p_1, ..., p_W, 0)$

While proving his uniqueness theorem Khinchin enlarged the second axiom. He supposed that all U states of the composite system Σ were equally probable, that is, $p_{ij} = 1/U$ for all i and j; whence he got

$$S(p) = \ln U \tag{4.4.4}$$

Besides, he supposed that U_i^{II} equally probable states of the subordinate subsystem Σ^{II} corresponded to each i-th state of the master subsystem Σ^I. Just this assumption is the most questionable. Indeed, the probability distribution for the subsystem Σ^{II} coupled with the master subsystem should be rather canonical distribution than equally probable one. Nevertheless, Khinchin took $Q(j|i) = 1/U_i^{II}$ for $i = 1, ..., W$ and $j \in U_i^{II}$; whence from Eqs. (4.4.1) and (4.4.3) he obtained

$$U_i^{II} = Up_i^I, \qquad S(Q(j|i) = \ln U_i^{II}, \qquad S(p^I|p^{II}) = \ln U + \sum_i p_i^I \ln p_i^I. \tag{4.4.5}$$

Substituting Eqs. (4.4.4), (4.4.5) into Eq. (4.4.2) Khinchin found the Gibbs–Shannon entropy for the master subsystem Σ^I as

$$S^{(G)}(p^I) = -\sum_i^W p_i^I \ln p_i^I.$$

The second axiom can be changed to the more weak form:

(2') The information entropy for independent subsystems Σ^I and Σ^{II} is additive:

$$S(p) = S(p^I) + S(p^{II}).$$

The set of axioms (1), (2'), (3) can be satisfied with both the Gibbs–Shannon entropy and the more general Renyi information entropy $S^{(R)}(p)$.

Abe [50] proposed a set of three axioms for the Tsallis entropy. The first and third of his axioms are copies of Khinchin's axioms (1) and (3). His second axiom is of crucial importance in justification of the Tsallis entropy. It is written as

$$S_q(p^I, p^{II}) = S_q(p^I) + S_q(p^I|p^{II}) + (1-q)S_q(p^I)S_q(p^I|p^{II}) \tag{4.4.6}$$

where

$$S(p^I|p^{II}) = \sum_i P_i^I S(Q|i)$$

and

$$P_i = \frac{p_i^q}{\sum_j p_j^q}$$

is the escort distribution associated with p_i [24].

On the basis of these three axioms the uniqueness theorem for the Tsallis entropy was proved [50]. There are two key points in his second axiom. First, the escort distribution is introduced into the Tsallis thermostatistics as a corner stone of its axiomatic foundation but not as one of auxiliary tools. Second, in the case of statistically independent subsystems Σ^I and Σ^{II} the entropy $S_q(p^I, p^{II})$ does not become a sum of two subsystem's entropies. Indeed, in this case the conditional entropy $S_q(p^I|p^{II})$ passes into the entropy of the subsystem Σ^{II} and the second axiom, Eq. (4.4.6) becomes

$$S_q(p^I, p^{II}) = S_q(p^I) + S_q(p^{II}) + (1-q)S_q(p^I)S_q(p^{II})$$

It means that this axiom introduces non-additivity or non-extensivity of the information entropy of a compound system even though its subsystems are statistically independent. Tsallis and his adherents enunciate this property as a crucial principle of their thermostatistics.

Another set of axioms for the Renyi entropy was proposed by Jizba and Arimitsu [51]:

1. For a given integer W and given $p = \{p_1, p_2, \ldots, p_W\}$ $(p_k \geq 0, \sum_k^W p_k = 1)$, $S^{(R)}(p)$ is a continuous with respect to all its arguments.
2. For a given integer W, $S^{(R)}(p_1, p_2, \ldots, p_W)$ takes its largest value for $p_k = 1/W$ $(k = 1, 2, \ldots, W)$ with the normalization $S^{(R)}\left(\frac{1}{2}, \frac{1}{2}\right) = 1$.
3. For a given $q > 0$; $S^{(R)}(A \cap B) = S^{(R)}(A) + S^{(R)}(B|A)$ with
$$S^{(R)}(B|A) = \phi^{-1}\left(\sum_k P_k(q)\phi(S^{(R)}(B|A = A_k))\right)$$
and $P_k(q) = p_k^q / \sum_k p_k^q$.
4. $\phi(x)$ is invertible and positive in $[0, \infty)$.
5. $S^{(R)}(p_1, p_2, \ldots, p_W, 0) = S^{(R)}(p_1, p_2, \ldots, p_W)$, i.e., adding an event of probability zero (impossible event) we do not gain any new information.

Note particularly the appearance of distribution $P(q)$ in the axiom (3). This is the same escort distribution that was introduced in the Abe's set of axioms. Note also the Kolmogorov-Nagumo generalized average in the axiom (3) for the conditional entropy. Jizba and Arimitsu chose the Kolmogorov-Nagumo function ϕ in the exponential form just as it was done in the original derivation of the Renyi entropy. As a result, they obtained their axiomatic foundation of same Renyi entropy, as it would be expected for such the choice of ϕ.

It should be observed that all above three sets of axioms differ mainly in the forms of conditional entropies. They appear as Procrustean Beds[1] of sorts each of which is fitted to select one information entropy and to exclude any other forms of it. Moreover, we have no *a priori* recipe to give one of them preference over another two. Such a controversy prompts us to recall the general notion of an axiom. According to Encyclopedia Britannica, *"Axiom in*

[1] After Procrustes, a mythical Greek giant who stretched or shortened captives to make them fit his beds.

logic, an indemonstrable first principle, rule, or maxim, that has found general acceptance or is thought worthy of common acceptance whether by virtue of a claim to intrinsic merit or on the basis of an appeal to self-evidence. An example would be: 'Nothing can both be and not be at the same time and in the same respect."' In this aspect, all above axioms related to conventional entropies can not be accepted as axioms. Indeed, each of them is not "an indemonstrable first principle, rule, or maxim, that has found general acceptance or is thought worthy of common acceptance whether by virtue of a claim to intrinsic merit or on the basis of an appeal to self-evidence" but man-made restriction fitted in well with one preferred kind of information entropy.

There is another set of five axioms that corresponds to the formal definition of the noun "axiom" by Shore and Johnson in 1980 [52, 53][2]; see also Uffink [12]. In their approach, it is assumed that one is looking for a procedure by which a prior probability distribution $\{u_i\}$ is changed into a posterior distribution $\{p_i\}$ when new information is taken into account. It is further subsumed that the procedure takes the form of maximizing some relative uncertainty expression of the form $F(u,p)$ under the constraint. However, the procedure is characterized by how the posterior depends on the prior distribution and the new information. MEP appears only as the special case where the prior distribution is uniform.

Technically, the problem is formulated as follows. A system has a set of W possible states with an unknown true distribution $\{p_i\}$. The prior distribution, $\{u_i\}$ represents a (subjective) estimate of $\{p_i\}$ before new information is given. In response to this information, the prior density $\{u_i\}$ is changed into a posterior density $\{p_i\}$. It is assumed that the inference rule yields this posterior $\{p_i\}$ as a function depending only on the prior $\{u_i\}$ and the constraint I, symbolically written as:

$$p = I \circ u$$

where $\circ$ is an 'updating operator'.

Shore and Johnson give five "consistency axioms" for this updating operator [52, 53].

1. *Uniqueness:* The result should be unique.
2. *Invariance:* The choice of coordinate system should not matter[3].
3. *System independence:* It should not matter whether one accounts for independent information about independent systems separately in terms of different densities or in terms of a joint density.
4. *Subset independence:* It should not matter whether one treats disjoint subsets of system states in terms of separate conditional densities or in terms of the full density.

[2] *Note added in proof:* Although their work has been unfortunately ignored so far, Abe and Bagci (Phys. Rev. **E71**, 016139 (2005)) has now published an account of these axiomatics.

[3] This axiom is important for continuous probability densities

5. In the absence of new information, we should not change the prior.

The above axioms do indeed seem reasonable for the outlined problem. The following theorem is proven (see [12]) on the base of these axioms.

Theorem 4.1. *An updating procedure satisfies the five consistency axioms above if and only if it is equivalent to the rule*

$$\text{Maximize } U_\eta(u,p) \text{ under the constraint } I$$

where

$$U_\eta(u,p) = \left(\sum_i^W \left(\frac{P_i}{u_i} \right)^{-\eta} P_i \right)^{1/\eta}, \qquad \eta < 1.$$

For the particular case of a homogeneous prior distribution $u_i = 1/W$ (for all i) this rule can be modified as

$$\text{Maximize any monotonous function } \Psi \text{ of}$$

$$U_\eta(p) = \left(\sum_i^W p_i^{1-\eta} \right)^{1/\eta}$$

under the constraints I.

The most evident choice of the monotonous function is $\Psi(U_\eta) = \ln U_\eta(p)$, that is the Renyi entropy $S_q^{(R)}$ for $q = 1 - \eta$. Such a choice of Ψ ensures a passage to the limit $S_q^{(R)} \to S^{(G)}$ when $q \to 1$. In the absence of any additional information I the operator o becomes the unit operator and the Renyi entropy for any q transforms into the Boltzmann entropy. Thus, the Shore-Johnson theorem provide quite conclusive foundation of the Renyi entropy as itself and the maximum entropy principle for it and in doing so it justifies the proposal stated in Sec. 2 that the Renyi entropy is maximal at a steady state of a complex system.

On the other hand, another choice of the monotonous function Ψ is possible that ensures the passage to the limiting case $S^{(G)}$ when $q \to 1$. It is $\Psi(U_\eta) = ([U_\eta(p)]^\eta - 1)/\eta$, that is the Tsallis entropy. In this aspect, the above axioms and theorem can be considered as an alternative foundation of the Tsallis entropy, as well. It should be noted here that if we impose the condition of an entropy additivity this form of the function Ψ is forbidden.

4.5 MEP for Renyi entropy

If the Renyi entropy $S^{(R)}$ is used instead of the Gibbs–Shannon entropy the equilibrium distribution must provide maximum of the functional

$$L_R(p) = \frac{1}{1-q} \ln \sum_i^W p_i^q - \beta \sum_i^W H_i p_i - \alpha \sum_i^W p_i,$$

where β and α are Lagrange multipliers. Note that, $L_R(p)$ passes to $L_G(p)$, Eq. (4.1.2), in the $q \to 1$ limit.

We equate a functional derivative of $L_R(p)$ to zero, then

$$\frac{\delta L_R(p)}{\delta p_i} = \frac{q}{1-q} \frac{p_i^{q-1}}{\sum_j p_j^q} - \beta H_i - \alpha = 0. \tag{4.5.1}$$

Multiplying this equation by p_i and summing up over i, with account of normalization condition $\sum_i p_i = 1$ we get

$$\alpha = \frac{q}{1-q} - \beta U. \tag{4.5.2}$$

Then, it follows from equation (4.5.1) that

$$p_i = \left(\sum_j^W p_j^q \left(1 - \beta \frac{q-1}{q} \Delta H_i \right) \right)^{1/(q-1)}, \qquad \Delta H_i = H_i - U.$$

Using once more the condition $\sum_i p_i = 1$ we get

$$\sum_j^W p_j^q = \left(\sum_i^W \left(1 - \beta \frac{q-1}{q} \Delta H_i \right)^{1/(q-1)} \right)^{-(q-1)}$$

and, finally,

$$p_i = p_i^{(RZ)} = Z_R^{-1} \left(1 - \beta \frac{q-1}{q} \Delta H_i \right)^{1/(q-1)} \tag{4.5.3}$$

$$Z_R = \sum_i \left(1 - \beta \frac{q-1}{q} \Delta H_i \right)^{1/(q-1)}$$

This is the Renyi distribution in the Z-form [54]. At $q \to 1$ the distribution $\{p_i^{(R)}\}$ becomes the Z-form of Gibbs canonical distribution, Eq. (4.1.3), in which the constant $\beta = 1/k_B T_0$ is the reciprocal of the temperature.

To obtain the S-form of the Renyi distribution we can find that for the Renyi distribution (4.5.3) $S^{(R)} = \ln Z_R$; hence

$$p_i^{(RS)} = e^{-S^{(R)}} \left(1 - \beta \frac{q-1}{q} \Delta H_i \right)^{1/(q-1)}.$$

At $q \to 1$ this distribution becomes the S-form of Gibbs canonical distribution, Eq. (4.1.4). An escort version of MEP for the Renyi entropy was discussed in our paper [55].

The thermodynamic Renyi entropy $\tilde{S}^{(R)}$ is defined as the Renyi information entropy for the Renyi distribution, just as in the Gibbs thermostatistics the thermodynamic entropy is the Gibbs-Shannon entropy for the Gibbs distribution. It can be represented as

$$\tilde{S}^{(R)} = k_B S_q^{(R)}(p^{(R)}) = \frac{k_B}{1-q} \ln \sum_i \frac{\bar{p}_i{}^q}{(\sum_i \bar{p}_i)^q} \qquad (4.5.4)$$

where the Boltzmann constant k_B is taken into account to ensure a right dimension of the thermodynamic entropy and

$$\bar{p}_i = \left(1 - \beta\frac{q-1}{q}\Delta H_i\right)^{1/(q-1)}$$

is the non-normalized Renyi distribution.

To define a thermodynamic temperature we use the Klausius relation between gains of the heat Q and thermodynamic entropy $\tilde{S}^{(R)}$, that is

$$\delta\tilde{S}^{(R)} = \frac{\delta Q}{T} \qquad (4.5.5)$$

Next we consider process in which the Hamiltonian H dependent on external parameters $a_1, a_2, ...a_K$ is varied. The state of the system described with the Renyi distribution is varied as well. Then, for the gain of the internal energy we get

$$\begin{aligned}\delta U &= \langle\sum_k \frac{\partial H}{\partial a_k}\delta a_k\rangle + \frac{\sum_i H_i \delta\bar{p}_i}{\sum_i \bar{p}_i} - \frac{U}{\sum_i \bar{p}_i}\sum_i \delta\bar{p}_i \\ &= -\delta A + \frac{1}{\sum_i \bar{p}_i}\sum_i \Delta H_i \delta\bar{p}_i\end{aligned}$$

were $\delta A = -\langle\sum_k(\partial H/\partial a_k)\delta a_k\rangle$ is the work produced by the system. Thus, for the heat gain we get

$$\delta Q = \delta U + \delta A = \frac{1}{\sum_i \bar{p}_i}\sum_i \Delta H_i \delta\bar{p}_i \qquad (4.5.6)$$

In the same manner we find the entropy gain

$$\begin{aligned}\delta\tilde{S}^{(R)} &= \frac{k_B q}{(1-q)\sum_i \bar{p}_i{}^q}\sum_i \bar{p}_i{}^{q-1}\delta\bar{p}_i - \frac{k_B q}{(1-q)\sum_i \bar{p}_i}\sum_i \delta\bar{p}_i \\ &= k_B\beta\frac{1}{\sum_i \bar{p}_i}\sum_i \Delta H_i \delta\bar{p}_i \qquad (4.5.7)\end{aligned}$$

We see from equations (4.5.6) and (4.5.7) that the Klausius relation (4.5.5) is true at any q, if we put $T = 1/k_B\beta$.

It is not difficult to check that the same value of the physical temperature is followed from the standard statistical definition of the temperature, that is

$$T = \left(\frac{\partial \tilde{S}^{(R)}}{\partial U} \right)^{-1}_{\{a_k\}} = \frac{1}{k_B \beta} \tag{4.5.8}$$

It will be shown below (Sec.4.8, Eq. (4.8.6)) that at least for the power-law Hamiltonian the Lagrange parameter β does not depend on q whence $\beta = \beta_0$. Therefore we can assert that the physical temperature concides with the usual thermodynamic temperature, that is $T = T_0$ where T_0 is the temperature of a heat bath.

By the way the zeroth law of thermodynamics is reinforced for complex systems being in contact with the entrostat.

4.6 MEP for Tsallis entropy

When the Tsallis entropy was used instead of the Gibbs–Shannon entropy, the steady state distribution was derived by Tsallis [33] in the form

$$p_i^{(Ts)} = \frac{(1 + \beta(1-q)H_i)^{1/(q-1)}}{\sum_i (1 + \beta(1-q)H_i)^{1/(q-1)}},$$

was also known as the 1st version of thermostatistics. It was noticed there that the parameter β was not a Lagrange multiplier because the starting functional was taken as

$$L_T(p) = -\frac{1}{1-q}\left(1 - \sum_i^W p_i^q\right) - \alpha\beta(q-1)\sum_i^W H_i p_i + \alpha \sum_i^W p_i. \tag{4.6.1}$$

Here, the question arises about forms of Lagrange multipliers $\alpha\beta(q-1)$ and $(+\alpha)$, but the main problem is that the functional $L_T(p)$ does not pass to the functional (4.1.2) when $q \to 1$ as the second term in (4.6.1) vanishes.

It seems reasonable to suppose that just this difficulty led to the introduction of the 3rd version of nonextensive thermodynamics [56] with the escort distribution $P_i = p_i^q / \sum_i p_i^q$. The consistency of the transition to the escort distribution is partly justified by the condition of conservation of a preassigned average value of the energy $U = \langle H \rangle_{\text{es}} \equiv \sum_i H_i P_i$, however other average values are to be calculated with the use of the same escort distribution also, that contradicts to the main principles of probability description.

For the generality sake, the resulted distribution of the Tsallis' 3rd version is reproduced here as

$$p_i^{(T3)} = Z_{T3}^{-1} \left(1 - \beta^*(1-q')(H_i - U)\right)^{1/1-q'}, \tag{4.6.2}$$

$$Z_{T3} = \sum_i (1 - \beta^*(1 - q')(H_i - U))^{1/1-q'}$$

where $\beta^* = \beta / \sum_j (p_j^{T3})^{q'}$ and β is the true Lagrange multiplier in the corresponding variational functional of the 3rd version.

It is shown below that careful solution of the variational problem for the 1st version of thermostatistics gives rise to a probability distribution which does not seem to be less acceptable then $p_i^{(T3)}$. If we take the functional

$$L_T(p) = \frac{1}{1-q}\left(\sum_i^W p_i^q - 1\right) - \beta \sum_i^W H_i p_i - \alpha \sum_i^W p_i,$$

and equate its functional derivative to zero we get

$$\frac{\delta L_T(p)}{\delta p_i} = \frac{q}{1-q} p_i^{q-1} - \beta H_i - \alpha = 0. \tag{4.6.3}$$

Multiplying this equation by p_i and summing up over i, with account of normalization condition $\sum_i p_i = 1$ we get $\alpha = (q/1-q)\sum_i p_i^q - \beta U$. Then, it follows from Eq. (4.6.3) that

$$\begin{aligned} p_i &= \left(\sum_j p_j^q - \beta \frac{q-1}{q} \Delta H_i\right)^{1/(q-1)} \\ &= \left(\sum_j p_j^q\right)^{1/(q-1)} \left(1 - \beta \frac{q-1}{q} \frac{\Delta H_i}{\sum_j p_j^q}\right)^{1/(q-1)} \end{aligned} \tag{4.6.4}$$

Using once more the condition $\sum_i p_i = 1$ we get

$$p_i = p_i^{TZ} = Z_{T1}^{-1} \left(1 - \beta \frac{q-1}{q} \frac{\Delta H_i}{\sum_j p_j^q}\right)^{1/(q-1)},$$

$$Z_{T1} = \sum_i \left(1 - \beta \frac{q-1}{q} \frac{\Delta H_i}{\sum_j p_j^q}\right)^{1/(q-1)},$$

our modification of the 1st version of the Tsallis distribution in the Z-form. It differs from the 3rd version, Eq. (4.6.2), by the signs before differences $q-1$ and does not invoke the escort distribution. In this respect it seems to be much more attractive then $p^{(T3)}$. At $q \to 1$ the distribution $p^{(TZ)}$ becomes the Z-form of the Gibbs canonical distribution, Eq. (4.1.3). It should be noted that both $p^{(T3)}$ and $p^{(TZ)}$ are explicitly self-referential in contrast to $p^{(RZ)}$.

To obtain the S-form of the modified 1st version of the Tsallis distribution we rewrite the sum $\sum_j p_j^q$ in the upper line of Eq. (4.6.4) in term of the Tsallis entropy. Then we get finally

$$p_i = p_i^{(TS)} = \left[1 - (q-1)\left(S_T - \frac{\beta}{q}U + \frac{\beta}{q}H_i\right)\right]^{1/(q-1)}.$$

At $q \to 1$ this distribution becomes the S-form of the Gibbs canonical distribution, Eq. (4.1.4).

Note that both Renyi and escort Tsallis (3$^{\rm rd}$ version) distributions are identical if $q' = 1/q$; in fact

$$1 - q' = \frac{q-1}{q}, \qquad \frac{q'}{1-q'} = \frac{1}{q-1}$$

and β^* is determined by the same second additional condition of MEP as well as β. Thus, the not always justifiable linearization of the logarithm in Renyi entropy and the questionable use of escort distribution leads to the same Renyi distribution when q and q' are considered as free parameters.

We conclude therefore that the numerous works (see [34]) confirming correspondence of the Tsallis escort distribution (with fitted q') with observed distributions in complex physical, biological, social and other systems, count more in favor of Renyi entropy rather than the nonextensive Tsallis entropy.

4.7 Small subsystem with fluctuating temperature

The problem to be solved for a unique definition of the Renyi distribution is determination of a value of the Renyi parameter q.

An excellent example of a solution to this problem for a physical non-Gibbsian system was presented by Wilk and Wlodarczyk [57]. They took into consideration fluctuations of both energy and temperature of a minor part of a large equilibrium system. This is a radical difference of their approach from the traditional Gibbs method in which temperature is a constant value characterizing the heat bath. As a result, their approach led (see [58]) to the Renyi distribution with the parameter q expressed via heat capacity C_V of the minor subsystem.

The approach by Wilk and Wlodarczyk was advanced by Beck [59] and Beck and Cohen [60] who offered for it a new apt term "superstatistics". In the frame of superstatistics, the parameter q is defined by physical properties of a system which can exchange energy and heat with a heat bath. As a result, $q \neq 1$ but $|q-1| \ll 1$, because of exchange entropy with an entropy bath is not taken into account by superstatistics explicitly. To clear up a physical sense of the Renyi distribution (4.5.3) and parameter q for such a subsystem we use here an approach proposed by Wilk and Wlodarchuk [57].

We will treat the subsystem as a small part of a larger equilibrium system that experiences thermal fluctuations of both energy and temperature. This is a radical difference of the suggested approach from the Gibbs approach traditionally employed in statistical physics, in which temperature is preassigned by a constant characterizing the heat bath.

In order to analyze the temperature fluctuations, we will invoke the Landau-Lifshitz theory of hydrodynamic fluctuations [61], in which the respective fluctuation contributions are added to regular flows of mass, momentum, and energy entering the set of hydrodynamic equations. No flows of mass and momentum are present in the case under consideration; however, an energy flux must be observed because of the temperature fluctuations. Then, the equation of conservation of energy density of the system $E(\mathbf{r}, t)$ takes the form

$$\frac{\partial E(\mathbf{r}, t)}{\partial t} = -\mathrm{div}(\mathbf{q}^R(\mathbf{r}, t) + \mathbf{q}^F(\mathbf{r}, t)) \tag{4.7.1}$$

where $\mathbf{q}^R(\mathbf{r}, t)$ describes a regular flow of energy density, and $\mathbf{q}^F(\mathbf{r}, t)$ represents the flow fluctuations.

We will single out in the system a subsystem of preassigned volume V, integrate Eq. (4.7.1) with respect to the volume V, and use the Gauss-Ostrogradsky formula to derive the equation of conservation for the energy of this subsystem,

$$\frac{d\bar{E}(t)}{dt} = -Q^R(t) - Q^F(t), \qquad Q^{R,F} = \int_A dA \cdot \mathbf{q}^{R,F}, \tag{4.7.2}$$

where the surface area A of the singled-out subsystem is introduced. We will further restrict ourselves to fluctuations of only one parameter, namely, temperature, and represent the energy $\bar{E}(t)$ in the form of $\bar{E}(t) = C_V T(t)$, where C_V is the heat capacity of the subsystem. In addition, the flux Q^R may be conveniently written in the form of the heat-transfer equation

$$Q^R(t) = -A\eta(T(t) - T_0)$$

where η is the heat-transfer coefficient, and T_0 is the average temperature of the system. Then, Eq. (4.7.2) takes the form

$$C_V \frac{dT(t)}{dt} = -A\eta(T(t) - T_0) - Q^F(t). \tag{4.7.3}$$

This equation is the Langevin equation for the temperature which characterizes the singled-out part of the system and fluctuates under the effect of a random energy flux $Q^F(t)$ through the boundary of the discontinuous system being treated.

In the nonequilibrium linear thermodynamics [62], the thermodynamic force conjugate to the flux Q^R is provided by the quantity

$$\left(\frac{1}{k_B T_0} - \frac{1}{k_B T(t)}\right) \simeq \frac{1}{k_B T_0^2}(T(t) - T_0).$$

rather than by the temperature difference.

Accordingly, the kinetic coefficient of the heat transfer equation must have the form $k_B T_0^2 A\eta$. Then, according to the Landau-Lifshitz theory of hydrodynamic fluctuations [61], in a linear approximation with respect to deviation from equilibrium, the stochastic properties of random flux have the form

$$\langle Q_l^F(t)\rangle = 0, \qquad \langle Q_l^F(t)Q_l^F(t')\rangle = 2k_B T_0^2 A\eta\delta(t-t')$$

The second one of these expressions indicates that, within the linear theory, $Q_l^F \propto T_0$. This fact suggests a simple way of including the nonlinearity by replacing $Q_l^F(t)$ by $Q^F(t) = T(t)\xi(t)$, where $\xi(t)$ is a random function of time satisfying the relations

$$\langle \xi(t)\rangle = 0, \qquad \langle \xi(t)\xi(t')\rangle = 2k_B A\eta\delta(t-t')$$

As a result, Eq. (4.7.3) takes the form of nonlinear stochastic Langevin equation,

$$\frac{dT(t)}{dt} = -\frac{1}{\tau}(T(t) - T_0) - \frac{1}{C_V}T(t)\xi(t), \tag{4.7.4}$$

where $\tau = C_V/(A\eta)$.

Corresponding to the derived stochastic Langevin equation with δ-correlated noise is the Fokker Planck kinetic equation for the temperature distribution function $f(T,t)$,

$$\frac{\partial f(T,t)}{\partial t} = -\frac{\partial}{\partial T}W_1(T)f(T,t) + \frac{1}{2}\frac{\partial^2}{\partial T^2}W_2(T)f(T,t), \tag{4.7.5}$$

The coefficients $W_1(T)$ and $W_2(T)$ of this equation are expressed in terms of the first $\langle T(t) - T(t+\tau)\rangle$ and second $\langle (T(t) - T(t+\tau))^2\rangle$ conditional moments of stochastic equation (4.7.4), which correspond to some preassigned value of $T(t)$. For a linear stochastic Langevin equation, these moments are determined quite simply (see, for example, [63]). For nonlinear equations of the type of Eq. (4.7.4), the solution to this problem is also known and used in various applications of the theory of random processes [64]. For the case treated by us, the coefficients of the Fokker Planck equation take the form

$$W_1(T) = -\frac{1}{\tau}(T - T_0) + k_B T\frac{1}{\tau^2 A\eta}, \tag{4.7.6}$$

$$W_2(T) = 2k_B T^2\frac{1}{\tau^2 A\eta}. \tag{4.7.7}$$

Because these coefficients are not explicitly dependent on time, a steady-state solution to Eq. (4.7.5) exists,

$$f(T) = \frac{K}{W_2(T)}\exp\left\{2\int^T \frac{W_1}{W_2}dT\right\} \tag{4.7.8}$$

The constant K will be determined from the normalization condition; as a result, the choice of the lower limit of integration in the exponent is arbitrary and may be omitted. We substitute expressions (4.7.6) and (4.7.7) into (4.7.8) to derive

$$f(T) = -\frac{K}{T}\exp\left\{\frac{\tau A\eta}{k_B}\int^T \frac{-T+T_0}{T^2}dT\right\},$$

whence follows

$$f(T) = -KT^{-1-\gamma} e^{-\gamma T_0/T}$$

where the dimensionless constant $\gamma = C_V/k_B$ is introduced. Note that the resultant steady-state solution does not depend on either the heat-transfer coefficient η or the surface area A. This fact suggests that the obtained distribution may be more general than the treated model of heat transfer. In what follows, we will be interested in the distribution function with respect to the quantity $\beta = 1/(k_B T)$, rather than in the distribution function with respect to the temperature. In view of the relation $d\beta = -dT/(k_B T^2)$, we derive

$$f(\beta) = K\beta^{\gamma-1} e^{-\gamma k_B T_0 \beta}$$

The constant K is determined from the normalization condition reduced to

$$\begin{aligned} K^{-1} &= \int_0^\infty \beta^{\gamma-1} e^{-\gamma k_B T_0 \beta} d\beta \\ &= (\gamma k_B T_0)^{-\gamma} \Gamma(\gamma), \end{aligned}$$

whence we finally derive

$$f(\beta) = \frac{(\gamma k_B T_0)^\gamma}{\Gamma(\gamma)} \beta^{\gamma-1} e^{-\gamma k_B T_0 \beta}.$$

This function may also be represented in the form of the distribution of the temperature ratio $u = k_B T_0 \beta = T_0/T$,

$$f(u) = \frac{\gamma^\gamma}{\Gamma(\gamma)} u^{\gamma-1} e^{-\gamma u}.$$

or of the distribution of the dimensionless quantity $z = \gamma k_B T_0 \beta = \gamma T_0/T$,

$$f(z) = \frac{1}{\Gamma(\gamma)} z^{\gamma-1} e^{-z}. \tag{4.7.9}$$

Therefore, the distribution function of the inverse temperature of the subsystem in the dimensionless form of (4.7.9) is a gamma distribution. In concrete calculations below, we will largely use $f(\beta)$ or $f(u)$; for brevity, we will refer to them as gamma distributions as well.

Note that, if the mean energy of the singled-out volume $\bar{E}_0 = C_V T_0$ is introduced, the expression for $f(\beta))$ takes the form

$$f(\beta) = \frac{(\gamma\beta/\beta_0)^\gamma}{\beta\Gamma(\gamma)} e^{-\beta\bar{E}_0}.$$

By its form, this expression is close to the Gibbs distribution; however, unlike the latter, it reflects the inclusion of the temperature fluctuation of the

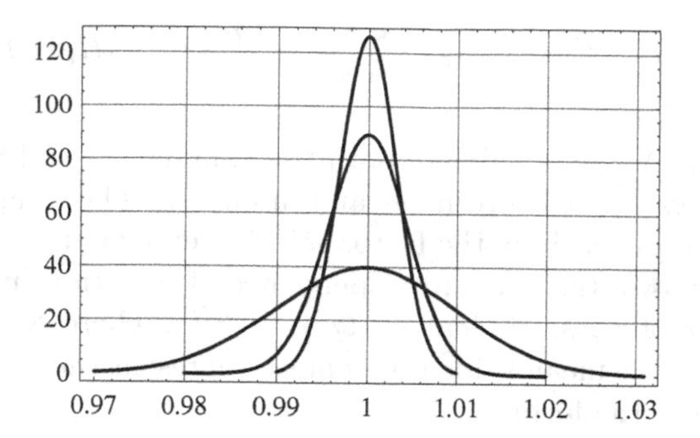

Fig. 4.2. Gamma distribution $f(u)$ of the temperature ratio $u = T_0/T$ for different values of the parameter $\gamma = C_V/k_B = 10^4$ (lower curve), $0.5 \cdot 10^5$ (middle curve), 10^5 (upper curve).

subsystem with the preassigned energy $\bar{E}_0$. Fig. 2 shows the form of gamma distribution $f(u)$ at $\gamma = 10^4$, $0.5 \cdot 10^5$, and 10^5. One can see that the dispersion of the inverse temperature of the subsystem decreases abruptly with increasing γ. However, small values of γ correspond to very small subsystems. Indeed, for an ideal monatomic gas at normal conditions, $\gamma = 3N/2$, where N is the number of particles in the volume; according to the Avogadro law, $N \simeq 0.3 \cdot 10^{20}\, V$. Then, for the singled-out volume with the characteristic size of the order of the free path length (about 10^{-5} cm), we have the value of $\gamma = 0.5 \cdot 10^5$. One can see in the figure that the temperature dispersion in this case is of the order of 0.005, which coincides with the result of the thermodynamic theory of fluctuations [65] $\delta T/T_0 \simeq (k_B/C_V)^{1/2}$. It appears to be more promising to apply these relations to heterogeneous systems in which the size of small particles (for example, atomic nucleus [66]) may not exceed several angstroms, and to low-temperature systems [67] with $C_V \to 0$ at $T_0 \to 0$

In order to describe a subsystem in contact with a large thermally equilibrium system (heat bath), the Gibbs canonical distribution is used in statistical physics (here and below, the factor G_i allowing for number of states of energy H_i will be omitted):

$$\rho_i = Q^{-1} e^{-\beta H_i}, \tag{4.7.10}$$

where H_i is the energy of the subsystem (the subscript i may indicate the number of discrete energy level or totality of the values of coordinates and momenta of molecules of the subsystem), and Q is the partition function. In so doing, the inverse temperature $\beta = 1/k_B T$ is taken to be known preassigned quantity. As was demonstrated above, the temperature may fluctuate. In view of this, the question arises as to how the Gibbs distribution is modified under the effect of temperature fluctuations. The answer to this question may be obtained by the way of averaging the Gibbs distribution (4.7.10) with the gamma-distribution for temperature T (or β). For further treatment, ρ_i may be conveniently represented in an equivalent form,

$$\rho_i = Q^{-1} e^{-\beta \Delta H_i}, \qquad Q = \sum_i e^{-\beta \Delta H_i}, \qquad \Delta H_i = H_i - \bar{H}$$

where the symbol $\sum_i$ may indicate both the summation and integration over a totality of the values of coordinates and momenta. The temperature dependence of ρ_i is defined both by the factor β in the exponent and, in the general case, by the unknown temperature dependence of partition function $Q(\beta)$ (in the simplest case of classical ideal gas $Q \propto \beta^{-3N/2}$, where N is the number of molecules). Using the mean value theorem we represent the Gibbs distribution averaged over β in the form

$$\bar{\rho}_i = \int_0^\infty d\beta f(\beta) \rho_i = \frac{1}{Q^*} \int_0^\infty d\beta f(\beta) e^{-\beta \Delta H_i},$$

where Q^* lies in the range of possible variation of $Q(\beta)$ from $Q(0)$ to $Q(\infty)$. From the conditions of normalization to unity of the distributions $f(\beta)$ and ρ, we have

$$\frac{1}{Q^*} \sum_i \int_0^\infty d\beta f(\beta) e^{-\beta \Delta H_i} = 1$$

whence we find

$$Q^* = \sum_i \int_0^\infty d\beta f(\beta) e^{-\beta \Delta H_i}.$$

Therefore, it is sufficient to calculate only the average value of the exponent,

$$\int_0^\infty d\beta f(\beta) e^{-\beta \Delta H_i} = \frac{(\gamma k_B T_0)^\gamma}{\Gamma(\gamma)} \beta^{\gamma-1} \int_0^\infty d\beta \beta^{\gamma-1} e^{-\beta(\gamma k_B T_0 + \Delta H_i)}$$
$$= \left(1 + \frac{\beta_0}{\gamma} \Delta H_i\right)^{-\gamma}.$$

Finally, the averaged Gibbs distribution takes the form

$$\bar{\rho}_i = \frac{\left(1 + \frac{\beta_0}{\gamma} \Delta H_i\right)^{-\gamma}}{\sum_i \left(1 + \frac{\beta_0}{\gamma} \Delta H_i\right)^{-\gamma}}. \tag{4.7.11}$$

In the $\gamma \to \infty$ limit corresponding to a high heat capacity of the singled-out subsystem, $\bar{\rho}_i$ goes to ρ_i.

The resulting equation for the modified Gibbs distribution is similar to the Renyi distribution $p_i^{(R)}$ in its structure. To identify $\bar{\rho}_i$ with p_i it is enough to present γ as $\gamma = (1-q)^{-1}$ when Eq. (4.7.11) takes the form

$$\bar{\rho}_i = \frac{[1 + \beta_0(1-q)\Delta H_i]^{-1/1-q}}{\sum_i [1 + \beta_0(1-q)\Delta H_i]^{-1/1-q}}.$$

The full identity of this expression with the probability p_i (4.5.3) ensuring the extremality of Renyi entropy enables one to take a new view of the physical meaning of the Renyi entropy and parameter

$$q = 1 - \frac{1}{\gamma} = \frac{C - k_B}{C}.$$

So, the Renyi parameter differs significantly from unity only in the case where the heat capacity of the singled out system is of the same order of magnitude as the Boltzmann constant k_B. This implies that the Renyi distribution for large systems in contact with a heat bath coincides practically with the Gibbs distribution. To obtain real description of complex self-organizing systems the superstatistics should be generalized with allowance made for entropy transfer due to contact with an entropy bath.

4.8 Power-law Hamiltonian

When $q \to 1$ the Renyi distribution $\{p_i^{(R)}\}$ becomes the Gibbs canonical distribution and $\beta/q \to \beta_0 = 1/k_BT$. Such behavior is not enough for unique determination of β. Indeed, in general, it may be an arbitrary function $\beta(q)$ which becomes β_0 in the limit $q \to 1$.

To find an explicit form of β, we return to the additional condition of the pre-assigned average energy $U = \sum_i H_i p_i$ and substitute there Renyi distribution (4.5.3). For the sake of simplicity, we will confine the discussion to the particular case of a power-law dependence of the Hamiltonian on a parameter x

$$H_i = Cx_i^\kappa.$$

This type of the Hamiltonian corresponds to an ideal gas model that is widely used in the Boltzmann-Gibbs thermostatistics and it seems reasonable to say that it may be useful in construction of thermostatistics of complex systems. Moreover, in most social, biological and humanitarian sciences the system parameter x can be considered (with $\kappa = 1$) as a kind of the Hamiltonian (e.g. the size of population of a country, effort of a word pronouncing, bank capital, etc.).

If the distribution $\{p_i\}$ allows for smoothing over the ranges much larger than the average distance $\Delta x_i = x_i - x_{i+1}$ without significant loss of information, we can pass from the discrete variable x_i to the continuous x. Then the condition (4.1.1) of a fixed average energy for the Renyi distribution becomes

$$Z^{-1} \int_0^\infty Cx^\kappa \left(1 - \beta \frac{q-1}{q} (Cx^\kappa - U)\right)^{1/(q-1)} dx = U \tag{4.8.1}$$

or

$$Z^{-1} \int_0^\infty C_u x^\kappa \left(1 - \beta U \frac{q-1}{q} (C_u x^\kappa - 1)\right)^{1/(q-1)} dx = 1, \tag{4.8.2}$$

where

$$Z = \int_0^\infty \left(1 - \beta U \frac{q-1}{q}(C_u x^\kappa - 1)\right)^{1/(q-1)} dx \tag{4.8.3}$$

and $C_u = C/U$. Both integrals in these equations may be calculated with the use of a tabulated [68] integral

$$I = \int_0^\infty \frac{x^{\mu-1}dx}{(a+bx^\nu)^\lambda} = \frac{1}{\nu a^\lambda}\left(\frac{a}{b}\right)^{\mu/\nu} \frac{\Gamma\left[\frac{\mu}{\nu}\right]\Gamma\left[\lambda - \frac{\mu}{\nu}\right]}{\Gamma[\lambda]}$$

under condition of convergence

$$0 < \frac{\mu}{\nu} < \lambda, \qquad \lambda > 1.$$

For the integrals in Eqs. (4.8.2) and (4.8.3) this condition becomes

$$0 < \frac{1+\kappa}{\kappa} < \frac{1}{1-q}. \tag{4.8.4}$$

Then, finally, we find from Eqs. (4.8.2), (4.8.3) with the use of the relation $\Gamma[1+z] = z\Gamma[z]$, that

$$\beta U = \frac{1}{\kappa} \qquad \text{for all } q. \tag{4.8.5}$$

Independence of this relation from q means that it is true, in particular, for the limit case $q = 1$ where the Gibbs distribution takes a place and, therefore,

$$\beta = \beta_0 \equiv 1/k_B T \qquad \text{for all } q. \tag{4.8.6}$$

When $H = p^2/2m$ (that is, $\kappa = 2$) we get from (4.8.5) and (4.8.6) that $U = \frac{1}{2}k_B T$, as would be expected for one-dimensional ideal gas. Additionally, the Lagrange parameter β can be eliminated from Renyi distribution (4.5.3) with the use of Eq. (4.8.5) and we have, alternatively,

$$p^{(R)}(x|q,\kappa) = Z^{-1}\left(1 - \frac{q-1}{\kappa q}(C_u x^\kappa - 1)\right)^{1/(q-1)} \tag{4.8.7}$$

or

$$p_i^R(q,\kappa) = Z^{-1}\left(1 - \frac{q-1}{\kappa q}(C_u x_i^\kappa - 1)\right)^{1/(q-1)}.$$

So, at least for power-law Hamiltonians, the Lagrange multiplier β does not depend on the Renyi parameter q and coincides with the Gibbs parameter $\beta_0 = 1/k_B T$ and can be eliminated with the use of the relation (4.8.5).

4.9 Transfer to Renyi thermostatistics as a phase transition

Thus, we have Gibbs and Renyi thermostatistics. Each of them provides an adequate description of corresponding class of systems and we need in a rigorous formulation of conditions of transfer from one thermostatistics to another. Indeed, the Renyi thermostatistics is more general and includes the Gibbs thermostatistics as a partial case of value $q = 1$ of the Renyi parameter.

Transfer from the Gibbs distribution describing a state of dynamic chaos [25] to power–law Renyi distributions that are characteristic for ordered self-organized systems [14] corresponds to an increase of an "order parameter" $\eta = 1 - q$ from zero at $q = 1$ up to $\eta_{\mathrm{max}} = 1 - q_{\mathrm{min}}$. As this takes place, the Gibbs–Shannon entropy passes into the Renyi entropy.

In accordance to the Landau theory [61] of phase transitions an entropy derivative with respect to the order parameter undergoes a jump at a point of the phase transition. In particular, when the order parameter is a temperature gain, the jump corresponds to a heat capacity jump that is characteristic for phase transitions of the second type.

Here we deal with the transfer from the Gibbs thermostatistics to the Renyi thermostatistics corresponding to non-zero values of the order parameter η. Let us consider a variation of the entropy at this transition. Substituting the Renyi distribution (4.5.3) into the Renyi entropy definition (4.2.3), we find the thermodynamic entropy in the Renyi thermostatistics as

$$\begin{aligned}\tilde{S}^{(R)} &= S_\eta^{(R)}(p_\eta^{(R)}) \\ &= k_B \ln \sum_i^W \left(1 + \beta \frac{\eta}{1-\eta} \Delta H_i\right)^{-1/\eta} = \ln Z_\eta^{(R)}.\end{aligned} \tag{4.9.1}$$

where, as before, the Boltzmann constant k_B is introduced.

When $\eta \to 0$, $(q \to 1)$, this entropy passes into thermodynamic entropy in the Gibbs thermostatistics

$$\tilde{S}^{(G)} = S^{(G)}(p^{(G)}) = k_B \ln \sum_i^W e^{-\beta \Delta H_i}. \tag{4.9.2}$$

Now it is not difficult to calculate the limiting value at $\eta \to 0$ of the derivative of the entropy difference $\Delta S = \tilde{S}^{(R)} - \tilde{S}^{(G)}$ with respect to η. We get

$$\lim_{\eta \to 0} \left(\frac{d\Delta S}{d\eta}\right) = \frac{k_B}{2} \beta^2 \sum_i^W p_i^{(G)} (\Delta H_i)^2$$

According to a fluctuation theory for the Gibbs equilibrium ensemble we have

$$\sum_i^W p_i^{(G)} (\Delta H_i)^2 = \frac{1}{k_B \beta^2} \frac{dU}{dT} = \frac{1}{k_B \beta^2} C_V$$

whence

$$\lim_{\eta \to 0} \left(\frac{d\Delta S}{d\eta} \right) = \frac{1}{2} C_V.$$

where C_V is the heat capacity at a constant volume.

Thus, the derivative of the entropy gain with respect to the order parameter exhibits the jump (equal to $C_V/2$) at $\eta = 0$. This permits us to consider the transfer to the Renyi thermostatistics as a peculiar kind of a phase transition into a more organized state. We can give this transition the name *entropic phase transition.*

A derivative of the entropy gain with respect to temperature can be calculated in the same manner. Hence

$$\lim_{\eta \to 0} \left\{ \left(\frac{d}{dT_0} S_\eta^{(R)}(p_\eta^{(R)}) \right) - \left(\frac{d}{dT_0} S^{(G)}(p^{(G)}) \right) \right\} = 0.$$

from which it follows that usual phase transition does not take place at $\eta = 0$.

As a result of the entropic phase transition the system passes into an ordered state with the order parameter $\eta \neq 0$. In contrast to the usual phase transition that take place at the temperature of phase transition, conditions of the entropic phase transition are likely to be determined partially for each concrete system. For example, a threshold of appearance of turbulence (see [69]) as an ordered structure is determined by a critical Reynolds number and an appearance of Benard cells is determined by a critical Reyleigh number (see [25]). Social, economic and biological systems are realized as a rule in ordered self-organized forms. This is the reason why power-law and closely related distributions are characteristic for them but not canonical Gibbs distribution.

The question of a definite value of the order parameter for different systems is treated next.

4.10 The most probable value of the Renyi parameter

The problem to be solved for a unique definition of Renyi distribution is the determination of a value of the Renyi parameter q. Some successes in this direction were achieved for particular cases of small subsystems with fluctuating temperature (so-called "superstatistics" discussed in Sec. 7), a set of independent harmonic oscillators [70, 71] and fractal systems [72]. On the other hand, there are many complex systems for which we have no information related to a source of fluctuations. In that cases the parameter q cannot be determined with the use of the superstatistics.

Here, we propose a further extension of MEP that consists of looking for a maximum of the Renyi entropy in a space of Renyi distributions with different values of q. Substitution of Renyi distribution $p^R(x|q,\kappa)$ into the definition of the Renyi entropy, Eq. (4.2.3), and variation of the q-parameter results in the picture of $S^{(R)}[p^{(R)}(x|q,\kappa)]$ as shown in Fig. 3 (left). It is seen

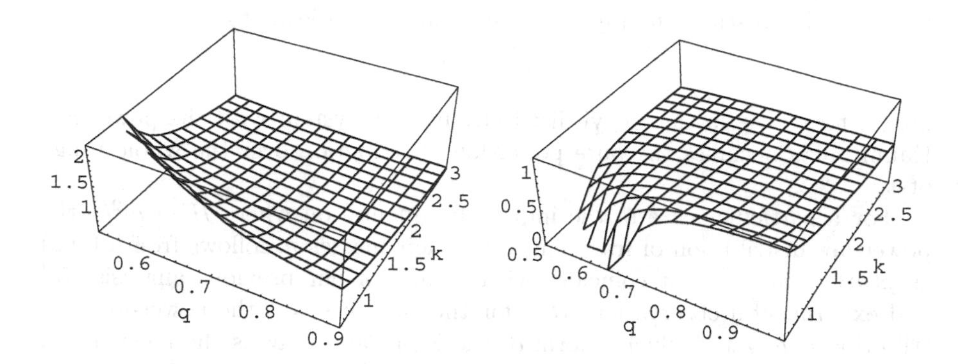

Fig. 4.3. The entropies $S_q^{(R)}[p^{(R)}(q,\kappa)]$ (left) and $S^{(G)}[p^{(R)}(q,\kappa)]$ (right) for the power-law Hamiltonian with the exponent κ within the range $3 > \kappa > 0.5$ and $q > 1/(1+\kappa)$.

that $S^{(R)}[p^{(R)}(x|q,\kappa)]$ attains its maximum at the minimal possible value of q which fulfills the inequality (4.8.4), that is,

$$q_{\min} = \frac{1}{1+\kappa}.$$

For $q < q_{\min}$, the integral (4.8.1) diverges and, therefore, Renyi distribution does not determine the average value $U = \langle H \rangle_p$, that is a violation of the second condition of MEP.

To check self-consistency of the proposed extension of MEP a similar procedure is applied to the Gibbs-Shannon entropy $S^{(G)}(p)$. Substituting $p = p^R(x|q,\kappa)$ we get the q-dependent function $S^{(G)}[p^{(R)}(x|q,\kappa)]$ illustrated in Fig. 3 (right). As would be expected, the Gibbs-Shannon entropy $S^{(G)}[p^{(R)}(x|q,\kappa)]$ attains its maximum value at $q = 1$ where $p^{(R)}(x|q,\kappa)$ becomes the Gibbs canonical distribution.

Hence, the maximum of the Renyi entropy is realized at $q = q_{\min}$ and is just the value of the Renyi parameter that should be used for the particular case of the power-law Hamiltonian when we have no additional information on behavior of the stochastic process under consideration. Recall that the Renyi entropy was derived here as a functional which attain its maximum value at the steady state of a complex system just as the Helmholtz free energy does for a thermodynamic system. (The kind of extremum, that is, maximum or minimum is determined by the sign definition for the constant λ). Then, a radically important conclusion follows from comparison of these two graphs:

In contrast to the Gibbs-Shannon entropy the Renyi entropy increases with increasing complexity ($\eta = 1 - q$), suggesting its consideration as a potential that allows evolution of the system to self-organization.

In support of such a conclusion, recall that a power-law distribution characteristic for self-organizing systems [14] is realized when the Renyi entropy is maximal. Substitution of $q = q_{\min}$ into Eq. (4.8.7) leads to

$$p = Z^{-1}x^{-(1+\kappa)};$$

hence, for $q = q_{\min}$ the Renyi distribution for a system with the power-law Hamiltonian becomes the pure power-low distribution over the whole range of x.

For a particular case of the impact fragmentation where $H \sim m^{2/3}$ the power-law distribution of fragments over their masses m follows from (4.9.2) as $p(m) \sim m^{-5/3}$ that coincides with results of our previous analysis [73] and experimental observations [74]; for the case of $\kappa = 1$ the power-low distribution is $p \sim x^{-2}$. Such a form of the Zipf-Pareto law is the most useful in social, biological and humanitarian sciences (see, e. g. [75, 76, 77]). The same exponent of power-low distribution was demonstrated [78] for energy spectra of particles from atmospheric cascades in cosmic ray physics and for distribution of users among the web sites [79].

It is necessary to notice here that inequalities (4.8.4) suggest in fact $q > q_{\min}$, that is, $q = q_{\min} + \epsilon$ where $\epsilon \ll 1$ is a positive infinitesimal value that should be a finite constant in physical realizations[4]. Taking into account that finite ϵ gives rise to Renyi distribution in the form

$$p^R(x) = Z^{-1}(C_u x)^{-K}\left[1 - \epsilon(\kappa+1)^2(1 - C_u x^{-\kappa})\right]^{-K/\kappa}$$

where we have used $K := (\kappa+1)(1+\epsilon(\kappa+1)/\kappa)$. For sufficiently large x this Renyi distribution passes to the power law distribution where all terms with ϵ can be neglected. On the other hand, for sufficiently small x, only the term $\epsilon(\kappa+1)^2 C_u x^{-\kappa}$ may be accounted for in the expression in the square brackets, hence

$$p^R(x)|_{x\ll 1} \sim (\epsilon(\kappa+1)^2)^{-(\kappa+1)/\kappa}.$$

This equation points to the fact that the asymptote to Renyi distribution for small $x's$ is a constant. Figure 4 shows the Renyi distribution over the whole range of x. Now there is no methods for a unique theoretical determination of ϵ, so it may be considered as a free parameter. It can be estimated for those experimental data where the head part preceding power law distribution is presented. As an example, for the probability distribution of connections in World Wide Web network [80] the parameter ϵ is estimated as $\sim 10^{-4}$.

4.11 Conclusive Notes on the Renyi entropy and Self-Organization

An equilibrium state of a thermodynamic system coupled with a heat bath is characterized by extremum (minimum) of the Helmholtz free energy that can

[4] The participant of the workshop Dr. Ramandeep S. Johal pointed to the fact that the Lagrange multiplier α, Eq. (4.5.2), tends to zero when $q = q_{\min}$. It is really so due to relation (4.8.5). For $q = q_{\min} + \epsilon$ we have $\alpha \propto \epsilon$.

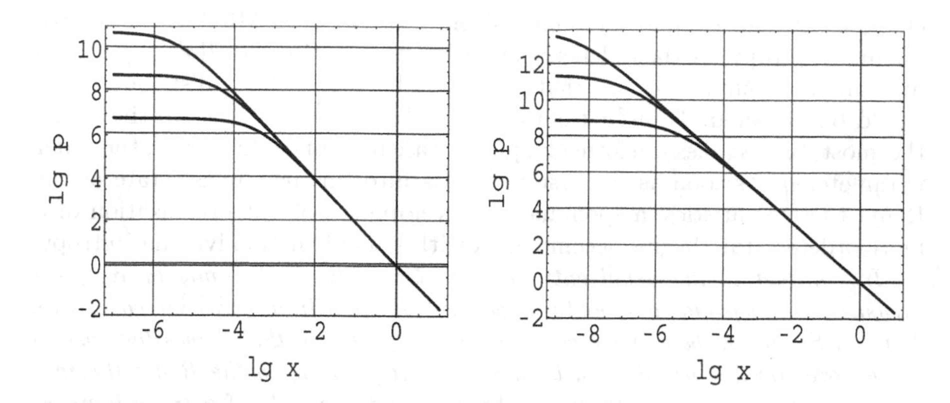

Fig. 4.4. The Renyi distributions (non-normalized) for the power-law Hamiltonian $H \sim x^{\kappa}$, $\kappa = 1$ (left) and $\kappa = 2/3$ (right) and different values $\epsilon = 10^{-6}$, 10^{-5}, 10^{-4} from upper to below in each graph.

be obtained [5] as a cumulant average of the Hamiltonian in contrast to the internal energy that is a simple mean of the Hamiltonian. It is known that the internal energy is not extremal for a system in contact with a heat bath. In a similar manner, the standard Gibbs-Shannon entropy for inhomogeneous distributions is derivable [16, 17] by simple averaging of the Boltzmann entropy for an uniform distribution. So, there is no reason to believe that the Gibbs-Shannon entropy for a complex system coupled with an entropy bath will be extremal at a steady state of the system. It is shown here that the procedure of cumulant averaging when applied to the Boltzmann entropy leads to the Renyi entropy. This provides reason enough to believe that the Renyi entropy should be extremal (maximal) at a steady state of a system being in contact with the entropy bath. Such a form of the information entropy and its maximality are justified by the Shore-Johnson theorem.

The MEP applied to the Renyi entropy gives rise to a Renyi distribution that depends on the Renyi parameter $q < 1$ and Lagrange multiplier β. It is shown here that for the particular case of a power-law Hamiltonian $H_i = Cx^{\kappa}$ the multiplier β does not depend on the Renyi parameter q and coincides with $\beta_0 = 1/k_B T_0$. Moreover, it can be expressed in terms of U and κ and thus eliminated completely from Renyi distribution function. In the absence of any additional information on the nature of a complex system, the q parameter can be determined with the further use of MEP in the space of the q-dependent Renyi distributions.

The q-dependent thermodynamic entropy in the Renyi thermostatistics is determined as the Renyi entropy for the Renyi distribution. Maximum maximorum of such an entropy is attained at minimal possible value $q = q_{\min} = 1/(1+\kappa)$, that is, at maximal possible value of an order parameter $\eta = 1 - q$. The Renyi distribution at such η becomes the power–law distribution that is characteristic for complex systems. When applying such MEP to the Gibbs-Shannon entropy for the q-dependent Renyi distribution, $S^{(G)}[p^R(x|q,\kappa)]$, its maximum is found at $q = 1$ that corresponds to the Gibbs distribution, as would be expected. Transfer from usual Gibbs thermostatistics to the Renyi thermostatistics takes the form of a phase transition of ordering with the order parameter η. As soon as the system passes into this new phase state of the Renyi thermostatistics, a spontaneous development of self–organization of a more ordered state begins accompanied with gain of thermodynamic entropy.

In contrast to the usual entropy, the Renyi thermodynamic entropy increases as the system complexity (departure of the η from zero) increases (see Fig. 3). So, it can be considered as a kind of potential that drives the system to self-organized state. Such a behavior of the entropy in the Renyi thermostatistics eliminates a contradiction between the principle of entropy increase and a system evolution to self-organization. Moreover, it may be supposed that biological evolution or development are governed by the extremal principle of the Renyi thermostatistics.

4.12 Thermodynamics of a Quantum Mechanical system in Coherent State

Considerable recent attention has been focused on coherent states [81] in the context of a new interpretation of quantum mechanics in which the crucial point is an interaction of the system with its surrounding. During the interaction process most of the states of the open quantum system becomes unstable, and, as a result, an arbitrary state of this system becomes a superposition of "selected states" that provides classical description. This phenomenon has come to be known as a decoherence.

An intrinsic connection between the decoherence phenomena and coherent states of a system was suggested in [82]. Moreover, it was noted there [82] that decoherence might produce coherent states. This was proved for the particular case of a system of harmonic oscillators in [83]. Analyzing a master equation for an open quantum system interacting with a thermostat Paz and Zurek [84] found that the coherent states were sampled as "selected states", when self-energy of the system was of the same order as an energy of its interaction with the thermostat. This process is accompanied by an increase of both the open system entropy and the thermostat entropy due to losing information about their states. If a total closed system was initially in a pure state (with the zero entropy) then the theorem states [85, 86, 87, 88] that the entropies produced in the open system and thermostat are equal.

At first glance there is a paradox of non-zero entropy and temperature of a coherent state which is a pure state. In our opinion, it is resolved by the fact that the coherent states are eigenstates of non-Hermitian operator, and hence values of observable variables (that are eigenvalues of Hermitian operators) remain indeterminate in the coherent state, among them an energy. Thus, it is found to be fruitful to consider a density matrix of the coherent state and determine entropy and temperature with the use of standard methods of statistical physics. Such a concept will be illustrated below (in Subsecs. 12.1 and 12.2) by the examples of well-known coherent states of open quantum systems, namely, a quantum harmonic oscillator and quantum scalar field in a vicinity of a static source.[5] This approach provides also fresh insight (in Subsec. 12.3) into the problem of the black hole entropy.

4.12.1 Entropy of an oscillator in a coherent state

The known isomorphism of Hilbert spaces of states of arbitrary quantum systems ensures that a space of states of any n-dimensional quantum system can be mapped on the space of states of n-dimensional harmonic oscillator. Owing to this fact, properties of particular model systems discussed below may be considered as sufficiently general.

For purposes of clarity we begin with the simplest system, that is the one-dimensional quantum harmonic oscillator. In our approach the oscillator in a coherent state is treated as an open system in a steady interaction with an external force center (e.g. atomic nucleus). The interaction with internal variables of the force center could ensure a modification of the Shrödinger equation such that it describes the decoherence of a harmonic oscillator and the passage of its state into a coherent state. Here, however we will confine our discussion to a prepared coherent oscillator state.

The Hamiltonian of the system is

$$H = \frac{1}{2m}(p^2 + m^2\omega^2 q^2),$$

where p, q are respective momentum and coordinate operators satisfying the commutation relation $[q, p] = i\hbar$. Coherent states are conveniently [90] discussed in terms of non-Hermitian creation and annihilation operators

$$a = \frac{m\omega q + ip}{(2m\hbar\omega)^{1/2}}, \qquad a^\dagger = \frac{m\omega q - ip}{(2m\hbar\omega)^{1/2}}, \tag{4.12.1}$$

satisfying the commutation relations $[a, a] = [a^\dagger, a^\dagger] = 0, [a, a^\dagger] = 1$. A coherent state can be defined as an eigenstate of the non-Hermitian operator a:

$$a|d\rangle = \left(\frac{m\omega}{2\hbar}\right)^{1/2} d|d\rangle. \tag{4.12.2}$$

where the eigenvalue d is a complex number.

The density matrix of the d-coherent state is $\rho^d = |d\rangle\langle d|$. To construct a thermodynamics of this state we need in a partition function of ρ^d. Since it is defined as a value ensuring the condition $\mathrm{Tr}[\rho^d] = 1$ we consider only diagonal

[5] This is an advanced and enlarged version of the paper [89].

elements of ρ^d. Since coherent states are not mutually orthogonal and form an overcomplete basis, we get in a coherent state representation

$$\rho^d_{ff} = |\langle f|d\rangle|^2 = \exp\left\{-\frac{m\omega|d|^2}{2\hbar}\right\}\exp\left\{\frac{m\omega}{2\hbar}\left(-|f|^2 + d^* f\right)\right\}.$$

The first multiplier does not depend on the index f and can be considered as a normalizing factor, that is, a reverse value of the partition function

$$Q^d_{\mathrm{osc}} = e^{\overline{n}_d}. \tag{4.12.3}$$

where $\overline{n}_d = m\omega|d|^2/(2\hbar)$ is the well-known [91] mean number of quanta in the state d. Considering that an energy representation is the most used in statistical mechanics, we demonstrate the same result in this representation:

$$\rho^d_{nn} = |\langle n|d\rangle|^2 = \frac{1}{n!}|\langle 0|a^n|d\rangle|^2 = \frac{1}{n!}\overline{n}^n_d|\langle 0|d\rangle|^2,$$

where the equation (4.12.2) and property $|n\rangle = (a^\dagger)^n|0\rangle/\sqrt{n!}$ are used. The partition function $Q^d_{\mathrm{osc}} = |\langle 0|d\rangle|^{-2}$ is determined with the use of a condition of completeness of the set $|n\rangle$, that is $\mathrm{Tr}[\rho^d] = \sum_n |\langle n|d\rangle|^2 = 1$, whence we get again Eq. (4.12.3). For the mean energy of the coherent $|d\rangle$-state of the oscillator we have

$$E^d_{\mathrm{osc}} = \langle d|H|d\rangle = \sum_n \rho^d_{nn}\langle n|H|n\rangle = \hbar\omega\left(\overline{n}_d + \frac{1}{2}\right).$$

According to one of basic principles of the statistical mechanics, all thermodynamic properties of a system can be found when the partition function $Q(\mathsf{T})$ is known. Really, in general, a characteristic thermodynamic function F is defined as

$$\mathsf{F} = -k_B\mathsf{T}\ln Q(\mathsf{T}) \tag{4.12.4}$$

and known as the Helmholtz free energy $F(T,V,N)$ for a canonical ensemble, or thermodynamic potential $J(T,V,\mu)$ for grand canonical ensemble, or Gibbs free energy $G(T,p,N)$ for $T-p$ ensemble. At any case the thermodynamic entropy is defined as

$$\mathsf{S} = -\frac{\partial\mathsf{F}}{\partial\mathsf{T}}. \tag{4.12.5}$$

This is just the key point of the proposed approach. Transition from the statistical mechanics to thermodynamics is performed here in Eq. (4.12.4) for the Helmholtz free energy only. The entropy is defined in terms of it with no use of microscopic interpretations of entropy. As for the temperature T, it should be determined on the base of thermodynamic relations between the entropy S and mean energy E, that is, an internal energy of the system under consideration. The difference of entropy in the two neighboring equilibrium states E and $\mathsf{E}+d\mathsf{E}$ (other thermodynamic parameters are fixed) in accordance with the first law of thermodynamics is

$$d\mathsf{S} = \frac{1}{\mathsf{T}} d\mathsf{E},$$

whence temperature is defined as

$$\mathsf{T} = \left(\frac{\partial \mathsf{S}}{\partial E}\right)^{-1}. \tag{4.12.6}$$

Thermodynamics of the $|d\rangle$-state of the oscillator is constructed in the same manner on the base of the partition function $Q^d_{\rm osc}$:

$$F^d_{\rm osc} = -k_B T_{\rm osc}\overline{n}_d, \tag{4.12.7a}$$

$$S^d_{\rm osc} = k_B\overline{n}_d + k_B T_{\rm osc}\frac{\partial \overline{n}_d}{\partial T_{\rm osc}}, \tag{4.12.7b}$$

$$T_{\rm osc} = \left(\frac{\partial S^d_{\rm osc}}{\partial E^d_{\rm osc}}\right)^{-1} \tag{4.12.7c}$$

In contrast to common thermodynamics where the temperature is the known parameter of a system state, here Eqs. (4.12.7*a-c*) are used as a self-consistent definition of the effective temperature of the coherent state. We assume first that $T_{\rm osc}$ is in the form $T_{\rm osc} = C|d|^q$, where C, q are nonzero constants to be found. Presenting $\overline{n}_d$ as $(T_{\rm osc})^{2/q} m\omega/(2\hbar C^{2/q})$ Eqs. (4.12.5-4.12.6) become

$$F^d_{\rm osc} = -k_B (T_{\rm osc})^{1+2/q}\frac{m\omega}{2\hbar C^{2/q}}, \tag{4.12.8a}$$

$$S^d_{\rm osc} = k_B\left(1 + \frac{2}{q}\right)\overline{n}_d, \tag{4.12.8b}$$

$$T_{\rm osc} = \frac{\hbar\omega}{k_B(1+2/q)}. \tag{4.12.8c}$$

From the obtained equation for $T_{\rm osc}$ we see that the assumed form $T_{\rm osc} = C|d|^q$ is invalid as the constant q is to be equal zero, and all Eqs. (4.12.8*a-c*) lose their meanings.

Another alternative form for the temperature is $T_{\rm osc} = C\omega^p$. Then $\overline{n}_d$ can be presented as $(T_{\rm osc})^{1/p} m|d|^2/(2\hbar C^{1/p})$ and Eqs. (4.12.5-4.12.6) become

$$F^d_{\rm osc} = -k_B (T_{\rm osc})^{1+1/p}\frac{m|d|^2}{2\hbar C^{1/p}},$$

$$S^d_{\rm osc} = k_B\left(1 + \frac{1}{p}\right)\overline{n}_d,$$

$$T_{\rm osc} = \frac{\hbar\omega}{k_B(1+1/p)},$$

whence the constants in the sought-for form of $T_{\rm osc}$ are $p = 1$, $C = \hbar/(2k_B)$. We can introduce an area $A_d = \pi|d|^2$ of a phase portrait (circle) of the corresponding classical oscillator in $(q,\, p/\omega)$ - phase plane, since $\langle q(t)\rangle_d = d\cos\omega t$

and $\langle p(t)\rangle_d = -d\omega \sin \omega t$. Then, $\overline{n}_d$, entropy and temperature of the coherent state are found as

$$\overline{n}_d = \frac{k_B T_{\mathrm{osc}} m A_d}{2\pi\hbar^2} = \frac{A_d}{2\pi l_0^2}, \tag{4.12.9a}$$

$$S^d_{\mathrm{osc}} = 2k_B \overline{n}_d = k_B \frac{A_d}{\pi l_0^2}, \tag{4.12.9b}$$

$$T_{\mathrm{osc}} = \frac{\hbar\omega}{2k_B}, \tag{4.12.9c}$$

where $l_0 = \sqrt{\hbar/(m\omega)}$ is the amplitude of zero-point oscillations. Hence, the entropy and temperature do not vanish even for a dynamical quantum system in the coherent state.

An alternative approach to definition of the temperature can be obtained with the use of a Wigner distribution function based on a coordinate representation of a wave function for the oscillator in the coherent state

$$W(p,q) = \frac{1}{\pi\hbar} \exp\left\{ -\frac{(p - \langle p(t)\rangle_d)^2/2m + m\omega^2(q - \langle q(t)\rangle_d)^2/2}{\hbar\omega/2} \right\},$$

where the expression in the numerator of the exponent represents an energy of fluctuations around a classical trajectory $(\langle p(t)\rangle_d, \langle q(t)\rangle_d)$. Thus, the Wigner function $W(p,q)$ can be considered as an equilibrium canonical distribution of a sort, where the factor $\hbar\omega/2$ presents an efficient temperature of quantum fluctuations $k_B T_W$, whence we get $T_W = \hbar\omega/(2k_B)$, that is, the same value T_{osc} as above. To clarify a physical sense of this temperature we consider a reduced Wigner distribution over coordinate

$$w(q) = \int W(p,q)\,dp = \sqrt{\frac{m\omega^2}{2\pi k_B T_{\mathrm{osc}}}} \exp\left\{ -\frac{m\omega^2 (q - \langle q(t)\rangle_d)^2}{2k_B T_{\mathrm{osc}}} \right\}, \tag{4.12.10}$$

that is the statistical distribution of q for the one-dimensional wave packet [90]. It is illustrated in Fig. 5.

On the other hand, the well-known Bloch formula (see, e.g. [65]) defines a probability distribution $b(q)$ for a harmonic oscillator interacting with a heat bath of a temperature T_{hb}:

$$b(q) = \sqrt{\frac{m\omega^2}{2\pi k_B T_{Bl}}} \exp\left\{ -\frac{m\omega^2 q^2}{2k_B T_{Bl}} \right\}, \qquad T_{Bl} = \frac{\hbar\omega}{2k_B} \coth \frac{\hbar\omega}{2k_B T_{hb}}.$$

At very low temperature of the heat bath, $k_B T_{hb} \ll \hbar\omega$, we get $T_{Bl} = T_0 \equiv \hbar\omega/(2k_B)$. The temperature T_0 does not depend on T_{hb} and determined by zeroth fluctuations entirely. It is remarkable that $T_0 = T_{\mathrm{osc}}$. It can be noted that the Bloch formula for this limit case is a particular case of the Wigner reduced function $w(q)$ when $\langle q\rangle_d = 0$. We can argue, therefore, that the temperature T_{osc} of a coherent state is of the same nature as T_0 and determined by the

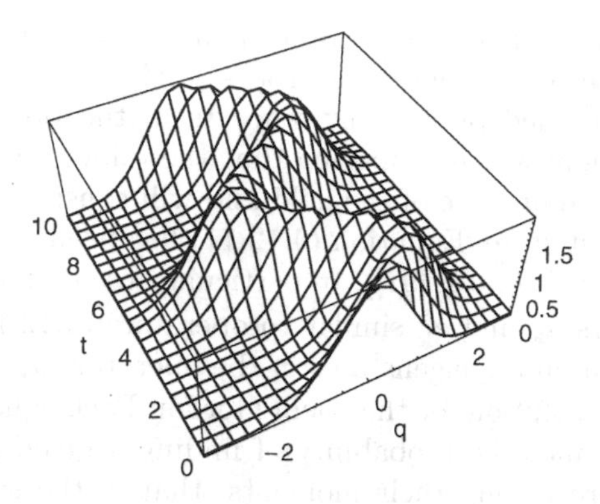

Fig. 4.5. The landscape of statistical distribution of the coordinate q for the one-dimensional wave packet of the coherent oscillator, that is, the reduced Wigner function, Eq. (4.12.10).

zeroth fluctuations as well. Just as T_0 does not change even when $T_{hb} \to 0$ the temperature $T_{\rm osc}$ is fixed also and can not be decreased without destroying the system.

Up to now we considered the thermodynamic definition of the entropy (4.12.5). Next we discuss the von Neumann statistical-mechanical entropy $S^{SM} = \mathrm{Tr}[\rho^d \ln \rho^d]$ of the coherent state $|d\rangle$. For a pure state it is to be zero. Indeed, we have the expansion $\ln \rho^d = -Q - Q^2/2 - Q^3/3 - Q^4/4 - ...$, where $Q = 1 - \rho^d$. Since the density matrix $\rho^d = |d\rangle\langle d|$ is a projection operator, that is, $(\rho^d)^2 = \rho^d$, we get $\rho^d Q = 0$ and $\rho^d \ln \rho^d = 0$. Therefore, the von Neumann entropy S^{SM} is equal to zero, as should be expected for the pure state. It would be different from zero for an ensemble of oscillators distributed over different coherent states. On contrary, the coherent state entropy $S^d_{\rm osc}$, Eq. (4.12.9c), corresponds to an indeterminancy of the single coherent state by itself. This indeterminancy is illustrated in Fig. 5 as a Gauss error curve of a frontal section of the landscape. Dispersion of this Gaussian distribution is determined by the temperature of zero-point oscillations $T_{\rm osc}$.

4.12.2 Entropy of scalar Klein-Gordon-Fock field

We now consider another example of an open quantum system, the scalar mesonic field $\phi(\mathbf{r}, t)$ described by the Klein-Gordon-Fock (KGF) equation with a static source term in the right hand side,

$$\left(\frac{1}{c^2}\frac{\partial^2}{\partial t^2} - \nabla^2 + \frac{m^2c^2}{\hbar^2}\right)\phi(\mathbf{r}, t) = g\varrho(\mathbf{r}). \qquad (4.12.11)$$

The density $\varrho(\mathbf{r})$ and intensity g of the source are given and fixed, which is natural if we treat (4.12.11) as a pure field problem and consider only processes outside the source. On the other hand, such a problem setting is an idealization because we disregard all processes that occur inside the source which is a real extended physical system. As in the previous example, we restrict the discussion by the supposing that these internal processes provide formatting of a coherent state of the field and will consider the field $\phi(\mathbf{r},t)$ in the resulted coherent state. Equation (4.12.11) has been analyzed in detail in quantum field theory (see, e.g. [91]). In particular it was shown that with non-Hermitian operators $a_{\mathbf{k}}$ and $a^{\dagger}_{\mathbf{k}}$ similar to operators (4.12.1), coherent states of the physical vacuum are eigenstates of the operators $a_{\mathbf{k}}$.

The well-known solution for the Klein-Gordon-Fock equation with a static source term determines the probability of finding a given number of quanta in the vacuum disregarding their moments, that is the sum of numbers of quanta with different moments. This probability is governed by the Poisson distribution

$$\rho_{nn} = Q^{-1}_{\mathrm{KGF}} \frac{\overline{n}^n}{n!}, \qquad n = \sum_i n_{\mathbf{k}_i}.$$

Hence we led to the same relations which were introduced above for the particular case of the harmonic oscillator. The partition function and entropy of this ensemble are

$$Q_{\mathrm{KGF}} = \sum_{n=0}^{\infty} \frac{\bar{n}^n}{n!} = e^{\bar{n}}, \tag{4.12.12a}$$

$$F_{\mathrm{KGF}} = -k_B T_{\mathrm{KGF}} \bar{n}, \tag{4.12.12b}$$

$$S_{\mathrm{KGF}} = 2k_B \bar{n}. \tag{4.12.12c}$$

An energy of virtual quanta is presented as the sum of energies of harmonic oscillators of the frequencies ω_i and mean occupation numbers $\bar{n}_{\mathbf{k}_i}$:

$$E_{\mathrm{KGF}} = \overline{n}\hbar\overline{\omega} + \frac{1}{2}\hbar\overline{\omega},$$

where $\overline{\omega} = \sum_i \omega_i \bar{n}_{\mathbf{k}_i}/\bar{n}$ is the mean frequency. Then, we find the thermodynamic temperature of the field as

$$T_{\mathrm{KGF}} = \frac{\hbar\overline{\omega}}{2k_B}.$$

Here, as in the previous example, the coherent state temperature is determined by the system properties and can not be decreased without destroying the coherent state.

To estimate $\bar{n}$, we consider the expectation value of the field potential of a spherical source of radius d, which can be represented [91] in the ground state as

$$\langle 0|\phi(\mathbf{r})|0\rangle|_{r>d} = g \int d^3r' \, \frac{e^{-|\mathbf{r}-\mathbf{r}'|/\lambda_c}}{|\mathbf{r}-\mathbf{r}'|} \varrho(r'), \tag{4.12.13}$$

where $\lambda_C = \hbar/(mc)$ is the Compton length for meson. Equation (4.12.13) implies that a cloud of virtual quanta envelops the source by a spherical layer of thickness λ_C, which do not depend on a size of the source. Therefore, $\bar{n}$ must be proportional to the volume of the layer $\lambda_C A_d$ (here $A_d = 4\pi d^2$ is the area of the layer surface and it is supposed that $d \gg \lambda_C$), divided by the quantum volume element $\sim \lambda_C^3$:

$$\bar{n} \sim \frac{A_d}{\lambda_C^2}.$$

Entropy (4.12.12c) then becomes

$$S_{\mathrm{KGF}} \sim k_B \frac{A_d}{\lambda_C^2}. \tag{4.12.14}$$

An important fact is that entropy (4.12.14) is proportional to the source surface area (cf. the equation (4.12.9c) for the harmonic oscillator). We have such a dependence because of the entropy results from stochastic processes occurring in the vacuum near the surface of the source. The contribution from the source itself into the entropy of the total system is not taken into account in accordance with the initial statement of the problem.

4.12.3 Black hole entropy

The entropy S_{BH} of a black hole is

$$S_{\mathrm{BH}} = k_B \frac{A_{\mathrm{BH}}}{4\lambda_P^2}, \tag{4.12.15}$$

where $A_{\mathrm{BH}} = 16\pi G^2 M_{\mathrm{BH}}^2/c^4$ is the horizon area, $\lambda_P = (\hbar G/c^3)^{1/2}$ the Planck length and M_{BH} the black hole mass. The entropy S_{BH} in the form (4.12.15) was first found by Bekenstein [92, 93] and Hawking [94] using purely thermodynamic arguments based on first and second laws of thermodynamics.

It is evident that the entropy (4.12.15) of a black hole S_{BH} is of the same form as the entropy S_{KGF}, Eq. (4.12.14). The proportionality of the entropy to the surface area of a black hole seemed paradoxical over a long time. This paradox was mostly resolved in [95, 96], where the contribution to S_{BH} coming only from virtual quantum modes which propagate in the immediate vicinity of the horizon surface was calculated. However, an ambiguity remains in the procedure for selecting such modes and we therefore think that the problem of justifying the expression (4.12.15) from the standpoint of statistical mechanics is still open, see for example the recent discussion in [97].

In this respect the review paper by Bekenstein [98] is worth mentioning, where a model quantization of the horizon area was proposed in the section with the ambitious title "Demystifying Black Hole's Entropy Proportionality to Area". There the horizon is formed by patches of equal area $\alpha\lambda_P^2$ which get added one at a time. Since the patches are all equivalent, each will have the same number of quantum states, say, κ. Therefore, the total number of quantum states of the horizon is

$$Q_{\rm BH} = \kappa^{\bar{n}_\alpha}, \qquad \bar{n}_\alpha = \frac{A_{\rm BH}}{\alpha \lambda_P^2}.$$

In essence Bekenstein considers his model construction as a microcanonical ensemble for the patches. As a result, he treats $Q_{\rm BH}$ as the thermodynamic weight of the system and defines the entropy of the horizon as the statistical (Boltzmann's) entropy for the microcanonical ensemble

$$S_{\rm BH}^B = k_B \ln Q_{\rm BH} = k_B \bar{n}_\alpha \ln \kappa.$$

An appearance of logarithmic function in an entropy expression is inevitable if one starts with Boltzmann's or von Neumann's statistical definitions of entropy. To exclude it, Bekenstein divided the obtained expression by the same logarithm function $\ln \kappa$ with the use of re-definition of the constant α:

$$\alpha = \gamma \ln \kappa$$

to obtain

$$Q_{\rm BH} = e^{\overline{n}_\gamma}, \qquad \overline{n}_\gamma = \frac{A_{\rm BH}}{\gamma \lambda_P^2},$$

The Bekenstein-Hawking formula (4.12.15) for the black hole entropy follows with $\gamma = 4$. Thus, the partition function, mean number of quanta, and entropy of a black hole are of the same form as the relevant values for coherent states of the harmonic oscillator or quantum field in a vicinity of the static source. Indeed, it was found [99, 100] that a strong gravitational field provides a decoherence of a system placed in the domain of this field. It is therefore quite natural to expect that a coherent state of virtual excitations of a vacuum is formed in a vicinity of the black hole horizon and, as a result, the black hole's entropy is the thermodynamic entropy of this coherent state. Then, according to the first law of thermodynamics, the black hole temperature is defined as (see Eq. (4.12.6))

$$T_{\rm BH} = \left(\frac{\partial S_{\rm BH}}{\partial E_{\rm BH}}\right)^{-1} = \frac{\hbar c^3}{8\pi k_B G M_{\rm BH}}, \qquad E_{\rm BH} = M_{\rm BH} c^2$$

and should be associated with zero-point fluctuations of vacuum, just as in the earlier examples of the harmonic oscillator and quantum mesonic field. Needless to say, any further rigorous treatment of the subject requires a solution of the problem of quantum field fluctuations of the physical vacuum in the strong gravitational field of a black hole, as was demonstrated above (Subsec. 12.2) for the mesonic field.

On behalf of the proposed approach we point to great unsolved problems in deriving the thermodynamic black hole's entropy (4.12.15) with the use of the standard Gibbs' equilibrium statistical ensemble and the von Neumann formula for entropy. Some years ago one could even read the assertion: " ··· it has been shown that the Bekenstein-Hawking entropy does not coincide with the statistical-mechanical entropy $S^{SM} = -\mathrm{Tr}(\rho \ln \rho)$ of a black hole" [101]. This point of view gained acceptance in recent years. In support of this assertion we present yet another quotation: "there are strong hints from black hole thermodynamics that even our present understanding of the meaning of the 'ordinary entropy' of matter is inadequate" [102]. Discussions are continuuing [97] on the question of whether S_{BH} is the entropy of only the surface system, and thus does not include an interior of black hole, or S_{BH} counts the total number of black hole's microstates including all configurations of interior, with all estimates of S_{BH} performed on the basis of statistical Boltzmann's or von Neumann's definitions of the entropy.

In my opinion, many problems of the black hole's entropy can be resolved if the statistical definitions of the entropy is abandoned in favor of thermodynamic definition, as it was demonstrated above for the particular case of the coherent state of the harmonic oscillator. I suppose that such an approach enhances the position of those who consider S_{BH} as a vacuum fluctuation entropy.

Acknowledgements.

It is a great pleasure to thank the organizers of this Meeting at Kanpur (India) for the exceptionally warm hospitality. I also acknowledge the warm, friendly, and illuminating discussions with S. Abe, R. Johal and A. Sengupta. It is pleasure to thank A. Vityazev for fruitful discussions and supporting this work of many years.

References

[1] Jaynes ET (1957) Phys Rev 106:620–630; 108:171–190
[2] Zubarev DN (1974) Nonequilibrium Statistical Thermodynamics. Consultant Bureau, New York
[3] Zubarev D, Morozov V, Röpke G (2001) Statistical Mechanics of Nonequilibrium Processes. Akademie Verlag, Berlin,
[4] http://astrosun.tn.cornell.edu/staff/loredo/bayes/
[5] Jaynes ET (1983) Probability, Statistics and Statistical Physics. Rosenkrantz (ed), Dordrecht, Reidel
[6] Penrose O (1979) Rep Progr Phys 42:1937–2006
[7] Lavis DA, Milligan PJ (1985) British Journal Philosophy Science 36:193–210
[8] Buck B, Macaulay VA (1991) Maximum Entropy in Action. Clarendon Press, Oxford

[9] Balian R (1991) From Microphysics to Macrophysics. Springer, Berlin
[10] Denbigh KG, Denbigh JS (1985) Entropy in its relation to incomplete knowledge. Cambridge University Press, Cambridge
[11] Dougherty JP (1993) Stud Hist Philos Mod Phys 24:843–866
[12] Uffink J (1995) Studies Hist Philos Mod Phys 26B:223–253
[13] Wheeler JA (1990) Information,Physics, Quantum: The Search for Links. In: Zurek WH (ed) Complexity, Entropy and the Physics of Information, SFI Studies in the Sciences of Complexity, vol VIII
[14] Bak P (1996) How nature works: the science of self-organized criticality. Copernicus, New York
[15] Montroll EW, Shlesinger MF (1983) Journ Stat Phys 32:209–230
[16] Haken H (1988) Information and Self-Organization. Springer, Berlin
[17] Nicolis GS (1986) Dynamics of Hierarchial Systems. An Evolutionary Approach Springer, Berlin
[5] Balescu R (1975) Equilibrium and nonequilibrium statistical mechanics. John Wiley and Sons Inc, New York
[19] Kadomtsev BB (1997) Dynamics and information. Published by Uspechi Fizicheskikh Nauk, Ìoscow (in Russian)
[20] Renyi A (1970) Probability theory. North-Holland, Amsterdam
[21] Selected papers by Alfred Renyi, Vol 2, (1976) Akademiai Kiado, Budapest
[22] Kolmogorov AN (1930) Atti R Accad Naz Lincei 12:388–396
[23] Nagumo M (1930) Japan J Math 7:71–79
[24] Beck C, Schlögl F (1993) Thermodynamics of Chaotic Systems. Cambridge Univ Press, Cambridge
[25] Klimontovich YuL (1994) Statistical Theory of Open Systems. Kluwer Academic Publishers, Dordrecht
[26] Aczel J, Daroczy Z (1975) On measures of information and their characterizations. Academic Press, London
[27] Arimitsu T, Arimitsu N (2002) Physica A 305:218–226
[28] Maassen H, Uffink JBM (1988) Phys Rev Lett 60:1103–1106
[29] Halsey TC, Jensen MH, Kadanoff LP, Procaccia I, Shraiman BI (1986) Phys Rev A 33:1141–1154
[30] Jensen MH, Kadanoff LP, Libchaber A, Procaccia I, Stavans J (1985) Phys Rev Lett 55:2798–2798
[31] Tomita K, Hata H, Horita T, Mori H, Morita T (1988) Prog Theor Phys 80:963–976
[32] Hentschel HGE, Procaccia I (1983) Physica D 8:435–458
[33] Tsallis C (1988) J Stat Phys 52:479–487
[34] http://tsallis.cat.cbrf.br/biblio.htm
[35] Anteneodo C, Tsallis C (2003) J Math Phys 44:5194–5209
[36] Kaniadakis G Scarfone AM (2004) Physica A 340:102–109
[37] Souza AMC, Tsallis C (2003) arXiv:cond-mat/0301304
[38] Abe S, Kaniadakis G, Scarfone AM (2004) arXiv:cond-mat/0401290
[39] Johal RS, Tirnakli U (2004) Physica A 331:487–496

[40] Abe S (2002) Phys Rev E 66:046134
[41] Leshe B (1982) J Stat Phys 27:419–422
[42] Bashkirov A (2005) Phys Rev E 72:028101; (2004) arXiv:cond-mat/0410667
[43] Abe S (2005) Phys Rev E 72:028102; (2004) arXiv:cond-mat/0412774
[44] Isihara A (1971) Statistical Physics. Academic Press, New York
[45] Bashkirov A (2004) Phys Rev Lett 93:13061
[46] Lesche B (2004) Phys Rev E 70:017102
[47] Bashkirov AG (2005) Comments on paper by B.Lesche (Phys.Rev. E 70, 017102 (2004)) "Renyi entropies and observables". arXiv:cond-math/0504103
[48] Shannon CE (1948) Bell Syst Techn Journ 27:379–423; 27:623–656
[49] Khinchin AI (1953) Uspekhi Mat Nauk 8:3–20; Khinchin AI (1957) Mathematical Foundations of Information Theory. Dover, New York
[50] Abe S (2000) Phys Lett A 271:74–79
[51] Jizba P, Arimitsu T (2001) AIP Conf Proc 597:341–354; cond–mat/0108184
[52] Shore JE, Johnson RW (1980) IEEE Trans Inform Theory IT-26:26–37
[53] Johnson RW, Shore JE (1983) IEEE Trans Inform Theory IT-29:942–943
[54] Bashkirov AG (2004) Physica A 340:153-162
[55] Bashkirov AG, Vityazev AV (2000) Physica A 277:136-145
[56] Tsallis C, Mendes RS, Plastino AR (1998) Physica A 261:534
[57] Wilk G, Wlodarczyk Z (2000) Phys Rev Lett 84:2770
[58] Bashkirov AG, Sukhanov AD (2002) Zh Eksp Teor Fiz 122:513–520 [(2002) JETP 95:440–446]
[59] Beck C (2001) Phys Rev Lett 87:18601
[60] Beck C, Cohen EGD (2003) Physica A 322:267–283
[61] Landau LD, Lifshitz EM (1980) Course of Theoretical Physics, Vol. 5: Statistical Physics. Pergamon, Oxford, Part 2.
[62] de Groot SR, Mazur P (1962) Nonequilibium Thermodynamics. North-Holland, Amsterdam
[63] Chandrasekhar S (1943) Stochastic Problems in Physics and Astronomy. American Inst. of Physics, New York
[64] Akhmanov SA, Dyakov YuE, Chirkin AS (1981) Introduction to Statistical Radio Physics and Optics. Nauka, Moscow (in Russian)
[65] Landau LD, Lifshitz EM (1980) Course of Theoretical Physics, Vol. 5: Statistical Physics. Pergamon, Oxford, Part 1
[66] Landau LD (1937) Zh. Eksp. Teor. Fiz. 7:819–832; Frenkel YaI (1950) Principles of the Nuclear Theory. Akad. Nauk SSSR, Moscow; Weisskopf V (1947) In Lecture Series in Nuclear Physics. US Govt Print Off, Washington
[67] Wu J, A. Widom A (1998) Phys Rev E 57:5178–5186
[68] Gradstein IS, Ryzhik IM (1994) Tables of Integrals, Summs, Series and Productions. 5th ed Acad Press Inc, New York

[69] Zubarev DN, Morozov VG, Troshkin OV (1992) Theor Math Phys 92:896–908
[70] Plastino AR, Plastino A (1994) Phys Lett A 193:251–257
[71] Plastino A, Plastino AR (1999) Brazil Journ Phys 29:50–72
[72] Lyra ML, Tsallis C (1998) Phys Rev Lett 80:53–56
[73] Bashkirov AG, Vityazev AV (1996) Planet Space Sci 44:909–916
[74] Fujiwara A, Kamimoto G, Tsukamoto A (1977) Icarus 31:277–293
[75] Zipf GK (1949) Human behavior and the principle of least efforts. Addison-Wesley, Cambridge
[76] Mandelbrot BB (1963) Journ. Business 36:394–426
[77] Price D (1963) Little Science, Big Science. Columbia Univ, New York
[78] Rybczyski M, Wlodarczyk Z, Wilk G (2001) Nucl Phys B (Proc Suppl) 97:81–95
[79] Adamic LA, Hubermann BA (2002) Glottometrics 3:143–151
[80] Wilk G, Wlodarczyk Z (2004) Acta Phys Polon B 35:871–883; cond-mat/0212056
[81] Peres A (1993) Quantum Theory: Concepts and Methods. Kluwer Academic Publ, Dordrecht
[82] Zurek WH, Habib S, Paz JP (1993) Phys Rev Lett 70:1187–1190
[83] Tegmark M, Shapiro HS (1994) Phys Rev E 50:2538–2547
[84] Paz JP, Zurek WH (1998) Quantum limit of decoherence: environment induces superselection of energy eigenstates. qu–ph/9811026
[85] Page DN (1993) Phys Rev Lett 71:1291–1294
[86] Elze H-T (1996) Phys Lett B 369:295–303
[87] Elze H-T (1997) Open quantum systems, entropy and chaos. qu–ph/9710063
[88] Kay BS (1998) Entropy defined, entropy encrease and decoherence understood, and some black–hole puzzles solved. arXiv:hep–th/9802172
[89] Bashkirov AG, Sukhanov AD (2002) Physica A 305:277–281
[90] Messiah A (1961) Quantum Mechanics. vol.1, North-Holland Publ Comp, Amsterdam
[91] Henley EM, Thirring W (1962) Elementary Quantum Field Theory. McGraw–Hill, New York
[92] Bekenstein JD (1972) Lett Nuovo Cimento 4:737–746
[93] Bekenstein JD (1973) Phys Rev D7:2333–2343
[94] Hawking SW (1975) Commun Math Phys 43:199–208
[95] Frolov V, Novikov I (1993) Phys Rev D 48:4545–4553
[96] Frolov V (1995) Phys Rev Lett 74:3319–3322
[97] Jacobson T, Marolf D, Rovelli C (2005) Black hole entropy: inside or out? hep–th/0501103
[98] Bekenstein JD (1997) Quantum black holes as atoms. gr–qc/9710076
[99] Hawking SW (1982) Commun Math Phys, 87:395–403
[100] Ellis J, Mohanty S, Nanopoulos DV (1989) Phys Lett B 221:113–117
[101] Frolov V (1994) arXiv:hep–th/9412211
[102] Wald RM (1999) Class Quant Grav 16A:177–185; gr-qc/9901033

5

Power Law and Tsallis Entropy: Network Traffic and Applications

Karmeshu[1] and Shachi Sharma[1,2]

[1]School of Computer and Systems Sciences, Jawaharlal Nehru University, New Delhi 110067, INDIA. [2]Centre for Development of Telematics, New Delhi 110030, INDIA.
E-Mail: [1]karmeshu@gmail.com, [2]shachi.sharma@rediffmail.com

Summary. A theoretical framework based on non-extensive Tsallis entropy is proposed to study the implication of long-range dependence in traffic process on network performance. Highlighting the salient features of Tsallis entropy, the axiomatic foundations of parametric entropies are also discussed. Possible application of non-extensive thermodynamics to study the macroscopic behavior of broadband network is outlined.

5.1 Introduction

The phenomenon of long-range dependence (LRD) seems to be ubiquitous. It has been observed in a variety of systems ranging from physical, engineering, biological and social systems [1-3]. The resulting correlation structure characterizing LRD decays in accordance with a power law indicating the presence of multifractal structure. There are numerous examples which illustrate power law like behavior in problems related to turbulence [4], DNA sequences [5], city populations [6], linguistics [7], fractal random walks [8], complex high energy processes [9], cosmic rays [10] and stochastic resonance [11]. Attempts have been made to capture power law using approaches based on Boltzmann Gibbs (BG) statistical mechanics and maximum entropy principle due to Jaynes [12] which seeks to maximize entropy subject to auxiliary conditions or moment constraints. The entropic form based on Shannon's measure of information [13] has successfully been employed in the context of BG statistical mechanics.

BG statistical mechanics describes systems in stationary states characterized by thermal equilibrium. Such systems are known to be ergodic and extensive [14]. Broadening the framework of BG statistical mechanics, Tsallis [15] proposed an entropic form to deal with systems which are non-extensive and non-ergodic. These systems are characterized by stationary states which are metastable.

As an extension of Shannon entropy, the proposed non-extensive parametric entropy by Tsallis [15] is defined as

Karmeshu and S. Sharma: *Power Law and Tsallis Entropy: Network Traffic and Applications*, StudFuzz **206**, 162–178 (2006)
www.springerlink.com

$$S_q = K\frac{1 - \sum_{i=1}^{w} p_i^q}{q-1} \tag{5.1.1a}$$

where w denotes the number of states, p_i represents the probability that the system is in state $i \in I = \{1, \cdots, w\}$, the parameter q measures the degree of non extensivity of the system and K is a constant. The non-extensive entropy S_q is concave for $q > 0$ and convex for $q < 0$. In the limit q tending to 1, Tsallis entropy reduces to Shannon entropy [13]

$$S_1 = \lim_{q \to 1} S_q = -K \sum_i p_i \ln p_i. \tag{5.1.1b}$$

which corresponds to entropy of the discrete random variate. It may be instructive to note that the case of continuous random variate is not on the same footing as that of discrete variate. As a limiting process, one can see from Eq. (5.1.1b), the entropy of continuous distribution becomes infinitely large and thus it cannot be used as a measure of uncertainty. However, the discrepancy can be reconciled if one takes the difference between two entropies with the same reference. To distinguish the case of continuous variate from that of discrete, the differential entropy of random variate X with probability density function $f(x)$ is defined as

$$h_B(X) = -\int_{-\infty}^{\infty} f(x) \ln f(x) dx \tag{5.1.2}$$

It is well-known that maximization of differential entropy Eq. (5.1.2) subject to appropriate constraints on the probability distribution may also yield distribution with a power tail such that $f(x) \sim Ax^{-v}$. For the purpose of illustration, one can obtain Cauchy as a maximum entropy probability distribution when the moment constraint on the auxiliary function $\phi(x) = \ln(a^2 + x^2)$ is given. Accordingly, the maximization of Eq. (5.1.2) subject to the following constraints

$$\int_{-\infty}^{\infty} f(x)dx = 1 \quad \text{and} \quad \int_{-\infty}^{\infty} \ln(a^2 + x^2) f(x) dx = \text{const}$$

yields the Cauchy distribution

$$f(x) = \frac{a}{\pi(a^2 + x^2)}, \qquad -\infty < x < \infty$$

Montroll and Shelsinger [16] have noted that any distribution with a power law behavior Ax^{-v} would require auxiliary function to behave asymptotically as $v \ln x$. They further observe "such a function has not been considered a natural one for use in auxiliary conditions. The general situation is even worse since one of the most natural long-tailed inverse power distributions that is connected with some physical model is the Lèvy distribution which is generally defined only through its Fourier integral representation." However, as has been demonstrated by Montroll and Shelsinger [16], in case of income distribution, the basic function which determines one's course of action, is the Bernoulli utility $U(x) = \ln(x/\bar{x})$. Employing maximum entropy formalism, subject to the constraints $E[U^2] = \text{const}$, they have introduced a model to indicate that Pareto-Lèvy tails can be derived from log-normal distribution.

Within the framework of maximization of Tsallis entropy subject to constraints, the power law behavior emerges in a natural way. Several researchers have successfully employed Tsallis entropy to capture power law behaviour in a variety of fields such as edge of chaos [17], Lèvy flights [18], earth-quakes [19], internet traffic [20]. Tsallis entropy framework could find useful applications in the field of teletraffic and internet engineering due to the fact that network traffic shows characteristic features which differ in fundamental ways from the conventional voice networks. It is interesting to note that data traffic measurements reveal the existence of long-range dependence having self-similar structure over a wide range of time scales. Based on simulation experiments with actual traces of ethernet LAN traffic, Erramilli, Narayan and Willinger [21] have shown that the effects of LRD dominate the behaviour of queuing system in terms of larger delays and buffer requirement. It is also observed that traffic in various packet networks such as LAN, WAN, WWW, SS7 and B-ISDN exhibits LRD [22-25]. The characteristic feature of such traffic is on account of burstiness with large variances. The new features in queuing system resulting from LRD traffic cannot be studied on the basis of traditional traffic models based on Markovian assumptions [26].

The purpose of this article is to highlight some of the salient features of non-extensive Tsallis entropy. The axiomatic foundations of various parametric entropies viz Renyi, Havrda-Charvat and Tsallis are underlined. We discuss the robustness of Jaynes entropy concentration theorem for both parametric and non-parametric entropies. The relevance of Tsallis entropic framework is brought out in the context of network traffic characterization as an alternative approach to model the phenomenon of LRD. This in turn helps to study its impact on network performance and gain a better insight into the quality of service (QofS) parameters. It is shown that the effect of LRD results in power law behavior of queue size as well as overflow probability. The usefulness of Tsallis statistics in connection with internet traffic is also discussed. The possibility of non-extensive thermodynamics for studying broadband network is suggested. The article ends with conclusion.

5.2 Tsallis Entropy and Some Other Parametric Entropies

5.2.1 Tsallis entropy and mutual information

Abe [27] has drawn attention to the attractive property of Tsallis entropy which can be uniquely identified by the principles of thermodynamics. Abe [27] observed that non-equilibrium states of such systems are occupied for significantly long periods with preserving scale invariant and hierarchical structures. A notable feature of the phase space is that it is generically inhomogeneous and additivity requirement may not be satisfied any more [28]. Thus for two independent systems one finds

$$S_q(A,B) = S_q(A) + S_q(B) + (1-q)S_q(A)S_q(B) \tag{5.2.1a}$$

In the limit $q \to 1$, Eq. (5.2.1a) reduces to the result due to Shannon

$$S_q(A,B) = S_q(A) + S_q(B) \tag{5.2.1b}$$

which states that sum of joint entropies is equal to the sum of marginal entropies.

An interesting observation is made by Abe [29] in connection with use of generalized differential operator introduced by Jackson (1909) [30, 31] in Tsallis entropy framework. The generalized differential operator is defined as

$$D_q f(x) = \frac{f(qx) - f(x)}{qx - x}$$

which in the limit $q \to 1$ reduces to

$$D_1 = \lim_{q \to 1} D_q = \frac{d}{dx},$$

Abe [28] notes

$$-K\left(D_q \sum_i p_i^q\right) = K\frac{1 - \sum_i p_i^q}{q-1} = S_q$$

whereas the result corresponding to Shannon entropy is recovered for $q = 1$, i.e.

$$-K\left(\frac{d}{dq}\sum_i p_i^q\right)_{q=1} = -K\sum_i p_i \ln p_i = S_1$$

Measures of divergence and mutual information: The divergence measure also referred to as Kullback-Leibler measure, denotes the distance between probability distributions $R = \{r_i\}$ and $P = \{p_i\}$ which in context of Shannon entropy reads as

$$D(R||P) = \sum_i r_i \ln \frac{r_i}{p_i} \tag{5.2.2}$$

The distance is not symmetric implying that $D(R||P) \neq D(P||R)$. The measure of divergence can be used to provide insight into Jaynes' entropy concentration theorem (JECT) which states that most of the probability distributions consistent with the given constraints will be centred around the probability distribution with the maximum entropy. Jaynes proved that for large number N of observations, the statistic $2N(S_{\max} - S)$ follows chi-square distribution where $S_{\max}$ is maximum entropy value and S is the entropy of any other probability distribution [32]. Shachi, Krishnamachari and Karmeshu [33] have investigated the validity of JECT in the context of parametric entropies, particularly the Tsallis entropy. The divergence measure for Tsallis entropy

$$D(R||P) = \frac{1}{q-1} \sum_i p_i \left[\left(\frac{r_i}{p_i} \right)^q - \frac{r_i}{p_i} \right]$$

which for $q \to 1$ yields Eq. (5.2.2). Shachi et. al. [33] find that JECT remains valid for Tsallis entropy and statistic $2N(S_{\max} - S)/q$ is asymptotically distributed as chi-square.

Recently an interesting application of Tsallis mutual information in conjunction with stochastic optimization approximation algorithm has been suggested in the area of image registration [34]. Tsallis mutual information can be expressed as

$$I(X:Y) = S_q(X) + S_q(Y) + (q-1)S_q(X)S_q(Y) - S_q(X,Y)$$

where $S_q(X,Y)$ is the entropy for the joint probability distribution of (X,Y). The authors [34] find that Tsallis entropy improves image registration accuracy and speed of convergence.

5.2.2 Some other parametric entropies

As early as 1967, Havrda and Charvat [35] proposed a non-additive measure of entropy of order α

$$S_\alpha^{HC}(P) = S_\alpha^{HC}(p_1, \cdots, p_n) = \frac{\sum_{i=1}^n p_i^\alpha - 1}{2^{1-\alpha} - 1}, \qquad \alpha \neq 1$$

which is almost similar to Tsallis entropy given in Eq. (5.1.1a). $S_\alpha^{HC}(P)$ reduces to Shannon entropy as α tends to 1. Here, we use the convention $0^\alpha = 0$, $(\alpha \neq 0)$. Though, Havrda and Charvat [35] obtained entropy measure similar to that of Tsallis [15], yet it remained obscured as the full import of its relevance could not be understood. It is Tsallis [36] who demonstrated the full relevance by generalizing BG statistical mechanics to non-extensive physical systems.

Another well-known additive parametric entropy is due to Renyi [37] involving only one parameter. Renyi's additive entropy of order α is defined as

$$S_\alpha^R(P) = S_\alpha^R(p_1, \cdots, p_n) = \frac{1}{1-\alpha} \ln \sum_{i=1}^n p_i^\alpha, \qquad \alpha > 0, \alpha \neq 1 \qquad (5.2.3)$$

which also gives Shannon entropy in the limit $\alpha \to 1$. An important finding is the connection between Havrda-Charvat entropy and Renyi entropy [38], i.e.

$$S_\alpha^{HC}(P) = \frac{2^{(1-\alpha)S_\alpha^R(P)} - 1}{2^{1-\alpha} - 1}, \qquad \alpha \neq 1 \tag{5.2.4}$$

It is obvious from Eq. (5.2.4) that Renyi and Tsallis entropies are also related to each other as a monotonic function. Accordingly, a pertinent question to be answered relates to appropriate choice of entropic measure. To this end, notion of stability can be useful for observable quantities to be experimentally reproducible [39].

5.2.3 Stability of Tsallis entropy

On the basis of stability consideration, Abe [39] has compared the three non-extensive entropies namely: Renyi entropy, Tsallis entropy and normalized Tsallis entropy, listed below

$$S_q^{(R)} = \frac{1}{1-q} \ln \sum_{i=1}^{n} p_i^q \tag{5.2.5a}$$

$$S_q^{(T)} = \frac{1}{1-q} \left(\sum_{i=1}^{n} p_i^q - 1 \right) \tag{5.2.5b}$$

$$S_q^{(NT)} = \frac{1}{1-q} \left(1 - \frac{1}{\sum_{i=1}^{n} p_i^q} \right) \tag{5.2.5c}$$

It is known from the principle of maximum entropy that all three entropies yield the q-exponential distribution. This raises an obvious question as to the choice of suitable criterion for selecting appropriate entropy form. Such a criterion is based on stability considerations which require small change under an arbitrary small deformation of the distribution. Using l^1 norm as a measure of size of deformation from $\{p_i\}_{i=1,\cdots}$ to $\{\acute{p}_i\}_{i=1,\cdots}$, i. e.

$$\|p - \acute{p}\|_1 = \sum_i |p_i - \acute{p}_i|,$$

Abe [39] established that only Tsallis entropy given by Eq. (5.2.5b) is stable under perturbations.

5.3 Axiomatic Foundations of Parametric Entropies

An area of investigation in vogue relates to characterization of entropy measure in terms of different sets of postulates. Notable attempts in the context of Shannon measure have been made by Khinchin [40] and Fadeev [41]. Subsequently, attempts have been made to characterize additive and non-additive measures of entropies due to Renyi and Havrda-Charvat [38]. Recently, Suyari [42] has generalized Shannon-Khinchin (GSK) axioms for non-extensive entropies. For the sake of completeness, we give the sets of postulates in the context of Shannon, Renyi, Havrda-Charvat and Tsallis entropies, with the following definition of state space

$$\triangle_n = \left\{ (p_1, \cdots, p_n) \colon (p_i \geq 0) \left(\sum_{i=1}^{n} p_i = 1 \right) \right\}. \tag{5.3.1}$$

5.3.1 Shannon-Khinchin axioms

We closely follow the article of Suyari [42] in this sub-section. The Shannon's measure of entropy S_1 is given by Eq. (5.1.1*b*). The Shannon-Khinchin axioms are:

1. *Continuity*: For any $n \in N$, the function S_1 is continuous with respect to $(p_1, \cdots, p_n) \in \triangle_n$.
2. *Maximality*: For given $n \in N$ and $(p_1, \cdots, p_n) \in \triangle_n$, the function S_1 takes its largest value for $p_i = 1/n$.
3. *Additivity*: If

$$p_{ij} \geq 0,\ p_i = \sum_{j=1}^{m_i} p_{ij}, \qquad (\forall i = 1, \cdots, n),\ (\forall j = 1, \cdots, m)$$

   then the following equality holds

$$S_1(p_{11}, \cdots, p_{nm_n}) = S_1(p_1, \cdots, p_n) + \sum_{i=1}^{n} p_i S_1 \left(\frac{p_{i1}}{p_i}, \cdots, \frac{p_{im_i}}{p_i} \right)$$

4. *Expandability*:

$$S_1(p_1, \cdots, p_n, 0) = S_1(p_1, \cdots, p_n)$$

The proofs of these axioms can be found in Khinchin [40].

5.3.2 Axioms for Renyi entropy

Renyi entropy as defined in Eq. (5.2.3) satisfies the following postulates [43]:

1. Symmetry: $S_\alpha^R(P)$ is a symmetric function of its variables for $n = 2, 3, \cdots$
2. Continuity: $S_\alpha^R(P)$ is a continuous function for p for $0 \leq p \leq 1$.
3. $S_\alpha^R\left(\frac{1}{2}, \frac{1}{2}\right) = 1$
4. Additivity: For two probability distributions $P = (p_1, \cdots, p_n)$ and $Q = (q_1, \cdots, q_n)$, the entropy of the combined distribution is equal to the sum of entropies of the individual distributions i.e.

$$S_\alpha^R(P * Q) = S_\alpha^R(P) + S_\alpha^R(Q)$$

5.3.3 Axioms for Havrda-Charvat entropy

Havrda-Charvat [35] introduced a set of postulates for non-additive structural α-entropy. Following Karmeshu and Pal [43], the postulates are:

1. $S_\alpha^{HC}(p_1, \cdots, p_n; \alpha)$ is continuous in the region $p_i \geq 0$, $\sum_i p_i = 1, \alpha > 0$
2. $S_\alpha^{HC}(1, \alpha) = 0, S_\alpha^{HC}(\frac{1}{2}, \frac{1}{2}; \alpha) = 1$
3. $S_\alpha^{HC}(p_1, \cdots, p_{i-1}, 0, p_{i+1}, \cdots, p_n; \alpha) = S_\alpha^{HC}(p_1, \cdots, p_{i-1}, p_{i+1}, \cdots, p_n; \alpha)$ for every $i = 1, 2, \cdots, n$
4. $$\begin{aligned} &S_\alpha^{HC}(p_1, \cdots, p_{i-1}, r_{i_1}, r_{i_2}, p_{i+1}, \cdots, p_n; \alpha) = \\ &\quad S_\alpha^{HC}(p_1, \cdots, p_{i-1}, p_{i+1}, \cdots, p_n; \alpha) + \alpha p_i^\alpha S_\alpha^{HC}\left(\frac{r_{i_1}}{p_i}, \frac{r_{i_2}}{p_i}; \alpha\right) \\ &\qquad\qquad \text{for every } r_{i_1} + r_{i_2} = p_i > 0, \; i = 1, \cdots, n \end{aligned}$$

Another attempt to characterize non-additive measure is due to Forte and Ng [44]. Following Mathai and Rathie [38], we present the axioms for Havrda-Charvat entropy:

1. *Symmetry*: $S_\alpha^{HC}(p_1, \cdots, p_6)$ is a symmetric function of its variables.
2. *Expansibility*:
$$S_\alpha^{HC}(p_1, \cdots, p_n, 0) = S_\alpha^{HC}(p_1, \cdots, p_n), \; (p_1, \cdots, p_n) \in \triangle_n, \; n \geq 2$$
3. *Branching*:
$$\begin{aligned} &S_\alpha^{HC}(p_1, \cdots, p_{n+1}) - S_\alpha^{HC}(p_1 + p_2, p_3, \cdots, p_{n+1}) > 0 \\ &\qquad \forall (p_1, \cdots, p_{n+1}) \in \triangle_{n+1}, \; p_1, p_2 > 0, \; n \geq 2 \end{aligned}$$
4. *Compositivity*:
$$\begin{aligned} &S_\alpha^{HC}(pp_1, pp_2, pp_3, pp_4, (1-p)q_1, (1-p)q_2) = \\ &\qquad \psi_{4,2}[S_\alpha^{HC}(p_1, p_2, p_3, p_4), S_\alpha^{HC}(q_1, q_2), p], \\ &\qquad \text{for all } (p_1, p_2, p_3, p_4) \in \triangle_4, \; (q_1, q_2) \in \triangle_2, \; p \in [0, 1] \text{ and} \\ &\qquad\qquad S_\alpha^{HC}(p_1, p_2) \text{ not constant in } \triangle_2 \end{aligned}$$
5. *Continuity*: $S_\alpha^{HC}(p_1, p_2, p_3)$ is continuous at the boundary points of $\triangle_3$
6. *Nullity*: $S_\alpha^{HC}(0, 1) = 0$
7. *Normalization* : $S_\alpha^{HC}(\frac{1}{2}, \frac{1}{2}) = 1$

5.3.4 Axioms for Tsallis entropy

Suyari [42] in a recent paper has proposed generalized Shannon-Khinchin axioms (1-4) for determining the function $S_q : \Delta_n \to R^+$ such that

$$S_q(p_1, \cdots, p_n) = \frac{1 - \sum_{i=1}^{n} p_i^q}{\phi(q)}$$

where $q \in R^+$ and $\phi(q)$ satisfies properties (i)–(iv)

(i) $\phi(q)$ is continuous and has same sign as $q-1$, i.e.,

$$\phi(q)(q-1) > 0, \qquad q \neq 1$$

(ii)

$$\lim_{q \to 1} \phi(q) = \phi(1) = 0, \qquad \phi(q) \neq 0, \; q \neq 1$$

(iii) there exists an interval $(a, b) \subset R^+$ such that $a < 1 < b$ and $\phi(q)$ is differentiable on the interval

$$(a, 1) \bigcup (1, b)$$

(iv) there exists a constant $k > 0$ such that

$$\lim_{q \to 1} \frac{d\phi(q)}{dq} = \frac{1}{k}$$

1. *Continuity*: S_q is continuous in Δ_n and $q \in R^+$
2. *Maximality*: for any $q \in R^+$, any $n \in N$ and any $(p_1, \cdots, p_n) \in \Delta_n$

$$S_q(p_1, \cdots, p_n) \le S_q\left(\frac{1}{n}, \cdots, \frac{1}{n}\right)$$

3. *Generalized Shannon additivity*: under the normalization constraint of probabilities, the following equality holds

$$S_q(p_1, \cdots, p_{nm_n} = S_q(p_1, \cdots, p_n) + \sum_{i=1}^{n} p_i^q S_q\left(\frac{p_{i1}}{p_i}, \cdots, \frac{p_{im_i}}{p_i}\right)$$

4. *Expandability*

$$S_q(p_1, \cdots, p_n, 0) = S_1(p_1, \cdots, p_n)$$

It would be worth examining the similarities and dissimilarities among the set of postulates for non-extensive entropies by Havrda-Charvat [35], Forte and Ng [44] and Suyari [42]. This will enable to identify the minimal set of postulates from physical considerations.

5.4 Tsallis Entropy and Network Traffic

The maximum entropy principle (MEP) as conceived by Jaynes [12] allows one to choose a probability distribution when some prior information about the system is prescribed. In the current setting, it is assumed that the kth moment of number of packets is known a priori. When Tsallis entropy as given in Eq. (5.1.1a) is maximized subject to k^{th} order moment constraint

$$\sum_i i^k p_i = A \tag{5.4.1a}$$

and normalization constraint

$$\sum_i p_i = 1 \tag{5.4.1b}$$

the queue length distribution of number of packets is obtained by employing Lagrange's method of undetermined multipliers. The Lagrangian function is given by

$$\phi_q = \frac{S_q}{K} - \alpha\left(1 - \sum_i p_i\right) + \alpha\beta(q-1)\left(A - \sum_i i^k p_i\right) \tag{5.4.2}$$

Here, α and β are Lagrange's parameters. By differentiating Eq. (5.4.2) with respect to p_i and making use of normalization constraint the queue length distribution is given by

$$p_i = Z^{-1}[1 + \beta(1-q)i^k]^{1/(q-1)} \tag{5.4.3}$$

where

$$Z = \sum_i [1 + \beta(1-q)i^k]^{1/(q-1)} \tag{5.4.4}$$

This result for $k = 1$ is similar to the one due to Tsallis [15]. In the limit q tends to 1, Eq. (5.4.3) becomes

$$p_i = \frac{\exp(-\beta i^k)}{\sum_i \exp(-\beta i^k)}$$

For large number of packets i in the network, Eq. (5.4.3) behave as

$$p_i \sim i^{k/(q-1)}$$

and mimicking power law behavior for $q < 1$. We next discuss cases for various values and ranges of k.

Case I: $0 < k < 1$. This case pertains to the fractional moments [45]. The packet networks traffic having long-range dependence may not have finite mean and variance. In such scenarios, some knowledge of finite fractional moment may be available. The range of entropic parameter q then becomes $1-k < q < 1$ since k is a fraction. This condition is required for the convergence of Z.

The mean number of packets in the network is given by

$$\overline{N} = Z^{-1} \sum_i i \left[1 + \beta(1-q)i^k\right]^{1/(q-1)} \tag{5.4.5}$$

which for finiteness requires $q > 1 - k/2$. It is easy to notice when q lies between $1-k$ and $1-k/2$, mean becomes infinite.

Case II $k = 1$. This case corresponds to availability of first moment as the constraint [46]. The corresponding probability distribution as given by Eq. (5.4.3) and Eq. (5.4.4) becomes

$$p_i = \frac{\left[\dfrac{1}{\beta(1-q)} + i\right]^{1/(q-1)}}{\zeta\left[\dfrac{1}{1-q}, \dfrac{1}{\beta(1-q)}\right]}$$

where $\zeta\left[\frac{1}{1-q}, \frac{1}{\beta(1-q)}\right]$ denotes the Hurwitz-Zeta function [47] defined by

$$\zeta\left[\frac{1}{1-q}, \frac{1}{\beta(1-q)}\right] = \sum_{i=0}^{\infty} \left[i + \frac{1}{\beta(1-q)}\right]^{-1/(q-1)}$$

The probability distribution of p_i is also known as the Zipf-Mandelbrot distribution [48].

The first moment of number of packets can also be expressed in terms of Hurwitz-Zeta function as

$$\overline{N} = \frac{\zeta\left[\dfrac{q}{1-q}, \dfrac{1}{\beta(1-q)}\right]}{\zeta\left[\dfrac{1}{1-q}, \dfrac{1}{\beta(1-q)}\right]} - \frac{1}{\beta(1-q)}, \qquad q > \frac{1}{2}$$

The range of entropic parameter now becomes $\frac{1}{2} < q < 1$.

Case III $k > 1$. This pertains to the case when higher order integral moments of number of packets is known to be finite. This available information is used as the constraint. The distribution of queue length is given by Eqs. (5.4.3) and (5.4.4), and for convergence we require $1-k < q < 1$. Since k is positive integer exceeding unity, q can be less than zero. The expression for the mean number of packets is same as given in Eq. (5.4.5) which for finiteness requires $1 - k^{-1} < q < 1$.

5.4.1 Performance Measures

One of the important QofS parameter is the overflow probability which denotes the probability of exceeding a given buffer size. This is often obtained from the tail of the probability of exceeding a threshold value in an infinite buffer system, given by

$$P(i > x) = Z^{-1} \sum_{i=x+1}^{\infty} \left[1 + \beta(1-q)i^k\right]^{1/(q-1)}$$

For large buffer size x, one finds

$$P(i > x) \sim x^{-k/(1-q)+1}$$

depicting again power law behaviour for $k/(1-q) > 1$.

Another QofS parameter of interest is the system utilization defined by

$$U = 1 - p_0 = 1 - \sum_i \left[1 + \beta(1-q)i^k\right]^{1/(q-1)}$$

5.4.2 Tsallis entropy and Escort distribution

In the context of non-extensive systems, it is suggested that the usage of probability $\{P_i\}$ defined as

$$P_i = \frac{p_i^q}{\sum_i p_i^q} \tag{5.4.6}$$

is more appropriate. Equation (5.4.6) is known as the Escort distribution of p_i of order q [28]. It has been argued that for power law distribution, mean values or q-expectation values formed with Escort distribution give more consistent results [49, 50]. Shachi and Karmeshu [51] observe that formalism based on q-expectation yields similar results as discussed in previous section except the admissible range of entropy parameter q. In the context of queueing problems, further information about system's utilization U_e is also known. The problem can be reformulated as,

$$\text{Max } S_q = \text{Max } K\frac{1 - \sum_i p_i^q}{q-1} \tag{5.4.7}$$

subject to

$$\sum_i p_i = 1, \qquad \sum_i i^k P_i = A_e \tag{5.4.8a}$$

$$\sum_i h(i)P_i = U_e, \quad \text{with} \quad h(i) = 0 \text{ if } i = 0$$
$$h(i) = 1 \text{ if } i = 1. \tag{5.4.8b}$$

The entropy optimization can be carried out using Langrange's method. This yields

$$p_0 = 1 - U_e, \quad i = 0$$

$$p_i = p_0 \lambda_1 \left[1 + \lambda_2 (q-1) i^k \right]^{1/(q-1)}, \quad i \neq 0 \tag{5.4.9}$$

where λ_1 and λ_2 are parameters to be calculated from Eq. (5.4.8a) and Eq. (5.4.8b). It is easy to see that Eq. (5.4.9) also asymptotically yields the well-known power law behavior for $q > 1$. The convergence of moment constraint given in Eq. (5.4.8a) requires $q < (k+1)$. Hence, the admissible range of q becomes $1 < q < (k+1)$.

5.5 Internet Traffic and Tsallis Entropy

There is evidence to the effect that Tsallis statistics describes the scale-invariant states of internet traffic [20]. Noting that internet provides an interesting example of self-organizing system, Abe and Suzuki [20] suggest that the actions of a large number of users can be understood within statistical mechanics framework. Based on data analysis of the echo experiment, the cumulative probability distribution of sparseness time interval in the internet is examined. They observe that the data is in accordance with the q-exponential distribution. The problem in a more generalized setting including k^{th} order moment can be stated mathematically as

$$\text{Max } S_q = \text{Max } K \frac{1 - \int_0^\infty p^q(x) dx}{q-1} \tag{5.5.1}$$

subject to

$$\int_0^\infty x^k p(x) dx = A \tag{5.5.2a}$$

$$\int_0^\infty p(x) dx = 1 \tag{5.5.2b}$$

Extremizing Eq. (5.5.1) subject to Eq. (5.5.2a) and Eq. (5.5.2b) results in

$$p(x) = Z^{-1} [1 + \beta(1-q) x^k]^{1/(q-1)}, \qquad x \geq 0 \tag{5.5.3}$$

for

$$Z = \int_0^\infty [1 + \beta(1-q) x^k]^{1/(q-1)} dx$$

where β is the Lagrange parameter and can be calculated from Eq. (5.5.2a), we found,

$$\beta = \frac{1}{A[q(1+k) - 1]} \tag{5.5.4}$$

Substituting Eq. (5.5.4) in Eq. (5.5.3), the resulting distribution of queue length reads as

$$p(x) = \frac{k\Gamma\left(\frac{1}{1-q}\right)\left[A\left(\frac{kq}{1-q}-1\right)+x^k\right]^{1/(q-1)}}{\Gamma\left(\frac{1}{k}\right)\Gamma\left(\frac{1}{1-q}-\frac{1}{k}\right)\left[A\left(\frac{kq}{1-q}-1\right)\right]^{1/(q-1)+1/k}}, \qquad x > 0 \tag{5.5.5}$$

which reduces to that of Abe and Suzuki [20] for $k = 1$. The moment generating function (MGF) is given by,

$$M_x(t) = \frac{\sum_{j=0}^{\infty}\frac{t^j}{j!}\left[A\left(\frac{kq}{1-q}-1\right)\right]^{(1+j)/k}\Gamma\left(\frac{1+j}{k}\right)\Gamma\left(\frac{1}{1-q}-\frac{1+j}{k}\right)}{\left[A\left(\frac{kq}{1-q}-1\right)\right]^{1/k}\Gamma\left(\frac{1}{k}\right)\Gamma\left(\frac{1}{1-q}-\frac{1}{k}\right)}$$

The effect of higher order moment (i.e $k > 1$) of sparseness time interval on the cumulative distribution function is worth examining.

5.6 Broadband Networks and Non-Extensive Thermodynamics

Broadband networks with large capacity of network components can be characterized as a dynamical system with complex nonlinear interactions among them. The understanding of the network performance has largely been dependent on the microscopic description of nodes and network dynamics. Due to unverifiable data, the microscopic description may sometimes not be reliable. Further, the presence of chaos in networks may render such data futile. In contrast to microscopic description Hui and Karasan [52] proposed a macroscopic description based on extensive thermodynamics employing scalability postulate for grade of service, bandwidth and buffer assignments and bandwidth demand. Earlier Benes [53] using Shannon's entropy framework proposed a theory for connecting networks and telephone theory.

We propose to extend the extensive thermodynamics framework to examine the role of non-extensivity in the field of broadband networks. It is widely known that characteristic features of traffic is long range dependent and exhibits the power law behaviour. Since Tsallis entropy captures such behaviour in a natural manner, it would be profitable to use non-extensive thermodynamics formalism to investigate broadband network behaviour. This will also give an insight into the dynamic routing of broadband traffic.

5.7 Conclusion

The article demonstrates the usefulness of non-extensive Tsallis entropic framework for capturing the phenomenon of long-range dependence in the context of broadband network traffic. It is noted that the effect of LRD results in the power law behavior of queue sizes in network. The apparent success of applicability of non-extensive statistical mechanics in widely diverse systems ranging from physical to social is noteworthy. This would suggest a deeper underlying mechanism as has been observed by Gellmann and Tsallis [14]. They observe: "An intriguing question that remains unanswered is: exactly what do all these systems have in common? One suspects, of course that the deep explanation must arise from microscopic dynamics. The various cases could all be associated with something like a scale-free dynamical occupancy of phase space, but this certainly deserves further investigation."

References

[1] V. Lathora, A. Rapisarda and S. Ruffo, Lyapunov Instablity and Finite Size Effects in a System with Long Range Forces, Phys. Rev. Lett. **80** (1998), 692–695.

[2] C. K. Peng, S .V. Buldyrev, A. L. Goldberger, S. Havlin, F. Sciotino, M. Simons and H.E. Stanley, Long Range Correlations in Nucleotide Sequences, Nature **356** (1992), 168–170.

[3] G. Malescio, N. V. Dokholyan, S. V. Buldyrev and H.E. Stanley, cond-mat/0005178.

[4] C. Beck, G. S. Lewis and H. L. Swinney, Measuring Nonextensitivity Parameters in a Turbulent Couette-Taylor flow, Phys. Rev. E **63**, 035303 (2001).

[5] Karmeshu and A. Krishnamchari, Sequence Variability and Long Range Dependence in DNA: An Information Theoretic Perspective in Neural Information Processing, Lecture Notes in Computer Science **3316** (2004), 1354–1361.

[6] A. H. Makse, S. Havlin and H. E. Stanley, Nature (London), **377**, (1995), 608–612.

[7] G. K. Zipf, *Selective Studies and the Principle of Relative Frequency in Language*, Harvard U. Press, Cambridge MA, 1937.

[8] A. Robledo, Renormalization Group, Entropy Optimization, and Nonextensivity at Criticality, Phys Rev. Lett. **83**, 2289 (1999).

[9] D. B. Walton and J. Rafelski, Equilibrium Distribution of Heavy Quarks in Fokker-Planck Dynamics, Phys Rev. Lett. **84** (2000), 31–34.

[10] C. Tsallis, J. C. Anjos and E.P. Borges, astro-ph/0203258.

[11] H. S. Wio and S. Bouzat, Brazil Journal of Physics, **29**, 136 (1999).

[12] E. Jaynes, Prior probabilities, IEEE Transactions on System Science Cybernetics **SSC-4** (1968), 227–241.

[13] C. E. Shannon, A Mathematical Theory of Communication, Bell System Technical Journal **27** (1948), 379–423.

[14] M. Gell-Mann and C. Tsallis, *Nonextensive Entropy Interdisciplinary Applications*-Preface, Oxford University Press, 2004.

[15] C. Tsallis, Possible Generalization of Boltzmann-Gibbs Statistics, J. Stat. Phys., **52** (12) (1988), 479–487.

[16] E. W. Montroll and M. F. Shlesinger, On the Wonderful World of Random Walks, *Nonequilibrium Phenomena from Stochastics to Hydrodynamics*, ed. J. L. Lebowitz and E. W. Montroll, North-Holland Physics Publishing, 1984.
[17] V. Latora, M. Baranger, A. Rapisarda and C. Tsallis, The Rate of Entropy Increase at the Edge of Chaos. Phys. Lett. A **273,** 90, (2000).
[18] S. Abe and A. K. Rajagopal, Rates of Convergence of Non-Extensive Statistical Distributions to Levy Distributions in Full and Half-Spaces. J. Phys. A **33,** (2000), 8723-8732.
[19] S. Abe and N. Suzuki, J. Geophys Res. **108**(B2) (2003) 2113.
[20] S. Abe and N. Suzuki, Itineration of the Internet over Non-equilibrium Stationary States in Tsallis Statistics. Phys. Rev. E **67,** 016106, (2003).
[21] A. Erramilli, O. Narayan, W. Willinger, Experimental Queueing Analysis with LRD Packet Traffic, IEEE/ACM Transactions on Networking **4** (2) (1996), 209-223.
[22] W. Leland, M. S. Taqqu, W.Willinger and D.V. Wilson, On the Self-Similar Nature of Ethernet Traffic, IEEE/ACM Transactions on Networking **2** (1) (1994), 1-15.
[23] Mark E. Crovella and Azer Bestavros, Self-similarity in World Wide Web Traffic: Evidence and Possible Causes, IEEE/ACM Transactions on Networking **5** (6) (1997), 835-846.
[24] D. Duffy, A. McIntosh, M. Rosenstein, W. Willinger, Statistical Analysis of CCSN/SS7 Traffic Data from Working CCS Subnetworks, IEEE Jounal on Selected Areas in Communication **12** (3) (1994), 544-551.
[25] William Stallings, *High Speed Networks and Internets: Performance and Quality of Service*, Second Edition, Pearson, 2002.
[26] Paxon and Floyd, Wide area Traffic : the Failure of Poisson Modelling, IEEE/ACM Transactions on Networking **3** (3) (1995), 226-244.
[27] S. Abe, Tsallis Entropy: How Unique? Preprint (2001).
[28] C. Beck and F. Schlogl, *Thermodynamics of Chotic Systems — An Introduction*, Cambridge University Press, 1993.
[29] S. Abe, A Note on the q-deformation Theoretic Aspect of the Genealized Entropies in Non-Extensive Physics, Phys Lett. A **224** (1997) 326.
[30] F. Jackson, Mess. Math, **38** (1909) 57.
[31] F. Jackson, Quart. J. Pure Appl. Math, **41** (1910) 1993.
[32] J. N. Kapur and H. K. Kesavan, *Entropy Optimization Principles with Application*, Academic Press Inc, London, 1992.
[33] Shachi Sharma, Krishanmachari and Karmeshu, Validity of Jaynes' Entropy Concentration Theorem: Tsallis and other Generalized Entropy Measures, Preprint (2005).
[34] S. Martin, G. Morison, W. Nailon and T. Durrani, Fast and Accurate Image Registration Using Tsallis Entropy and Simultaneous Perturbation Stochastic Approximation, Electonics Letters, **40**, 10, (2004).
[35] J. Havrda and F. Charvat, Quantifiation Method of Classification Processes: Concept of Structural α-entropy, Kybernetika **3** (1967), 30–35.

[36] C. Tsallis, Non-extensive Statistical Mechanics: Construction and Physical Interpretation, in *Nonextensive Entropy Interdisciplinary Applications* ed. M. Gellmann and C. Tsallis, Oxford University Press, 2004, 1–52.
[37] A. Renyi, On Measures of Entropy and Information, Proceedings Fourth Berkeley Symp. Math. Statist and Prob., University of California Press **1** (1961), 547–561.
[38] A. M. Mathai and P. N. Rathie, *Basic Concepts in Information Theory and Statistics*, John Wiley, 1975.
[39] S. Abe, Stability of Tsallis Entropy and Instabilities of Renyi and Normalized Tsallis Entropies: A Basis for q-exponential Distributions, Phys. Rev. E **66**, 046134, (2002).
[40] A. I. Khinchin, *Mathematical Foundations of Information Theory*, Dover Publications, 1957.
[41] D. K. Fadeev, On the Concept of Entropy of a Finite Probabilistic Scheme (Russian), Uspeki Mat. Nauk **11** (1956), 227–231.
[42] Hiroki Suyari, Generalization of Shannon-Khinchin Axioms to Nonextensive Systems and the Uniqueness Theorem for the Nonextensive Entropy, IEEE Transactions on Information Theory, **50** (8) (2004), 1783–1787.
[43] Karmeshu and N. R. Pal, Uncertainty, Entropy and Maximum Entropy Principle — An Overview, in *Entropy Measures, Maximum Entropy Principle and Emerging Applications*, ed. Karmeshu, Springer-Verlag, 2003.
[44] B. Forte and C. T. Ng, On a Characterization of the Entropies of Degree β, Utilitas Mathematica **4** (1973), 193–205.
[45] Karmeshu and Shachi Sharma, Queue Length Distribution of Network Packet Traffic: Tsallis Entropy Maximization with Fractional Moments, IEEE Communiation Letters, To Appear 2005.
[46] Karmeshu and Shachi Sharma, Long Tail Behavior of Queue Lengths in Broadband Networks: Tsallis Entropy Framework, Preprint (2005).
[47] T. M. Apostol, *Introduction to Analytic Number Theory*, Springer-Verlag, New York, 1976.
[48] B. Mandelbrot, *The Fractal Geometry of Nature*, Freeman, San Francisco, 1983.
[49] S. Abe and A. K. Rajgopal, Microcanonical Foundation for Systems with Power-Law Distributions. J. Phys. A **33**, 8733–8738 (2000).
[50] S. Abe and G. B. Bagci, Necessity of q-expectation Value in Nonextensive Statistical Mechanics. Phys. Rev. E **71**, 016139, (2005).
[51] Shachi Sharma and Karmeshu, Asymptotic Power Law Characteristics in Queueing Systems — Using Maximum Tsallis Entropy Principle, Preprint (2005).
[52] J. Y. Hui and E. Karasan, A Thermodynamic Theory of Broadband Networks with Application to Dynamic Routing, IEEE Journal of Selected Areas in Communications **13**, (1995), 991-1003.
[53] V. E. Benes, *Mathematical Theory of Connecting Networks and Telephone Traffic*, New York, Academic, 1965.

6

The Role of Chaos and Resonances in Brownian Motion

John Realpe[1] and Gonzalo Ordonez[2]

[1] Departamento de Física, Universidad del Valle, A.A. 25360, Cali, COLOMBIA,
[2]The University of Texas at Austin, University Station, Austin, TX 78712 USA, and Butler University, Indianapolis, IN 46208, USA
E-Mail:[1]jrealpe@calima.univalle.edu.co [2]gordonez@butler.edu

Summary. We study two points of view regarding the origin of irreversible processes. One is the "chaotic hypothesis" that says that irreversible processes are rooted in the randomness generated by chaotic dynamics. The second point of view, put forward by Prigogine's school, is that irreversibility is rooted in non-integrable dynamics, as defined by Poincaré. Non-integrability is associated with resonances. We consider a simple model of Brownian motion, a harmonic oscillator (particle) coupled to lattice vibration modes (field). We compute numerically the "(ϵ, τ) entropy", which indicates how random are the trajectories and how close they are to Brownian trajectories. We show that (1) to obtain trajectories close to Brownian motion it is necessary to have a resonance between the particle and the lattice, which allows the transfer of information from the lattice to the particle. This resonance makes the system non-integrable in the sense of Poincaré. (2) For random initial conditions, chaos seems to play a secondary role in the Brownian motion, as the entropy is similar for both chaotic and non-chaotic dynamics. In contrast, if the initial conditions are not random, chaos plays a crucial role, as it leads to the thermalization of the lattice, which then induces the Brownian motion of the particle through resonance.

6.1 Introduction

In this work we study the relationship that some features of mechanical systems (such as Poincaré resonances and chaos) keep with irreversible phenomena, specifically Brownian motion, a topic which has woken up great interest over the last years. We center on two points of view regarding the origin of irreversible processes. One is the "chaotic hypothesis" that says that irreversible processes are rooted in the randomness generated by chaotic dynamics [1, 2]. The second point of view, put forward by Prigogine's school, is that irreversibility is rooted in non-integrable dynamics, as defined by Poincaré [3, 4, 5, 6, 7, 8, 9].

Poincaré's non-integrability occurs when the perturbation expansions of the invariants of motion diverge due to resonances. Chaotic systems are

J. Realpe and G. Ordonez: *The Role of Chaos and Resonances in Brownian Motion*, StudFuzz **206**, 179–206 (2006)
www.springerlink.com

non-integrable in Poincaré's sense, but systems that are non-integrable in Poincaré's sense are not necessarily chaotic. Chaos is a sufficient but not necessary condition for Poincaré's non-integrability.

The chaotic hypothesis may be regarded as an extension of the ergodic hypothesis to systems out of equilibrium which are in a stationary state. In fact, the condition of Anosov implies the existence of an invariant measure μ univocally determined, known as the SRB distribution (after Sinai, Ruelle and Bowen) such that, *for almost all* the points $P \in \Gamma$ of the phase space Γ and *all* the observables f, the following equality holds [10, 11]:

$$\lim_{T\to\infty} \frac{1}{T}\int_0^T f(S^t P_0)\mathrm{d}t = \int_\Gamma f(P)\mathrm{d}\mu(P)$$

where S^t is the dynamical evolution group. Concerning to the chaotic hypothesis, Gaspard *et. al.* [12] have carried out an experiment measuring 145 612 successive positions of a Brownian particle at regular time intervals $\tau = 1/60$ s in order to calculate the ϵ-entropy of the so-obtained time series. On the basis of this analysis they claim to have found experimental evidence of chaos on the microscopic level, a conclusion which has been subject to many discussions (see e.g. Dettmann and Cohen [13]).

In this paper we will focus on Brownian random motion. We will investigate the roles that chaos and resonance (thus Poincaré's non-integrability) play in the generation of Brownian motion. We will study a simple model, a harmonic oscillator (Brownian particle) coupled to a lattice. The lattice vibration modes have a non-linear coupling as in Fermi-Pasta-Ulam's model, which may lead to chaos depending on the coupling strength. In order to identify the Brownian motion we will compute numerically the "pattern entropy" and the "(ϵ, τ)-entropy," which give a measure of the degree of randomness (or information content) of the motion of the particle. We start by giving a review of the ideas of Prigogine's school concerning irreversibility. In subsequent chapters we present a review on entropy and information and we give our numerical results.

6.2 The Prigogine school

Irreversibility of the physical processes in Nature is characterized by statistical entropy. For systems in equilibrium this magnitude is expressed as a functional of the probability distribution in phase space, $S = -k_B \int \rho \log \rho \, \mathrm{d}\Gamma$. Nonetheless, because of Liouville's theorem this quantity remains constant along the evolution toward equilibrium, in contradiction with the second law of thermodynamics.

The proposal of Prigogine's school is to introduce a change of representation through a transformation operator Λ, such that for

$$\tilde{\rho} = \Lambda \rho \tag{6.2.1}$$

the quantity $\tilde{S} = -k_B \int \tilde{\rho} \log \tilde{\rho}\, \mathrm{d}\Gamma$ does increase monotonically with time. In the Λ representation time-symmetry is broken. The transformed distribution $\tilde{\rho}$ follows an irreversible dynamics associated with noise and fluctuations. The existence of Λ imposes some restrictions over dynamics. Misra [3] showed that a necessary condition is that the system is mixing, while a sufficient condition is that it is a K-system.[1] Nevertheless, it is possible to establish the existence of Λ for systems that display weaker instabilities, as that associated with the existence of Poincaré resonances in non-chaotic systems, provided there is a continuous spectrum of frequencies and we restrict to a suitable set of initial conditions. The transformation Λ is non-unitary. However, it is invertible. This allows us to go back to the original representation. As we discuss now, this transformation is directly connected to the problem of integrability of dynamics, as studied by Poincaré (for more details see [3, 4, 7]).

The time evolution of ensembles is given by Liouville's equation

$$i\frac{\partial \rho}{\partial t} = L_H \rho$$

where $L_H = \{H, \}$ is the Liouville operator. We consider a system that can be described by a free motion component L_0 and an interaction λL_V, where λ is the interaction strength. We have $L_H = L_0 + \lambda L_V$. If the system is integrable in the sense of Poincaré, then we can construct by perturbation expansion in λ a canonical transformation U such that $\bar{L}_0 = U L_H U^{-1}$ has the same form as L_0. We obtain a dynamics of free (although renormalized) particles. U maps unperturbed invariants to perturbed invariants of motion. With $\bar{\rho} = U\rho$ the transformed Liouville equation is

$$i\frac{\partial \bar{\rho}}{\partial t} = \bar{L}_0 \bar{\rho}$$

Changing ρ to $\bar{\rho}$ allows us to integrate the equations of motion. Defining a suitable inner product between dynamical variables A and ensembles ρ such as

$$\langle A|\rho\rangle = \int d\Gamma\, A^* \rho$$

where $d\Gamma$ is the phase space volume element, we can define the Hermitian conjugate operator,

$$\langle A|U\rho\rangle = \langle U^\dagger A|\rho\rangle$$

and show that U is unitary, $U^{-1} = U^\dagger$. Moreover U is time reversal invariant. This means that dynamics is equivalent to a time reversible dynamics of free particles through a unitary transformation.

[1] Mixing implies that under the dynamics an evolving subset A_t becomes uniformly distributed on the complete phase space S. Thus for a fixed subset B the measures satisfy $\lim_{t\to\infty} \mu(A_t \cap B)/\mu(B) = \mu(A_0)/\mu(S)$ [5]. K-systems are systems with positive Kolmogorov-Sinai entropy (or KS-entropy). For systems without escape this is equivalent to the requirement that the system is chaotic [14].

The perturbation expansion of U has terms of the form

$$\lambda L_V \frac{1}{\omega - L_0}, \ \lambda L_V \frac{1}{\omega - L_0} \lambda L_V \frac{1}{\omega' - L_0}, \cdots$$

Here the parameters ω, ω' are characteristic frequencies of the system. For integrable systems in Poincaré's sense, the denominators are non-vanishing. The physical meaning of this is that there are no resonances in the system. As a result, each term of the perturbation expansion of U is well defined.

For non-integrable systems there appear resonances, and the denominators may vanish, giving divergences. The perturbation expansion of U loses its meaning. We can however regularize the denominators by interpreting them as distributions

$$\frac{1}{\omega - L_0} \Rightarrow \frac{1}{\pm i\epsilon + \omega - L_0}$$

where ϵ is an infinitesimal. Then we have

$$\frac{1}{\pm i\epsilon + \omega - L_0} = \mathcal{P}\frac{1}{\omega - L_0} \mp \pi i \delta(\omega - L_0)$$

where $\mathcal{P}$ is the principal part and δ is Dirac's delta function. The regularization of the denominators leads to the transformation Λ, which is no longer unitary and which breaks time-reversal invariance. Time reversal invariance is broken when we fix the sign of $i\epsilon$. For $\tilde{\rho} = \Lambda\rho$ the Liouville equation is transformed to

$$i\frac{\partial \tilde{\rho}}{\partial t} = \Theta \tilde{\rho}, \qquad \text{with} \quad \Theta = \Lambda L_H \Lambda^{-1} \tag{6.2.2}$$

The transformed Liouville operator Θ takes the form of a kinetic collision operator. For example for the Friedrichs model of Brownian motion we consider below, Θ is a Fokker-Planck operator [6]. Eq. (6.2.2) describes a Markov process that breaks time-symmetry. An interesting property of Λ is that it is non-distributive with respect to multiplication. This introduces fluctuations, which are equivalent to the fluctuations associated with stochastic processes, e.g., white noise in Brownian motion.

The main point of Prigogine's school is that irreversibility and fluctuations emerge from reversible, deterministic dynamics when the dynamics is non-integrable in Poincaré's sense. One cannot construct a transformation U that allows the integration of the equations of motion, but one can construct a transformation Λ that reveals the irreversible, fluctuating dynamics that is hidden in the usual formulation of dynamics. The invertible transformation Λ maps Hamilton's equations to stochastic irreversible equations.

Poincaré's work was based on the three-body problem, which was one of the first examples of chaotic systems. Prigogine and coworkers have shown that Poincaré's resonances appear not only in chaotic systems with few degrees of freedom, but also in large systems with many degrees of freedom (not necessarily chaotic) where there is a continuum of energies.

6.3 Mechanical model for Brownian motion

We investigate the evolution of a classical system that consists of a harmonic oscillator (which we will call "particle") and a lattice in one-dimensional space, represented as a chain of oscillators. The particle corresponds to an impurity within the lattice that weakly interacts with the lattice atoms through harmonic potentials. The coupling between the impurity and the atoms is assumed to be much weaker than the coupling between the lattice atoms.

We consider two cases:

- Friedrichs model: the oscillators of the chain interact among themselves via a nearest-neighbor harmonic potential in which case the interaction can be eliminated with a transformation to normal modes.
- Friedrichs-Fermi-Pasta-Ulam (FFPU) model: the chain of oscillators includes nearest-neighbor nonlinear interactions (anharmonic potential) which are not eliminated in the normal-mode representation.

The Friedrichs model gives linear equations of motion, so all the Lyapunov exponents are equal to zero. However, this system may present a Poincaré resonance between the particle and the lattice modes that leads to Brownian motion in the thermodynamic limit.[2]. The existence of the thermodynamic limit requires that initial phases of the lattice modes are distributed at random [15]. In terms of ensembles, this means that the initial ensemble is independent of the phases of the lattice modes. The initial condition contains a large amount of information. As we shall show, the resonance allows the transfer of information from the lattice to the particle, giving the time evolution of the particle a flavor of stochasticity.

The FFPU model, on the contrary, is non-linear and may show chaotic behavior with sensitivity to initial conditions, though it does not constitute an Anosov system. Chaotic dynamics can generate a lot of information by extracting digits from initial conditions represented by irrational numbers. This mechanism works even if we have few degrees of freedom. Thus the FFPU model has two sources of randomness, namely: chaotic dynamics and the large amount of degrees of freedom that are randomly distributed at $t = 0$.

The question we want to investigate is to what extent these two sources contribute to the Brownian motion.

6.3.1 Friedrichs model

First, let us consider a lattice with N harmonic oscillators and denote displacement from equilibrium positions as $x_1, \ldots, x_N$ and respective momenta as $p_1, \ldots, p_N$. The Hamiltonian of this system is given by

[2] Total number of particles N and volume V go to infinity, keeping N/V finite.

$$\mathcal{H}_0^{\mathrm{latt}} = \frac{1}{2}\sum_{n=1}^{N} p_n + \frac{\Omega^2}{2}\sum_{n=1}^{N}(x_{n+1} - x_n)^2, \tag{6.3.1}$$

where Ω^2 characterizes the harmonic coupling intensity among nearest neighbors. Periodic boundary conditions are assumed. Introducing the *modes*

$$q_k = \frac{1}{\sqrt{2N\omega_k}}\sum_{n=1}^{N}\left(\omega_k x_n \mathrm{e}^{-i\,2\pi nk/N} + i\,p_n \mathrm{e}^{i\,2\pi nk/N}\right), \tag{6.3.2}$$

with

$$\omega_k^2 = \Omega^2\left[2\sin\left(\frac{k\pi}{N}\right)\right]^2,$$

$\mathcal{H}_0^{\mathrm{latt}}$ takes the "diagonal" form

$$\mathcal{H}_0^{\mathrm{latt}} = \sum_{k=1}^{N} \omega_k q_k q_k^*. \tag{6.3.3}$$

For N large the dispersion relation is (with $L = N/\Omega \sim$ size of the box)

$$\omega_k = \left|\frac{2\pi k}{L}\right|$$

with $k \in \mathbb{Z}$. As L and N go to infinity, ω_k becomes a continuous variable over the positive real line, thus

$$\frac{2\pi}{L}\sum_k \xrightarrow[L\to\infty]{} \int \mathrm{d}\omega_k \qquad \text{and} \qquad \frac{L}{2\pi}\delta_{k0} \xrightarrow[L\to\infty]{} \delta(\omega_k),$$

As described above, the Friedrichs model consists of a harmonic oscillator, the particle, coupled to a scalar field, which here is the field of lattice vibrations. Introducing an additional mode corresponding to the particle

$$q_0 = \frac{1}{\sqrt{2\omega_0}}(\omega_0 x_0 + i\,p_0), \tag{6.3.4}$$

the Hamiltonian of the Friedrichs model may be written as follows

$$\mathcal{H}^F = \omega_0 q_0\, q_0^* + \mathcal{H}_0^{\mathrm{latt}} + \lambda\mathcal{V}, \tag{6.3.5}$$

where $\lambda\mathcal{V}$ represents the linear part of the interaction between particle and lattice. Since we assume the coupling is weak we neglect the terms quadratic in the field modes in the interaction $\mathcal{V}$. In modes representation $\mathcal{V}$ takes the form

$$\mathcal{V} = \sum_k (V_k q_0^*\, q_k + V_k^* q_0\, q_k^*).$$

To guarantee existence of the continuous limit $L \to \infty$, we must have $V_k \sim \mathcal{O}(L^{-1/2})$, in other words

$$V_k = \sqrt{\frac{2\pi}{L}}\, v_k, \qquad \text{with} \qquad v_k \sim \mathcal{O}(L^0).$$

When there is a continuous spectrum of frequencies ω_k, there can appear a resonance between the particle frequency ω_0 and the frequencies ω_k. Resonance means that there is an $\omega_k \in \sigma(H)$ such that $\omega_0 = \omega_k$, where $\sigma(H)$ stands for the continuous spectrum of H. As a result of this resonance, the system can become non-integrable in the sense of Poincaré. This means that there is no perturbed action of the particle obtained by a unitary transformation that is expandable in a power series in λ (for more details see [16, 7]). Because of the linearity of Hamilton's equations for the Friedrichs model, the time-dependent normal mode of the particle takes the linear form

$$q_0(t) = f_{00}(t) q_0(0) + \sum_k f_{0k}(t) q_k(0), \tag{6.3.6}$$

from which we get

$$\frac{\mathrm{d}}{\mathrm{d}t} q_0(t) = -i\, z_0(t) q_0(t) + R(t), \tag{6.3.7}$$

with

$$z_0(t) = i \frac{\partial}{\partial t} \ln f_{00}(t), \tag{6.3.8}$$

$$R(t) = \sum_k h_k(t) q_k(0), \tag{6.3.9}$$

$$h_k(t) = \dot{f}_{0k} + i\, z_0(t) f_{0k}(t). \tag{6.3.10}$$

We will consider the thermodynamic limit, where we have $L \to \infty$ with the condition that $\langle q_k q_k^* \rangle \sim \mathcal{O}(L^0)$ in such a way that the average energy per field mode is finite (independent of L) and non-vanishing as $L \to \infty$. To satisfy this condition, the initial field modes $q_k(0)$ must have random phases [15]. If the phases were not random, then the displacement or velocity of the particle would diverge as $L \to \infty$.

As a result of the randomness in the $q_k(0)$ modes, $R(t)$ is an erratic function and it plays the role of noise. In general, this noise keeps memory,

$$\langle R^*(t) R(t') \rangle \neq 0, \qquad \text{for} \qquad t \neq t',$$

where $\langle \cdot \rangle$ denotes statistical average over a Gibbs ensemble. For time scales $t > 0$ of the order of the relaxation time of the oscillator, $z_0(t)$ approaches the constant value z_0, which is an energy pole of the resolvent operator. If there is a resonance between the particle and the field, then z_0 is a complex number $z_0 = \omega_0 - i\gamma$, which gives the shifted frequency and the damping rate of the oscillator. Under these conditions the function $R(t)$ behaves as a complex Gaussian white noise $\hat{R}(t)$ satisfying

$$\langle \hat{R}^*(t)\hat{R}(t')\rangle = \hat{R}_c^2\delta(t-t'),$$

with $\hat{R}_c^2 = 2\gamma\langle q_k^* q_k\rangle$.

The replacement of $z_0(t)$ by z_0 and $R(t)$ by $\hat{R}(t)$ is traditionally associated with the "Markovian approximation." However, as shown in [6] one can obtain the exact stochastic Langevin equation ($t > 0$)

$$\frac{\mathrm{d}}{\mathrm{d}t}\hat{q}_0(t) = -iz_0\hat{q}_0(t) + \hat{R}(t), \tag{6.3.11}$$

that is equivalent to the time evolution of the transformed variable $\Lambda^\dagger q_0$. Here Λ is the transformation discussed in section 6.2. The autocorrelation function for $\hat{q}_0(t)$ is given by

$$\langle \hat{q}_0(t)\hat{q}_0^*(t)\rangle = \hat{q}_0(0)\hat{q}_0^*(0)\mathrm{e}^{-2\gamma t} + \frac{\hat{R}_c^2}{2\gamma}(1-\mathrm{e}^{-2\gamma t}), \tag{6.3.12}$$

which is characteristic of a Markov process.

To see the connection between resonance and irreversibility, consider the lowest order approximation of γ in a perturbation expansion in powers of λ. This corresponds to Fermi's Golden Rule,

$$\gamma = \lambda^2|v_k|^2\delta(\omega_0 - \omega_k) + \mathcal{O}(\lambda^4)$$

If there is resonance between the particle and the lattice, the Dirac delta-function is non vanishing, and hence γ is non-vanishing. If there is no resonance, then $\omega_0 \neq \omega_k$ for all $\omega_k \in \sigma(H)$, and γ vanishes. There is no damping and no approach to equilibrium. In short, we see that the appearance of Brownian motion is closely connected to the existence of a resonance between the particle and the lattice, and consequently, it is closely connected to Poincaré's non-integrability.

6.3.2 FFPU model

The well-known Fermi-Pasta-Ulam (FPU) model consists on a chain of anharmonic oscillators, whose Hamiltonian is given by

$$\mathcal{H}^{FPU} = \mathcal{H}_0^{\text{latt}} + g\mathcal{V}^{\text{latt}},$$

where $\mathcal{H}_0^{\text{latt}}$ has been defined in (6.3.1) and g is the intensity of interaction due to potential $\mathcal{V}^{\text{latt}}$ which, in terms of the coordinates x_n and p_n, may be written as

$$\mathcal{V}^{\text{latt}} = \sum_{n=1}^{N} (x_{n+1} - x_n)^4.$$

It's a well-known fact that this system may present chaotic behavior for high enough energy values [17].

The FFPU model is a fusion of the Friedrichs and FPU models. The Hamiltonian of the model is

$$\mathcal{H}^{FFPU} = \mathcal{H}^{F} + g\mathcal{V}^{\text{latt}}.$$

Using x_n and p_n coordinates for field oscillators and q_0 for the particle, $\mathcal{H}^{FFPU}$ reads

$$\begin{aligned}\mathcal{H}^{FFPU} &= \omega_0\, q_0^*\, q_0 + \frac{1}{2}\sum_{n=1}^{N} p_n^2 + \frac{\Omega^2}{2}\sum_{n=1}^{N}(x_{n+1} - x_n)^2 + g\sum_{n=1}^{N}(x_{n+1} - x_n)^4 \\ &+ \lambda\left\{\sum_{n=1}^{N}(f_n x_n - i\, g_n p_n)\, q_0 + \text{c.c.}\right\};\end{aligned} \tag{6.3.13}$$

here c.c. denotes complex conjugate of the previous term. The terms f_n and g_n stand for the sums

$$f_n = \sum_{k=1}^{N} \sqrt{\frac{\omega_k}{2N}} V_k \mathrm{e}^{-i\,2\pi kn/N},$$

$$g_n = \sum_{k=1}^{N} \frac{1}{\sqrt{2N\omega_k}} V_k \mathrm{e}^{i\,2\pi kn/N};$$

for $g = 0$ the Friedrichs model is recovered.

6.4 Entropy and information

In the English language there exists less probability of hearing words starting with the letter Z than starting with letter C, for the same reason, in a crossword puzzle, the letter Z gives us more information than letter C. This simple example contains the fundamental idea behind information theory and of entropy as a quantification of it.

Let (X, Ω, μ) be a probability space where X is the sample space, Ω a σ-algebra of subsets of X and μ a probability measure on Ω. Let $\imath$ be the amount of information gained to know that event $A \in \Omega$ has occurred; such a measurement must satisfy the following conditions [18]:

1. $\imath(\mu(A))$: the amount of information depends on the degree of uncertainty.

2. $\imath \geq 0$: there is no loss of information by knowing something new.

3. $\imath(\mu(A)) = 0$ if $\mu(A) = 1$: there is no gain of information by knowing something completely certain.

4. $\imath(\mu(A)) \uparrow$ if $\mu(A) \downarrow$: for more certainty there is less information to transmit.

5. $\imath(\mu(A \cap B)) = \imath(\mu(A)) + \imath(\mu(B))$: the information gained to know that two independent events $A, B \in \Omega$ occurred is equal to the sum of the information gained to know, independently, that each one has occurred.

The function

$$\imath(\mu(A)) = -\log(\mu(A)),$$

satisfies all these requirements. There is full arbitrariness in the choice of the basis of the logarithm. We will use here logarithm in base 10. In the theory of communication [19, 20] it is common to talk about a source of information emitting bi-infinite sequences

$$\cdots \omega_{-T} \cdots \omega_0 \omega_1 \cdots \omega_T \cdots$$

of characters (letters) from a set $\alpha = \{1, \ldots, \mathcal{L}\}$ (alphabet) that here we assumed finite. Each one of these letters is considered as a random variable since if there were no uncertainty there would be no information to transmit. In this way the sequence turns out to be a stochastic process defined by the distributions $\mu(\omega_0 \cdots \omega_{T-1})$ which, for each value of T, yield the probability to obtain a sequence containing the so called T-word $\omega_0 \cdots \omega_{T-1}$. The source is defined by the alphabet and the probability measure μ together. The average information or *block entropy* corresponding to the T-words is given by

$$H_T = - \sum_{\omega_0 \cdots \omega_{T-1}} \mu(\omega_0 \cdots \omega_{T-1}) \log \mu(\omega_0 \cdots \omega_{T-1}). \tag{6.4.1}$$

For stationary sources H_T does not depend on the place on the sequences where we choose the T-words so that, on average, the amount of information per symbol emitted by the source is H_T/T. The limit

$$h = \lim_{T \to \infty} \frac{H_T}{T}, \tag{6.4.2}$$

is known as the entropy of the source. If the T-words are equally probable we have $\mu(\omega_0 \cdots \omega_{T-1}) = \mathcal{L}^{-T}$ where $\mathcal{L}$ is the number of letters in the alphabet, such a way that $H_T = T \log \mathcal{L}$ and $h = \log \mathcal{L}$, hence the total number of T-words can be expressed as $\eta = \mathcal{L}^T = \mathrm{e}^{Th}$.

For a nonuniform distribution, the interpretation of h is provided by the following (see e.g. [20])

Theorem: For any stationary and ergodic source the following equality holds

$$\lim_{T\to\infty} \int_{\text{sequences}} \left| \frac{\log \mu(\omega_0 \cdots \omega_{T-1})}{T} + h \right| \mathrm{d}\mu(\omega_0 \cdots \omega_{T-1}) = 0,$$

except perhaps for a set of zero measure.

That is, for any ϵ, $\delta > 0$ and T large enough, the set of the T-words can be divided in two groups such that:

1. For any set of the first set we have

$$\left| \frac{\log \mu(\omega_0 \cdots \omega_{T-1})}{T} + h \right| < \epsilon.$$

2. The sum of the probabilities of all the T-words of the second set is less than δ.

Thus in the first group we have

$$\mu(\omega_0 \cdots \omega_{T-1}) \approx \mathrm{e}^{-hT},$$

and then the number of T-words in this group is $\eta_{eff} \approx \mathrm{e}^{hT}$, whereas the total number of T-words is $\eta = \mathcal{L}^T = \mathrm{e}^{T \log \mathcal{L}}$. Thus, for $h < \log \mathcal{L}$ the first group contains just a fraction of the total number of T-words. In this way, h characterizes the number of T-words typically observable (of high probability) which are those that practically are detected in experiments. Despite h is statistical in nature as it refers to the source, if the source is ergodic h can be obtained from a unique sequence, provided that it is long enough and chosen from the first group as described above. In this sense it is meaningful to talk about the entropy of a (typical) sequence of an ergodic source. Moreover, a simplified version of Shannon's first theorem [19] states that the T-words emitted from a source with entropy h and an alphabet of $\mathcal{L}$ letters, can be reconstructed (on average) from m-words of an alphabet with the same number of letters, where $m \geq h/\log \mathcal{L}T$. Thus, $h/\log \mathcal{L}$ is the maximum compression rate that can be reached.

Therefore, the entropy characterizes the complexity of a source of information from two points of view. First, it gives an estimate of how many T-words are really representative for a given value of T; a source with a value of entropy greater than another one can be considered more complex, i.e. less predictable, than the latter. Second, it gives the maximum compression rate theoretically possible (on average) for a typical sequence of the source. A source allows more compression than another one with a greater value of h and can be considered less complex than the latter since it contains more recurrence patterns and then is more predictable. For instance, a source whose typical sequences are of the form

$$\cdots\cdots 10110010011000000011 \cdots\cdots,$$

can be considered more complex than another whose typical sequences are of the form

$$\cdots\cdots 10101010101010101010\cdots\cdots .$$

6.4.1 (ϵ, τ)-entropy

Up to now we have discussed discrete process. In the continuous case (e.g. withe noise) entropy may become infinite since information is obtained from arbitrarily small scales of observation. To deal with this situation, the dependence of entropy on spatial and temporal scales of observation (ϵ and τ, respectively) is taken into account [1, 21]. The behavior of the divergence of the entropy as $\epsilon \to 0$ and $\tau \to 0$ characterizes the stochastic process. With this aim, the signal $X(t)$ is discretized in time intervals $\Delta t = \tau$ obtaining a new random variable which approximates the signal, namely

$$\mathbf{X} = \{X(t_0), \ldots, X(t_0 + T\tau - \tau)\}.$$

where T is the number of points in the time series. For the particle lattice models we consider in this paper, $X(t) = q_0(t)$ is the trajectory of the particle mode. We focus on a coarse grained description relative to a partition

$$\Delta = \{C_1, \ldots, C_{\mathcal{L}}\}$$

in one-to-one correspondence with an alphabet $\alpha = \{1, \ldots, \mathcal{L}\}$.

Let $\mu(\omega_0 \cdots \omega_{T-1})$ be the probability to visit successively the cells indexed by $\omega_0, \ldots, \omega_{T-1}$ ($\omega_j \in \alpha$). Using Eqs. (6.4.1) and (6.4.2) we get the entropy relative to the partition Δ and time scale τ, $h(\Delta, \tau)$.

The entropy so defined depends on the chosen partition; to avoid this ambiguity, the (ϵ, τ)-entropy either is defined either as

$$\underline{h}(\epsilon, \tau) = \inf_{\Delta:\mathrm{diam}(\Delta)\leq\epsilon} h(\Delta, \tau)$$

or

$$\overline{h}(\epsilon, \tau) = \sup_{\Delta:\mathrm{diam}(\Delta)\geq\epsilon} h(\Delta, \tau),$$

where $\mathrm{diam}(\Delta)$ denotes the maximum diameter of the cells over the partition.

6.4.2 Cohen-Procaccia method

Cohen and Procaccia [22] proposed a numerical method for the evaluation of the (ϵ, τ)-entropy of any time series. The calculation of $h(\epsilon, \tau)$ involves the study of the asymptotic behavior of the term H_T/T which can be rewritten as

$$\frac{H_T}{T} = \frac{1}{T}\sum_{k=2}^{T}(H_k - H_{k-1}) + \frac{H_1}{T} = \frac{1}{T}\sum_{k=1}^{T} A_k,$$

where

$$A_k = \begin{cases} H_k - H_{k-1} & \text{if } k \geq 2, \\ H_1 & \text{if } k = 1. \end{cases}$$

It can be shown that $A_k \geq 0$ and that $A_k \leq A_{k-1}$ and hence, there must exist a number A such that

$$\lim_{k\to\infty} A_k = A.$$

Hence for T and n large enough, with n fixed, we have

$$\frac{1}{T}\sum_{k=1}^{T} A_k \approx \frac{1}{T}\sum_{k=1}^{n} A_k + \frac{T-n}{T}A \xrightarrow[T\to\infty]{} A,$$

thus

$$h = \lim_{T\to\infty} \frac{H_T}{T\tau} = \frac{1}{\tau}\lim_{T\to\infty}(H_T - H_{T-1}). \tag{6.4.3}$$

Furthermore

$$\frac{H_T}{T} = \frac{1}{T}\sum_{k=1}^{T} A_k \geq A_T,$$

such a way that

$$|A_T - A| \leq \left|\frac{H_T}{T} - A\right|,$$

which shows that A_T converge faster than H_T/T resulting more efficient to work with the right hand side of (6.4.3).

Let $\{X_n\}_{n=1}^{T}$ be the time series (TS) to be studied and $\Delta = \{C_i\}_i$ a uniform partition of the phase space in cubic cells with sides of length ϵ. Let T_i be the number of points of the TS within the cell C_i. The points of the TS can also be indexed as X_i^k, where $1 < k < T_i$. The distance between two points, X_r and X_p, of the TS is defined as $\rho_1(X_p, X_r) = |X_p - X_r|$. Based on the the law of large numbers we expect that $p_i \approx T_i/T$, where p_i is the probability to find a point in the cell C_i. Then H_1 can be calculated as

$$H_1 \approx -\sum_i \frac{T_i}{T}\log\frac{T_i}{T} = -\frac{1}{T}\sum_i T_i \log R^1_{Z_i}\left(\frac{\epsilon}{2}\right), \tag{6.4.4}$$

where Z_i denotes the center of the cell C_i and

$$R_Z^1(\epsilon) = \frac{\#\text{ of trajectories } X_r \text{ such that } \rho_1(X_r, Z)) < \epsilon}{T},$$

Assume that

$$\left(\prod_{k=1} R^1_{X_i^k} \left(\frac{\epsilon}{2} \right) \right)^{1/T_i} = R^1_{Z_i} \left(\frac{\epsilon}{2} \right), \tag{6.4.5}$$

whose meaning is that the value of $R_Z(\epsilon/2)$ calculated at the center of the cell ($Z = Z_i$) coincides with the geometric mean of the values of $R_Z(\epsilon)$ calculated at each point of this cell ($Z = X_i^k$), which seems to make sense if the distribution is smooth and ϵ is small enough.

Introducing (6.4.5) into (6.4.4) we obtain

$$H_1 \approx -\frac{1}{T} \sum_i \sum_{k=1}^{T_i} \log R^1_{X_i^k} \left(\frac{\epsilon}{2} \right) = -\frac{1}{T} \sum_n \log R^1_{X_n} \left(\frac{\epsilon}{2} \right) = -\left\langle \log R^1 \left(\frac{\epsilon}{2} \right) \right\rangle,$$

where $\langle \cdots \rangle$ denotes average over all the points of the TS.

Similar considerations show that

$$H_l = -\left\langle \log R^l \left(\frac{\epsilon}{2} \right) \right\rangle, \tag{6.4.6}$$

with

$$R_Z^l(\epsilon) = \frac{\#\text{ of trajectories} X_r \text{ such that } \rho_l(X_r, Z)) < \epsilon}{T - l + 1}, \tag{6.4.7}$$

and

$$\rho_l(X_p, X_r) = \max\{|X_p - X_r|, \ldots, |X_{p+l-1} - X_{r+l-1}|\}. \tag{6.4.8}$$

In order to evaluate the entropy we choose $N_R \ll T$ random reference trajectories of length l and count the total number of trajectories within a ρ_l-distance of length ϵ. The block entropy is then

$$H_l(\epsilon, \tau) = -\frac{1}{N_R} \sum_{r=1}^{N_R} \log R^l_{X_r}$$

In our model the trajectories of the particle have an oscillating character. If there are no interactions, the trajectories in phase space (with scaled variables) are on a circle centered at the origin. Different sections of the circle are not really different trajectories. To take this into consideration, we will work with the *pattern entropy* $H(\epsilon, \tau, l)$, rather than the block entropy. The only difference between the pattern entropy and the block entropy lies in the definition of the distance between two trajectories. For the pattern entropy the distance is defined as

$$\rho_l(X_p, X_r) = \max\{(|(X_p - X_{p+a}) - (X_r - X_{r+a})|\}_{a=1\cdots l-1} \qquad (6.4.9)$$

instead of Eq. (6.4.8). This distance measures how similar two trajectories are, independent of their position or orientation. It is expected that in a random process $H(\epsilon, \tau, l)$ will grow linearly with l for $T \to \infty$ and $l \to \infty$. Indeed, the information content of random "words" of length l is proportional to l. Moreover, the probability of finding two identical words is an inverse power of the length of the words.

Let us then define the $h(\epsilon, \tau, l)$ entropy as

$$h(\epsilon, \tau, l) = \frac{H(\epsilon, \tau, l) - H(\epsilon, \tau, l-1)}{\tau}$$

This measures the rate of increase of information content as we increase the length of the words. i.e., the degree of randomness of the trajectory. The (ϵ, τ)-entropy is obtained as

$$h(\epsilon, \tau) = \lim_{T,l \to \infty} h(\epsilon, \tau, l)$$

In practice, we have a limitation in the maximum number of data T. The pattern entropy cannot grow past the maximum value $H(\epsilon, \tau, l) = \log T$. We will find though that for certain parameter values $H(\epsilon, \tau, l)$ grows fairly linearly with l and hence $h(\epsilon, \tau, l)$ is fairly constant with respect to l. For an Ornstein-Uhlenbeck process (Brownian motion), the set of curves for different values of τ form an envelope that behaves as [1, 21]

$$h_{env}(\epsilon) \sim \frac{1}{\epsilon^2}.$$

So the slope of the envelope of a plot of $\log h$ vs. $\log \epsilon$ for different values of τ should have a slope of -2 for Brownian motion.

6.5 Results

The fourth order Runge-Kutta method has been implemented in order to solve the equations of motion corresponding to Hamiltonian (6.3.13). The initial modes q_k, have been chosen as

$$|q_k(0)|^2 = \frac{k_B T}{\omega_k}, \qquad k = 1, \ldots, N-1$$

with random phases. We note that the mode q_k with $k = N$ does not contribute to the total energy. Moreover it is proportional to the total momentum, which we set equal to zero. Hence we exclude it in the numerical simulations.

The computer simulations were done with a system consisting of 1000 field modes. For the interaction in the Hamiltonian we chose the function

$$v(k) = \frac{\sqrt{2\omega_k}}{1 + (\omega_k/\omega_M)^2}$$

with $\omega_M = 2$. The size of the one-dimensional box containing the system was chosen as $L = 250$, and we imposed periodic boundary conditions on the field. We chose $\beta = 1/(k_B T) = 1.5$. The coupling constant of the particle-lattice interaction was set to $\lambda = 0.07$. The simulations were run from $t = 0$ to $t = 200$, taking 20000 time steps. For the frequency of the oscillator we chose either $\omega_0 = -1$ (non-resonant case, namely $\omega_0 \notin \sigma(H)$) or $\omega_0 = 1$ (resonant case, namely $\omega_0 \in \sigma(H)$). The coupling constant of the non-linear interaction between lattice modes was set to either $g = 0$ (non-chaotic case, Friedrichs model) or $g = 1$ (chaotic case, FFPU model).

In short, we consider the following four Cases:

1. No resonance, no chaos (Friedrichs model with $\omega_0 = -1$ and $g = 0$)
2. Resonance, no chaos (Friedrichs model with $\omega_0 = 1$ and $g = 0$)
3. No resonance, chaos (FFPU model with $\omega_0 = -1$ and $g = 1$)
4. Resonance, chaos (FFPU model with $\omega_0 = 1$ and $g = 1$)

Fig. 6.1 shows the evolution of the distance between two points of the full phase space (including all the field modes) initially close ($d(t) \sim 10^{-8}$ units), for the case with resonance and chaos.

As can be seen, the trajectories present exponential divergence characteristic of chaos. The non-resonant, chaotic case presents a similar behavior, while the non-chaotic cases (either resonant or non-resonant) present no exponential divergence of trajectories, as expected (see Fig. 6.2). Below we show examples of trajectories for each one of these cases (Figs. 6.3-6.6). As we see there is a striking difference between the cases where there is resonance and there is no resonance, regardless of the presence or absence of chaos. For the non-resonant cases the trajectories follow a cyclic pattern. For the resonant cases the trajectories are erratic.

To see this more in detail we consider the pattern entropy. Figs. 6.7-6.9 show the pattern entropy $H(\epsilon, \tau, l)$ as a function of the trajectory length l, with $\tau = 36$, $\epsilon = 0.45 \times 1.116^s$ and $s = 1, \cdots, 12$. The different curves correspond to different values of ϵ starting with $s = 1$ for the top curve. The number of reference trajectories is 50. Since there are 20000 time steps, taking $\tau = 36$ gives a small enough time interval to calculate the entropy.

For the cases with no resonance (Figs. 6.7, 6.8) there is a flat region in the pattern entropy. This means that no information is generated by the corresponding trajectories. In contrast, for the cases with resonance, we see that there is a region of approximately linear increase of the pattern entropy. This corresponds to generation of information in the trajectory of the particle. This is the information transferred from the lattice to the particle. In Fig. 6.9 we compare the chaotic and non-chaotic resonant cases. We see that in both

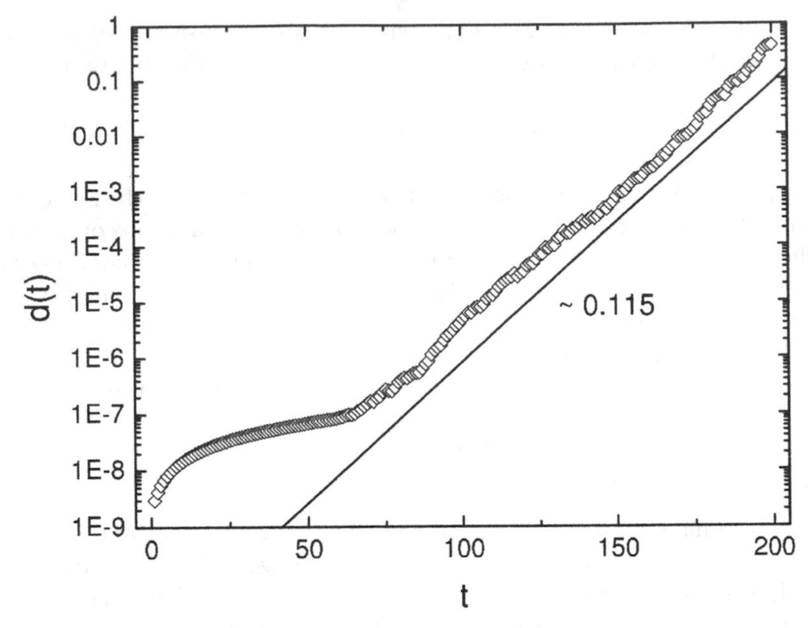

Fig. 6.1. Distance in phase space between two initially close points vs. time for the resonant chaotic case (Case 4). The non-resonant chaotic case (Case 3) gives a similar result.

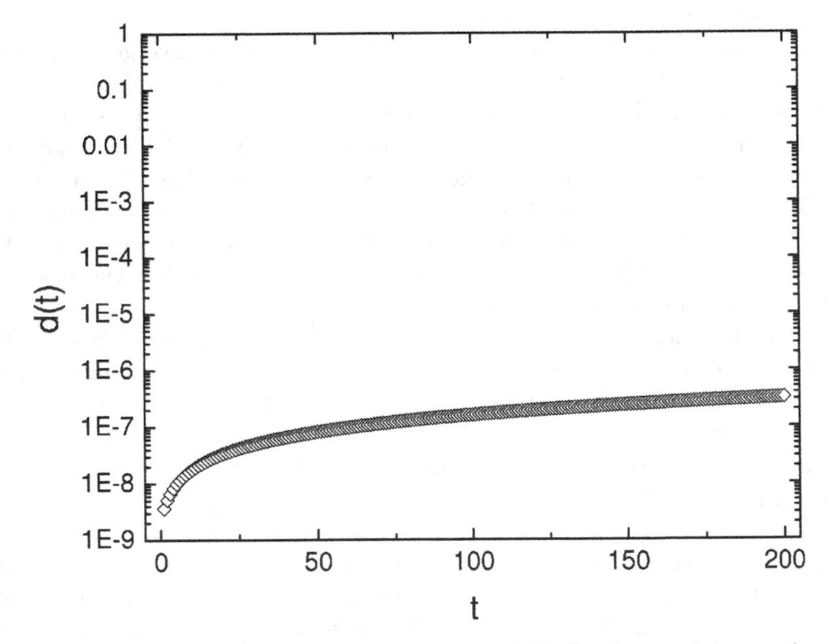

Fig. 6.2. Distance in phase space between two initially close points vs. time for the resonant non-chaotic case (Case 2). The non-resonant non-chaotic case (Case 1) gives a similar result.

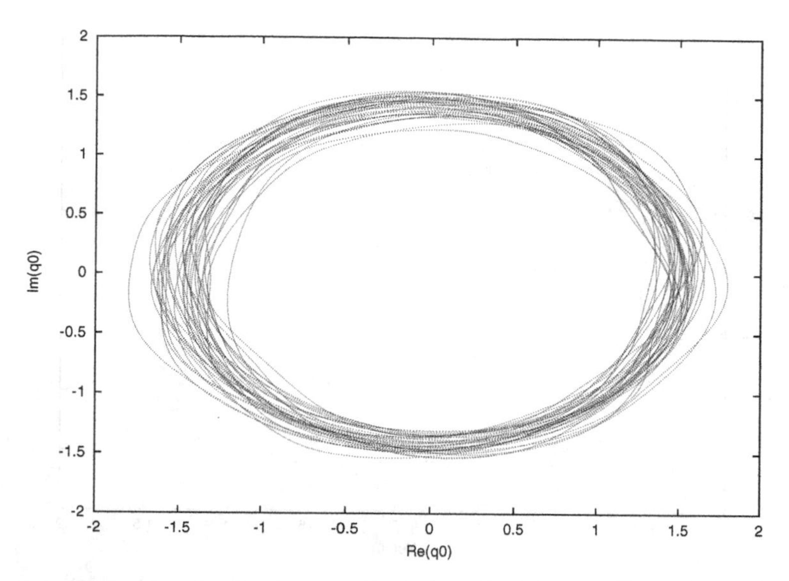

Fig. 6.3. Trajectory for the non-resonant, non-chaotic case (Case 1). The x and y axis are the real and imaginary part of the mode q_0, and are proportional to the oscillator's position and momentum, respectively.

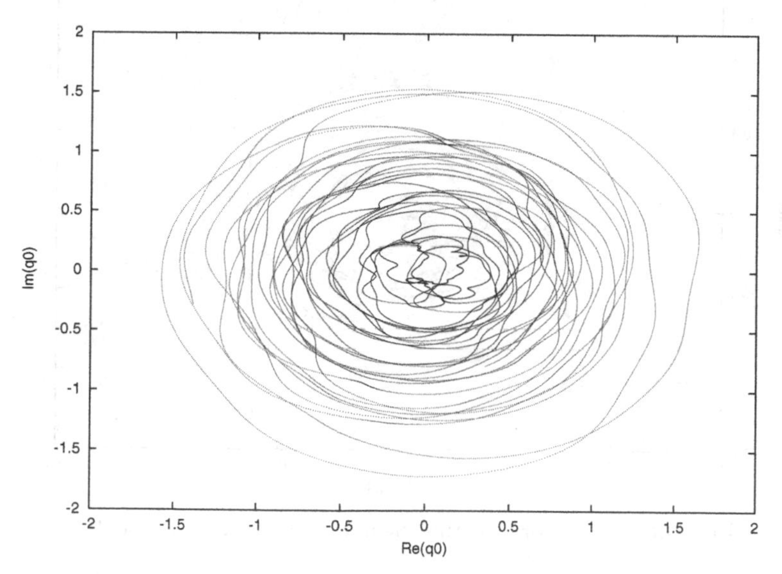

Fig. 6.4. Trajectory for the resonant, non-chaotic case (Case 2).

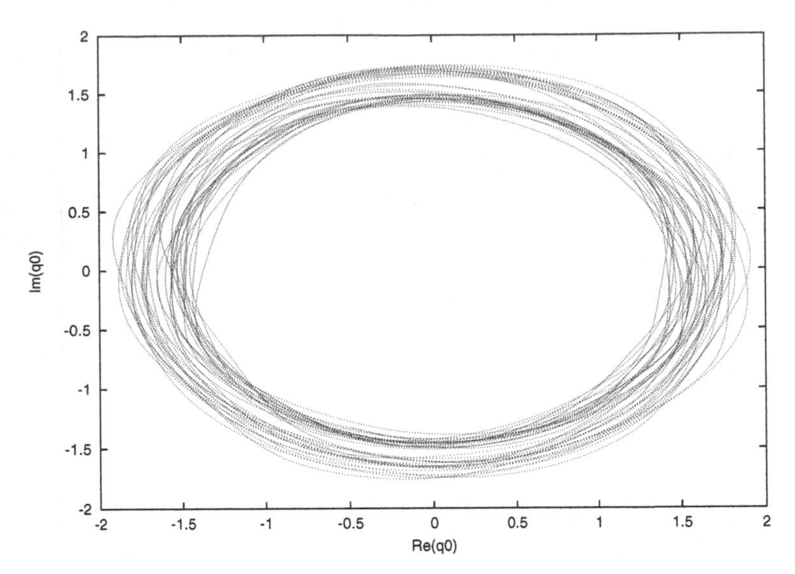

Fig. 6.5. Trajectory for the non-resonant, chaotic case (Case 3).

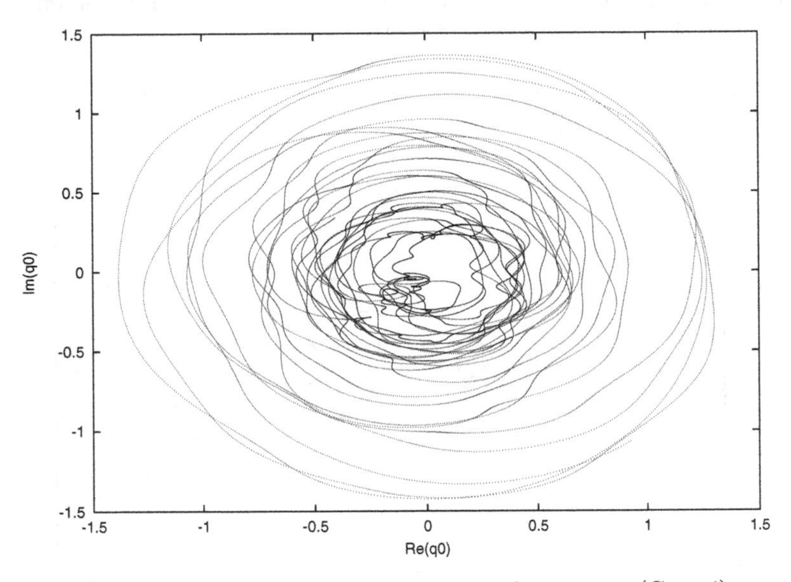

Fig. 6.6. Trajectory for the resonant, chaotic case (Case 4).

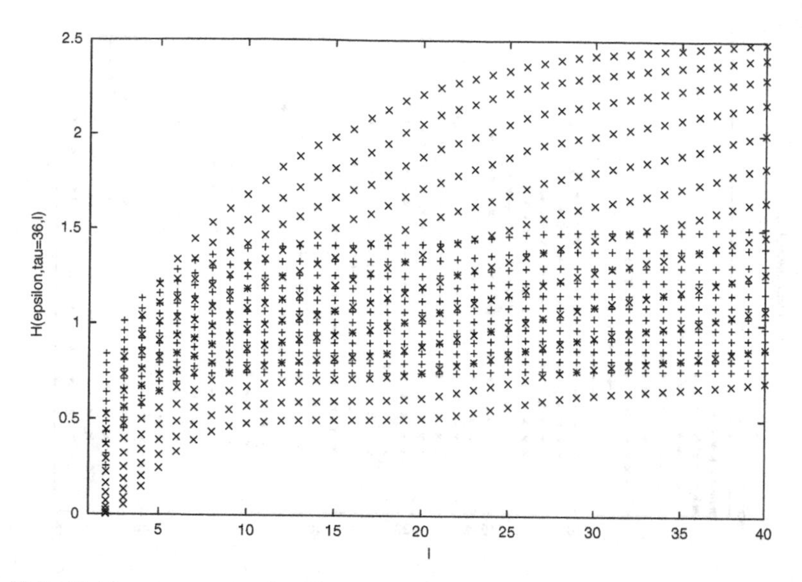

Fig. 6.7. Pattern entropies for the non-resonant non-chaotic case (Case 1, +) and the resonant, non-chaotic case (Case 2, ×).

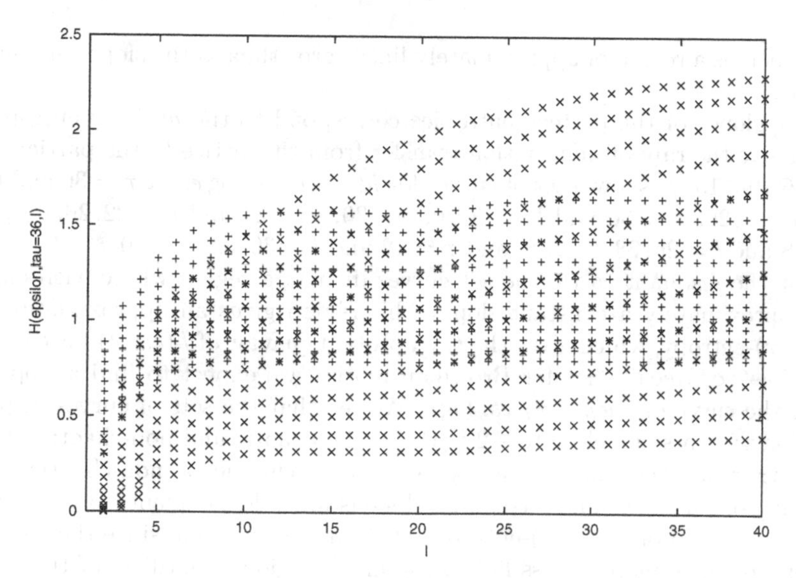

Fig. 6.8. Pattern entropies for the non-resonant chaotic case (Case 3, +) and the resonant, chaotic case (Case 4, ×).

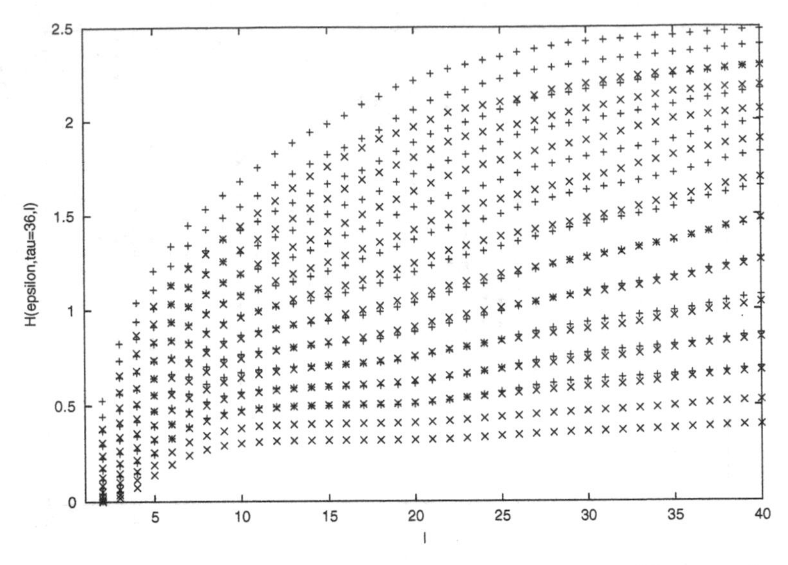

Fig. 6.9. Pattern entropies for the resonant non-chaotic case (Case 2, +) and the resonant, chaotic case (Case 4, $\times$).

cases there is a region of approximately linear growth, and the slopes are quite similar.

The slopes of the pattern entropies correspond to the $h(\epsilon, \tau, l)$ entropies, which give the rate of information transfer from the lattice to the particle. In Figs. 6.10-6.13 we show a set of curves $\log h(\epsilon, \tau, l)$ vs. $\log \epsilon$, for $\tau = 36$ and $l = 14, 16, \cdots, 26$, $\tau = 30$ and $l = 18, 20, \cdots, 30$, $\tau = 24$ and $l = 22, 26, \cdots, 34$, $\tau = 18$ and $l = 26, 28, \cdots, 38$, and $\tau = 12$ and $l = 30, 32, \cdots, 40$.

For the resonant cases, the set curves display a region where each curve has approximately a constant slope. This region gives an approximation of the (ϵ, τ)-entropy, which gives the asymptotic behavior of $h(\epsilon, \tau, l)$ for $l \to \infty$, and $T \to \infty$. Recall that for Brownian motion we expect to find a slope of -2 in the curve of $\log h(\epsilon, \tau)$ vs. $\log \epsilon$. We see that the sets of curves display a region with a slope close to -2 both for the chaotic and non-chaotic cases. In contrast for the non-resonant cases (either chaotic or non-chaotic) any grouping of curves around a common slope is much less evident, so we cannot see a clear trend of the (ϵ, τ)-entropy. This can be expected, since the pattern entropy remains more or less flat in the linear region, regardless of the value of ϵ (see Figs. 6.7 and 6.8).

6.6 Thermalization of the lattice

So far we discussed results where the phases of the initial modes were completely random numbers between 0 and 2π. Now we choose the initial con-

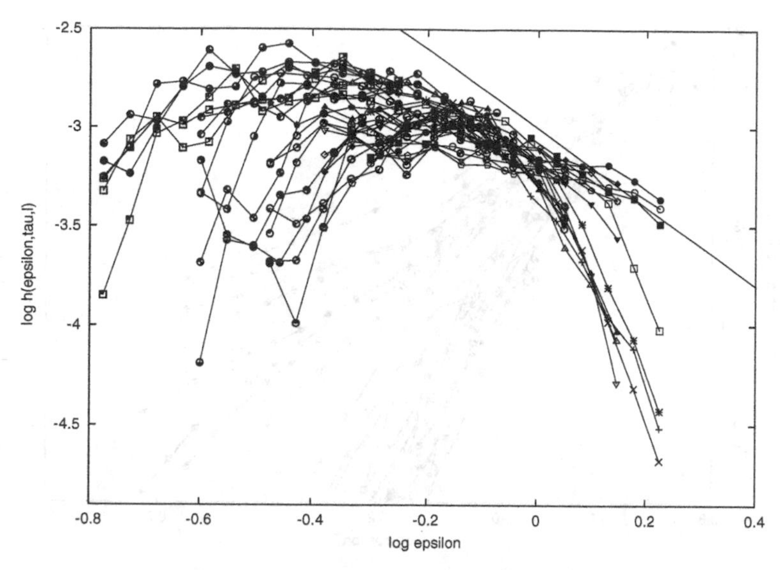

Fig. 6.10. $h(\epsilon, \tau, l)$ entropies for the resonant, non-chaotic case (Case 2). The line in the upper right corner has a slope of -2.

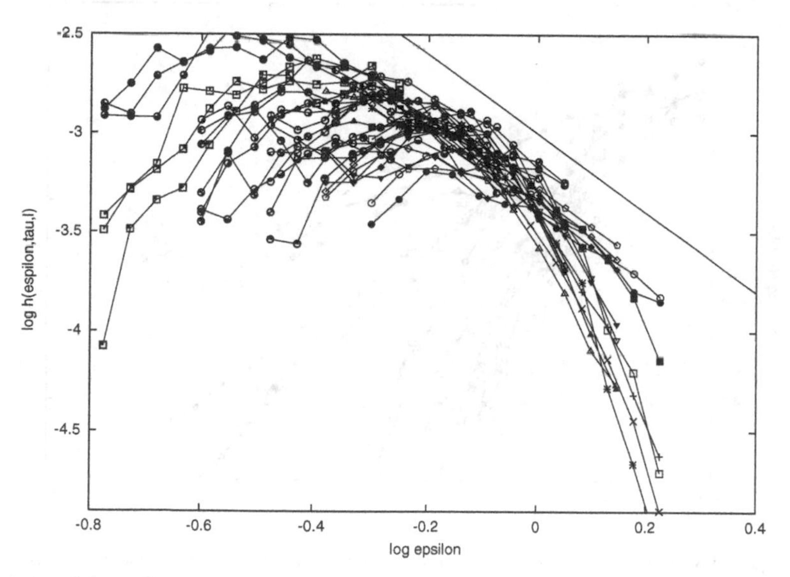

Fig. 6.11. $h(\epsilon, \tau, l)$ entropies for the resonant, chaotic case (Case 4). The line in the upper right corner has a slope of -2

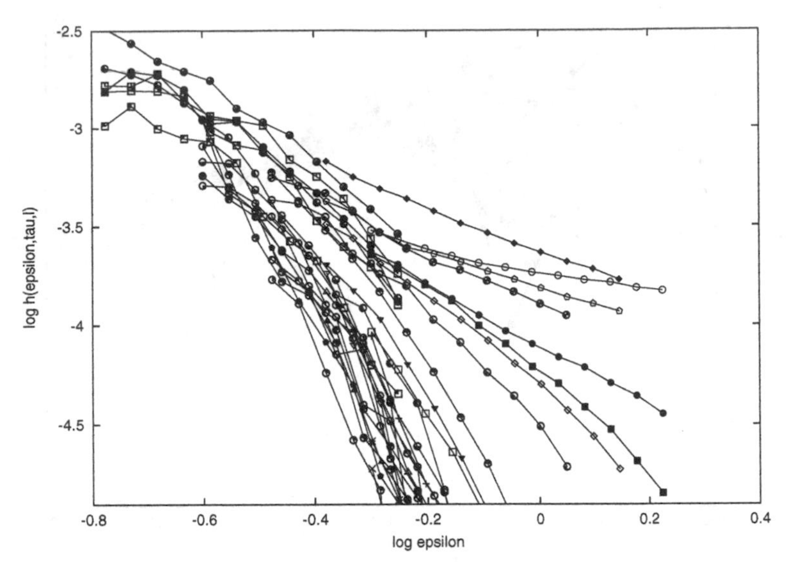

Fig. 6.12. $h(\epsilon, \tau, l)$ entropies for the non-resonant, non-chaotic case (Case 1).

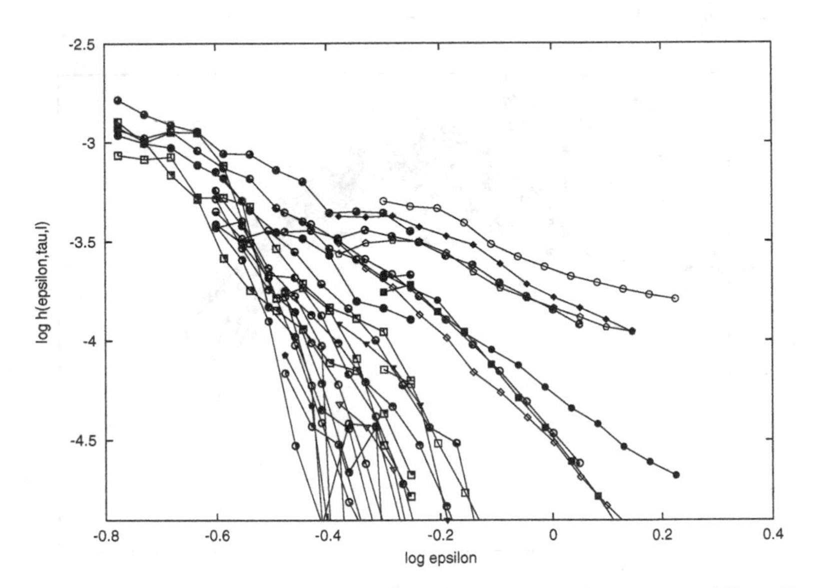

Fig. 6.13. $h(\epsilon, \tau, l)$ entropies for the non-resonant, chaotic case (Case 3).

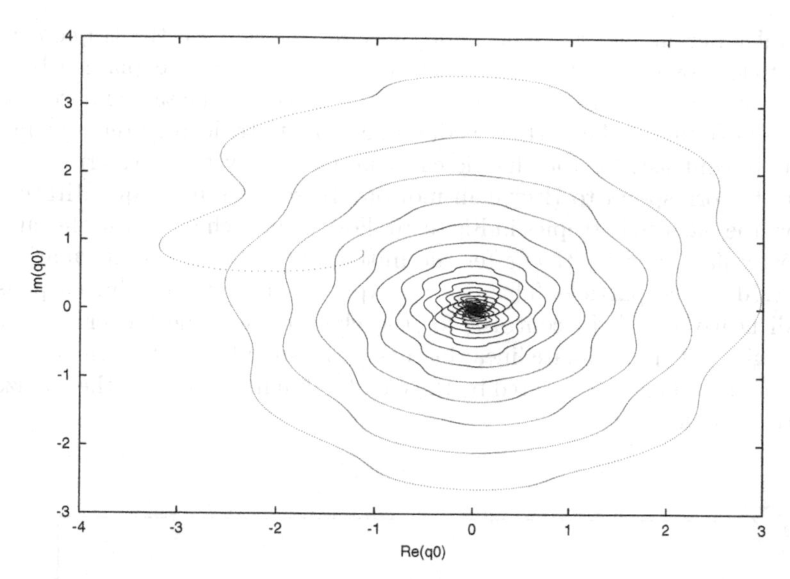

Fig. 6.14. Trajectory for the resonant, non-chaotic case (Case 2).

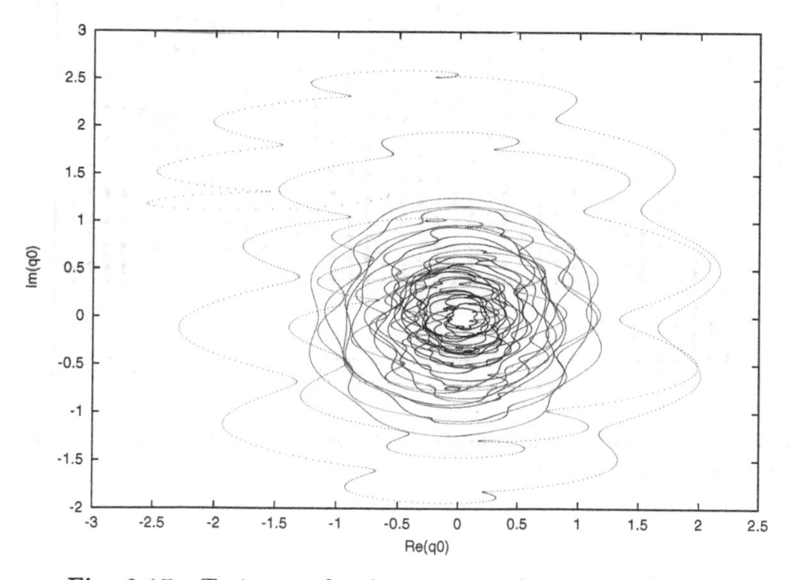

Fig. 6.15. Trajectory for the resonant, chaotic case (Case 4).

ditions in such a way that the initial phases are nearly identical for all the lattice modes. We choose the initial phases as random numbers between 0 and 10^{-3}.

In Figs. 6.14 and 6.15 we show trajectories obtained for the resonant case with and without chaos, respectively. We see that now chaos does introduce a big difference in the trajectory. For the non-chaotic case the trajectory of the particle shows an initial large excitation, which can be explained by the initial excitation from the lattice modes having virtually the same phase. After the initial excitation the particle is damped, but its trajectory remains fairly regular. In contrast, for the chaotic case the trajectory becomes erratic, and appears to correspond to Brownian motion. To see this more quantitatively, we show the pattern entropies in Fig. 6.16. For the non-chaotic case the curves are mostly flat, showing that after the initial stage almost no information is transferred to the particle. This can be expected, because the initial phases were all nearly equal. In contrast, for the chaotic case, the pattern entropy shows regions of approximate linear increase corresponding to the transmission of information. This is generated by the chaotic dynamics, which "thermalizes" the lattice modes.

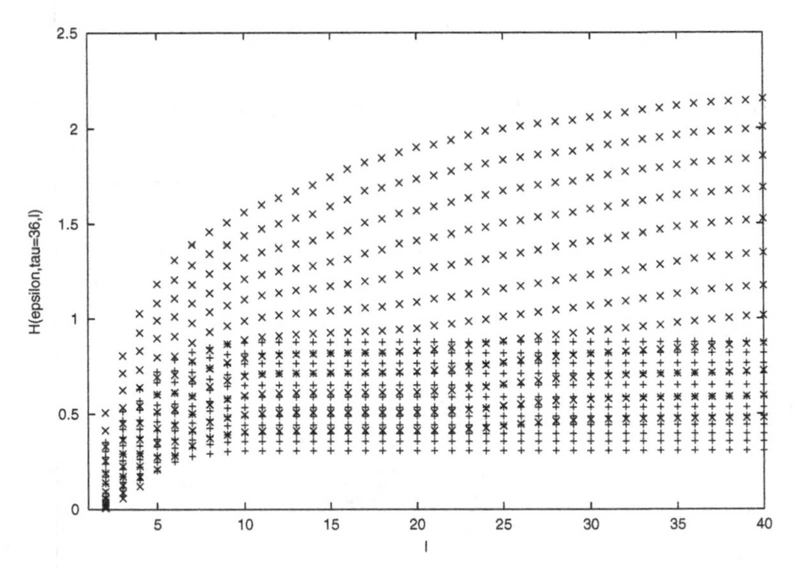

Fig. 6.16. Pattern entropies for the resonant non-chaotic case (Case 2, +) and the resonant, chaotic case (Case 4, ×).

In Fig. 6.17 we show the $h(\epsilon, \tau, l)$ entropy for the non- chaotic case. The curves are totally different from the curves we obtained before, when the initial phases of the lattice modes were random. Most of the curves fall below a truncation value of $h = 10^{-7}$, because the slopes of the pattern entropy are nearly zero. In contrast, for the chaotic case (Fig. 6.18) the curves are quite similar to the curves obtained with random initial conditions, and they are consistent with the slope -2 associated with Brownian motion. For the non-resonant cases (either non-chaotic or chaotic) there is no big difference

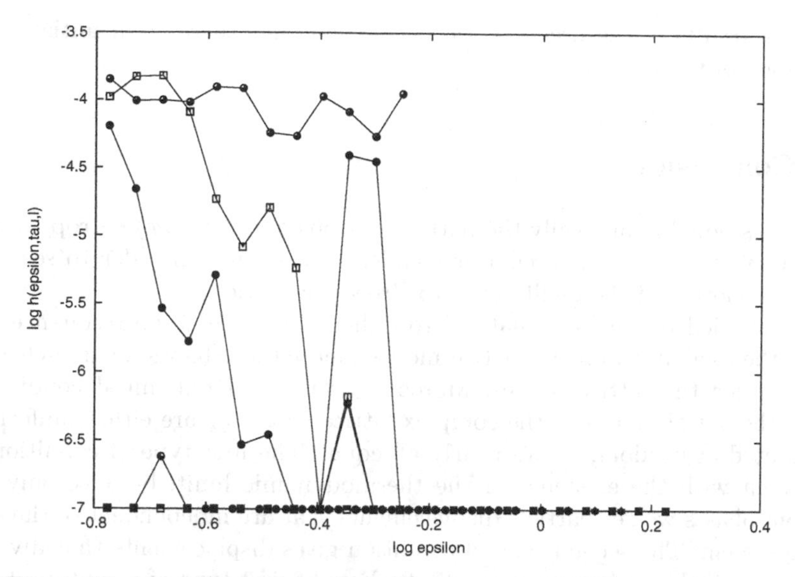

Fig. 6.17. $h(\epsilon, \tau, l)$ entropies for the resonant, non-chaotic case (Case 2).

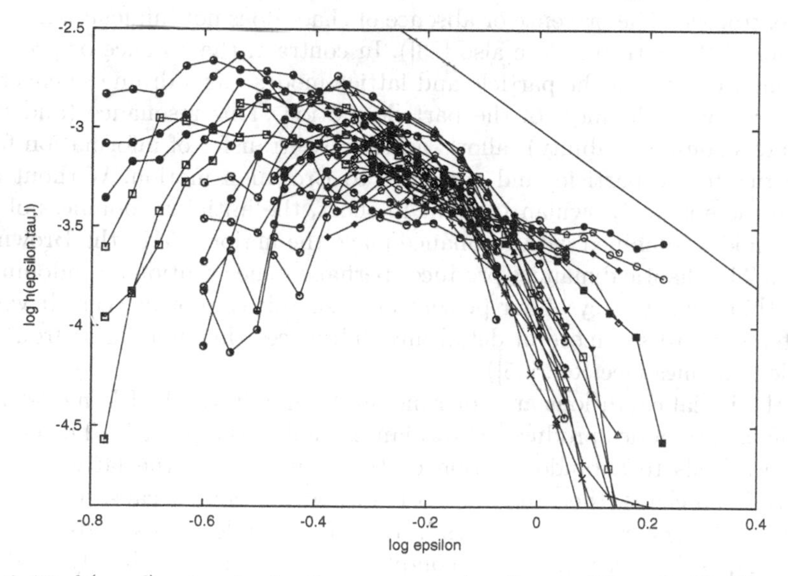

Fig. 6.18. $h(\epsilon, \tau, l)$ entropies for the resonant, chaotic case (Case 4). The line in the upper right corner has a slope of -2

with the results of the previous section, except that there is an initial large excitation of the particle.

6.7 Conclusion

We have studied numerically the pattern entropy and the (ϵ, τ)-entropy of the trajectory of a harmonic oscillator coupled to a lattice, in order to see how close the motion of the oscillator is to Brownian motion.

We studied parameters that led to either resonance or no resonance between the oscillator and the lattice modes, and either chaotic or non-chaotic behavior for the lattice modes. Moreover, we considered initial conditions where the initial phases of the complex lattice modes q_k are either uniformly distributed at random, or are nearly all equal. The first type of condition is consistent with the existence of the thermodynamic limit, because only for random phases we get particle displacements that are independent of the size of the system. The second type of condition gives displacements that diverge as the size of the system goes to infinity. For the first type of condition, both chaotic and non-chaotic lattice modes produce similar pattern entropies and (ϵ, τ)-entropies. The presence or absence of chaos does not influence the gross behavior of the entropies (see also [23]). In contrast, the absence or presence of resonance between the particle and lattice modes has a dramatic effect on the degree of randomness of the particle motion. This resonance (and thus Poincaré's non-integrability) allows an effective transfer of information from the lattice to the particle, and gives rise to Brownian motion. Without resonance there is no Brownian motion. In short, the initial randomness of the lattice modes combined with resonance plays the major role in the Brownian motion. The chaotic dynamics produces perhaps some additional randomness, but within the accuracy of our present analysis this cannot be seen. It would be interesting to see more in detail any differences that may arise from the chaotic dynamics (see Ref. [25]).

If the initial conditions are not random, then only the FFPU model with chaotic lattice modes produces Brownian motion of the particle. The chaotic dynamics leads to a randomization or thermalization of the lattice modes. If there is no chaos, then there is no Brownian motion, because there is no mechanism to generate randomness in the lattice modes. This leads to an interesting question: were the initial conditions of the Universe already random? If this is the case, then it would seem that the chaotic dynamics has played a secondary role in the generation of kinetic processes such as Brownian motion. On the contrary, if the initial conditions were not random, then chaotic dynamics has been necessary to randomize the positions or velocities in many-particle systems. In either case, the existence of resonances is essential to have kinetic processes. Resonance associated with Poincaré's non-integrability allows an effective "communication" between different degrees of freedom, and leads to the appearance of basic irreversible processes.

References

[1] Gaspard, P., *Chaos, Scattering and Statistical Mechanics* (Cambridge, Cambridge University Press, 1998).
[2] Gallavotti, G., Chaos **8**, 384 (1998).
[3] I. Prigogine, *From being to becoming* (Freeman, New York, 1980).
[4] T. Petrosky and I. Prigogine, Physica A **147**, 439 (1988).
[5] R. Balescu *Equilibrium and nonequilibrium statistical mechanics*, (John Wiley & Sons, 1975).
[6] S. Kim and G. Ordonez, Phys. Rev. E **67**, 056117 (2003).
[7] G. Ordonez, T. Petrosky and I. Prigogine, Phys. Rev. A **63**, 052106 (2001).
[8] T. Petrosky, G. Ordonez, and I. Prigogine, Phys. Rev. A **68**, 022107 (2003).
[9] B. A. Tay and G. Ordonez, Phys. Rev. E **73**, 016120 (2006).
[10] Sinai, Y. G., Russ. Math. Surveys **27**, 21 (1972).
[11] Bowen, R. and Ruelle, D., Invent. Math. **2**, 181 (1975).
[12] Gaspard, P. et al, Nature **394**, 865 (1998).
[13] Dettmann, C. P. and Cohen, E. G. D., J. Stat. Phys. **101**, 775 (2000).
[14] van Beijeren, H. arXiv:cond-mat/0407730 (2004).
[15] T. Petrosky and I. Prigogine, Chaos, Soliton and Fractals **11**, 373 (2000).
[16] T. Petrosky, I. Prigogine and S. Tasaki, Physica A **173**, 175 (1991).
[17] Livi, R. and Ruffo, S., J. Phys. A **19**, 2033 (1986).
[18] Chernov, N. and Markarian, R., *Introduction to the Ergodic Theory of Chaotic Billiards*, 2nd Edition (Rio de Janeiro, IMPA, 2003).
[19] Shannon, C. E. and Weaver, W., *The Mathematical Theory of Communication* (Urbana, The University of Illinois Press, 1962).
[20] Khintchine, A. I., *Mathematical Foundations of Information Theory* (New York, Dover Publications, 1957).
[21] Gaspard, P. and Wang, X.-J., Phys. Rep. **235** (6), 321 (1993).
[22] Cohen, A., Procaccia, I., Phys. Rev. A **31**, 1872 (1985).
[23] Dettmann, C. P. and Cohen, E. G. D, and van Beijeren, H., Nature **401** 875 (1999).
[24] Alekseev, V. M. and Yakobson, M. V., Phys. Rep. **75** (5), 287 (1981).
[25] Romero-Bastida, M. and Braun, E., Phys. Rev. E. **65** 036228 (2002).

7

Models of Finite Bath and Generalised Thermodynamics

Ramandeep S. Johal

Department of Physics, Lyallpur Khalsa College, Jalandhar-144001, INDIA
E-Mail: johal_jld@dataone.in

Summary. We consider the approach of a sample system in contact with a heat bath environment, to arrive at equilibrium distributions for the energies of the sample system. In canonical ensemble, we assume the size of the bath to be infinite. The real baths or environments are expected to have a finite size. Different conditions on the bath properties yield different equilibirum distributions of the sample system. Explicitly, the gaussian ensemble and q-exponential distributions are discussed. The various thermodynamic quantities like entropy and free energy are nonadditive in these formalisms. We also present a new model which can be seen as an intermediate case of the above two scenarios. The connection between noadditivity in these models with the deformed numbers in the context of q-analysis is also highlighted.

7.1 Introduction

The development of statistical mechanics based on ensemble theory is founded on the postulate of "equal *a priori* probabilities", which is assumed to apply to all microstates consistent with the given macrostate of an isolated system [1]. The corresponding statistical ensemble is the so-called microcanonical ensemble. A representative system in this ensemble has all "mechanical" variables such as energy E, volume V, magnetization M etc., fixed. For convenience in calculations, other ensembles are used which invariably suppose the existence of a subsidiary system or reservoir in contact with the actual system. For instance, in the canonical ensemble the walls of the system permit an exchange of energy with the reservoir while in the grand canonical ensemble, both energy and matter can be exchanged. In general, the different ensembles are constructed by allowing one or more mechanical variables to fluctuate. The exchange of each of these variables is controlled by a parameter which is a characteristic of the reservoir. For instance, in the case of the canonical ensemble, this parameter is precisely the temperature of the reservoir and determines the mean energy of the system. Actually, this is adequate when the reservoir is a very large system that can exchange arbitrary amounts of energy, without modification of its intensive properties. In practical situations, this is

R.S. Johal: *Models of Finite Bath and Generalised Thermodynamics*, StudFuzz **206**, 207–217 (2006)
www.springerlink.com

not always the case. However, very few studies have been devoted to analyse the consequences of possible deviations from these ideal reservoir properties.

In this contribution, we discuss the derivation of equilibrium distributions in the context of statistical mechanics, by considering the contact of the sample system with a finite heat bath. We discuss two previously studied models of finite bath; section (7.2) summarises the standard text-book derivation of the canonical ensemble. Section (7.3) is devoted to the so-called gaussian ensemble. Section (7.4) discusses the origin of q-exponential distributions from the specific properties of a finite bath. Then in section (7.5), we propose a novel model with specific bath properties as an intermediate case to the above two models. The last section is devoted to the connection between q-analysis and nonadditive statistical mechanics.

7.2 Derivation of canonical ensemble from microcanonical ensemble

Consider a system 1, exchange energy with an environment (system 2), with which it is in thermal contact. The energy E of the total system (system 1 + system 2) is being held fixed. Thus when system 1 has energy E_1, the environment has energy $E_2 = E - E_1$. According to the postulate of *apriori* equiprobabilities, each accessible microstate of the total system is equally probable. The probability denoted by $p(E_1)$, of finding system 1 in a microstate with energy E_1 is thus directly proportional to the number of microstates available to the environment at the corresponding energy E_2. Therefore,

$$p_1(E_1) = \frac{\Omega_2(E - E_1)}{\Omega_{1+2}(E)}, \tag{7.2.1}$$

where $\Omega_{1+2}(E)$ is the total number of states available for the set $1 + 2$. We define the entropy of the reservoir as: $S_2(E_2) = \ln \Omega_2(E_2)$, where we choose the entropy units so that Boltzmann's constant, $k_B = 1$. Thus equivalently, the probability for a microstate of system 1, is given by

$$p_1(E_1) = \frac{e^{S_2(E-E_1)}}{\Omega_{1+2}(E)}. \tag{7.2.2}$$

In the following, we present the so called small-E_1 derivation of the canonical ensemble. Thus expanding $S_2(E_2)$ around the energy value $E_2 = E - U$, where U is the equilibrium value for energy of system 1, we can write

$$S_2(E_2) = S_2(E - U) + \left.\frac{dS_2}{dE_2}\right|_{E-U} (U - E_1). \tag{7.2.3}$$

Using Eq. (7.2.3) in (7.2.2), we get

$$p(E_1) \propto \exp(-\beta E_1), \tag{7.2.4}$$

where we have defined the inverse temperature at equilibrium of the bath or environment, as

$$\beta = \frac{dS_2}{dE_2}. \tag{7.2.5}$$

The lesson is that if we look at a small subsystem of a much larger microcanonical system, then the distribution of the smaller system appears to be exponentially distributed. The same conclusion can be derived if we assume that the environment is so large that its temperature is constant, i.e. its heat capacity is infinite so that any finite exchange of energy does not change its temperature. However, realistically speaking, it may be the case that the bath is not infinitely large or we look at the subsystem of the larger system which is after all not that small as compared to the rest of the system. Then it seems reasonable to include the higher order terms in the expansion of S_2. Different approaches have been suggested to model finite size of the baths. Below, we discuss a couple of them.

7.3 Gaussian ensemble

This generalisation was proposed by Hetherington and Challa [2, 3, 4]. Later, it was extended and studied from different aspects in [5, 6]. It was introduced so that it is equivalent to the canonical ensemble in the limit of large systems, except in the energy range of a first-order transition. Interestingly, it enables a smooth interpolation between the microcanonical and the canonical ensembles. Taking into account these features, Challa and Hetherington [3, 4] showed the interest of this ensemble for Monte Carlo simulation studies of phase transitions. They demonstrated a significant reduction in computer time (compared to standard simulations in the canonical ensemble) and, its adequacy for distinguishing second-order from first-order transitions. More recently, Costeniuc et. al [6] have proposed generalised canonical ensembles and shown their equivalence in the thermodynamic limit, to the microcanonical ensemble. The gaussian ensemble forms an important particular case of this class of ensembles.

Gaussian ensemble is defined [5] if, in the expansion of S_2, we also keep the second order term. Thus

$$S_2(E-E_1)=S_2(E-U)+\left.\frac{dS_2}{dE_2}\right|_{E-U}(U-E_1)+\frac{1}{2!}\left.\frac{d^2S_2}{dE_2^2}\right|_{E-U}(U-E_1)^2+\mathcal{O}(U-E_1)^3. \tag{7.3.1}$$

where apart from inverse temperature β, we also define the second order derivative of the entropy as

$$\left.\frac{d^2S_2}{dE_2^2}\right|_{E-U} = -2\gamma. \tag{7.3.2}$$

Then substituting (7.3.1) in (7.2.2) and denoting the energies of the microstates of the sample by ϵ_i $(i = 1, ..., M)$, we obtain

$$p_i = \frac{1}{Z_G}\exp[-\beta\epsilon_i - \gamma(\epsilon_i - U)^2], \tag{7.3.3}$$

where the normalization constant Z_G is given by

$$Z_G = \sum_{i=1}^{M} \exp[-\beta\epsilon_i - \gamma(\epsilon_i - U)^2]. \tag{7.3.4}$$

Note that U is the mean energy and must be obtained self-consistently from the equation:

$$UZ_G = \sum_{i=1}^{M} \epsilon_i \exp[-\beta\epsilon_i - \gamma(\epsilon_i - U)^2]. \tag{7.3.5}$$

Eqs. (7.3.3), (7.3.4) and (7.3.5) reduce to the standard canonical ensemble definitions when $\gamma = 0$. Therefore, it is natural to relate the parameter γ with the finite size of the reservoir.

We remark that the above distribution within the gaussian ensemble can also be derived from a maximum statistical entropy principle. Thus we maximize the standard Gibbs-Boltzmann-Shannon entropy given by

$$S_G = -\sum_{i=1}^{M} p_i \ln p_i, \tag{7.3.6}$$

subject to the constraints of normalisation of the probability, the given mean value of the energy and the fixed value of the fluctuations, respectively as

$$\sum_{i=1}^{M} p_i = 1, \tag{7.3.7}$$

$$\langle\epsilon_i\rangle \equiv \sum_{i=1}^{M} \epsilon_i p_i = U, \tag{7.3.8}$$

$$\langle(\epsilon_i - U)^2\rangle \equiv \sum_{i=1}^{M} (\epsilon_i - U)^2 p_i = W. \tag{7.3.9}$$

Then the maximization procedure is done by introducing the Lagrange multipiers λ, β and γ for the respective constraints, and maximizing the following functional $\mathcal{L}$:

$$\begin{aligned}\mathcal{L} = &-\sum_i p_i \ln p_i - \lambda\left(\sum_i p_i - 1\right) \\ &-\beta\left(\sum_i \epsilon_i p_i - U\right) - \gamma\left(\sum_i (\epsilon_i - U)^2 p_i - W\right).\end{aligned} \tag{7.3.10}$$

By requiring the condition:

$$\frac{\partial \mathcal{L}}{\partial p_i} = 0, \tag{7.3.11}$$

it is easy to see that the optimum form of the probability distribution is given by the expression in Eq. (7.3.3). Therefore β and γ, within this context, are simply Lagrange multipliers that allow to fix, self-consistently, a mean value of the energy $U = \langle \epsilon_i \rangle$ and a specific value of the variance $W = \langle (\epsilon_i - U)^2 \rangle$.

To indicate the modification incurred in the standard thermodynamic relations, we define a thermodynamic potential $\Phi(\beta, \gamma)$ as

$$\Phi(\beta, \gamma) = \ln Z_G. \tag{7.3.12}$$

By differentiating Eq. (7.3.4), it can be straightforwardly obtained that:

$$-\left(\frac{\partial \Phi}{\partial \beta}\right)_\gamma = U(\beta, \gamma), \tag{7.3.13}$$

$$-\left(\frac{\partial \Phi}{\partial \gamma}\right)_\beta = W(\beta, \gamma). \tag{7.3.14}$$

The second derivative renders:

$$-\left(\frac{\partial^2 \Phi}{\partial \beta^2}\right)_\gamma = -\left(\frac{\partial U}{\partial \beta}\right)_\gamma = \frac{1}{W^{-1}(\beta, \gamma) - 2\gamma}, \tag{7.3.15}$$

which represents a generalization of the standard formula for energy fluctuations in the canonical ensemble. It is natural to define the extended heat capacity as:

$$\mathcal{C} \equiv -\beta^2 \left(\frac{\partial U}{\partial \beta}\right)_\gamma = \frac{\beta^2 W}{1 - 2\gamma W}. \tag{7.3.16}$$

This equation is the same that was already derived in Ref. [4]. Note that, contrary to what happens in the standard canonical ensemble, the positivity of the fluctuations W does not guarantee the positivity of $\mathcal{C}$.

For $\gamma \to 0$, it is seen that the relations (7.3.13) and (7.3.15) go to the corresponding relations for the case of canonical ensemble. Also in this limit, from Eqs. (7.3.14) and (7.3.15) we get an interesting relation given by:

$$\lim_{\gamma \to 0} \left(\frac{\partial \Phi}{\partial \gamma}\right)_\beta = \left(\frac{\partial^2 \Phi}{\partial \beta^2}\right)_\gamma, \tag{7.3.17}$$

which resembles in form with a diffusion equation.

The entropy S_G as given by (7.3.6) is the inverse Legendre transform of $\Phi(\beta, \gamma)$, and can be expressed as:

$$S_G(U, W) = \beta U + \gamma W + \Phi, \tag{7.3.18}$$

whereby S_G is a function of the specified values of the constraints i.e. U and W. Therefore we have the following thermodynamic relations

$$\left(\frac{\partial S_G}{\partial U}\right)_W = \beta, \tag{7.3.19}$$

$$\left(\frac{\partial S_G}{\partial W}\right)_U = \gamma. \tag{7.3.20}$$

Also note that the thermodynamic potential $\Phi = \ln Z_G$ is in general, nonadditive with respect to two subsystems with additive hamiltonians. Thus assume that a composite system hamiltonian can be written as $H = H_1 + H_2$, where H_i are the hamiltonian of the subsystems. Then the average energy of the total system is $U_{1+2} = U_1 + U_2$. The potential is given as

$$\Phi_{1+2}(\beta,\gamma) = \Phi_1(\beta,\gamma) + \Phi_2(\beta,\gamma) - \ln\left\langle e^{2\gamma(H_1-U_1)(H_2-U_2)}\right\rangle. \tag{7.3.21}$$

7.4 q-exponential distributions and model of finite heat bath

q-exponential distributions are the central predictions of the generalized statistical mechanics proposed by Tsallis [7]. These distributions have been considered as model distributions to describe various complex systems at their stationary states [8, 9, 10]. The general form of such distributions is given by $p(x) \sim e_q(x)$, where the q-exponential is defined as $e_q(x) = [1+(1-q)x]^{1/(1-q)}$. This function goes to the usual $\exp(x)$ function for $q \to 1$. For definiteness, we restrict to the range $0 < q < 1$. For our purpose, we rewrite the q-exponential as

$$e_q(x) = \exp\left[\frac{\ln[1+(1-q)x]}{(1-q)}\right], \tag{7.4.1}$$

and expand the ln function using the series $\ln[1+y] = y - \frac{y^2}{2} + \frac{y^3}{3} - \ldots$, provided that $-1 < y \le 1$. Thus we can write

$$e_q(x) = \exp\left[\sum_{n=1}^{\infty}\frac{1}{n}\{-(1-q)\}^{n-1}x^n\right], \tag{7.4.2}$$

for $-1 < (1-q)x \le 1$.

Now we show that the q-exponential distributions can be obtained by imposing the following requirements on the first and second derivatives of the entropy of the bath (see also [11]):

$$\frac{dS_2}{dE_2} = \beta(E_2); \quad \frac{d^2S_2}{dE_2^2} = -(1-q)\beta^2(E_2). \tag{7.4.3}$$

In general, for all integer values of n

$$\frac{d^nS_2}{dE_2^n} = (n-1)!(-(1-q))^{n-1}\beta^n(E_2). \tag{7.4.4}$$

Now the entropy of the bath expanded around the equilibrium value of the bath energy, is given by

$$S_2(E - E_1) = S_2(E - U) + \sum_{n=1}^{\infty} \frac{1}{n!} \left. \frac{d^n S_2}{dE_2^n} \right|_{E-U} (U - E_1)^n. \tag{7.4.5}$$

On applying Eq. (7.4.4) for the case of equilibrium, we can write

$$S_2(E - E_1) = S_2(E - U) + \sum_{n=1}^{\infty} \frac{1}{n} (-(1 - q))^{n-1} \beta^n (U - E_1)^n, \tag{7.4.6}$$

where note that β is given by its value at equilibrium. The equilibrium probability distribution is then given from (7.2.2) as

$$p(E_1) \sim \exp \left[\sum_{n=1}^{\infty} \frac{1}{n} (-(1 - q))^{n-1} \beta^n (U - E_1)^n \right], \tag{7.4.7}$$

which is in the form of q-exponential distribution, Eq. (7.4.2).

7.5 A new model for finite bath

In the above sections, we have encountered two possible models motivated from the fact that the environment or the bath with which the sample system exchanges energy, may be finite in size, in contrast to the infinite size idealisation assumed in the derivation of the canonical ensemble. We note that different conditions modeling the bath, can in principle, give rise to different equilirium distributions for the energies of the sample system. In this manner one may hope to arrive at distributions more general than the expoential, by route of ensemble theory or more precisely, by considering thermal contact of system with a specifically modelled finite bath. We obtained two distributons above: q-exponential and the distribution of the gaussian ensemble. It may be remarked that one reason for intense interest in q-exponential distributions is due to the fact that they can mimic power-law type distributions.

The conditions yielding the gaussian ensemble and the q-exponential ensemble are Eqs. (7.3.2) (together with the definition of inverse temperature) and (7.4.3), respectively. In the following, we propose a new set of conditions, which may be seen as a case intermediate to the above two scenarios. We define our new model of the finite bath, by specifying

$$\frac{dS_2}{dE_2} = \beta(E_2); \quad \frac{d^2 S_2}{dE_2^2} = r\beta(E_2), \tag{7.5.1}$$

where r is a parameter, independent of energy. Clearly, $r = 0$ implies canonical ensemble. In fact, using the above conditions, all derivatives of S_2 in this model, can be expressed in a compact form as

$$\frac{d^n S_2}{d{E_2}^n} = r^{(n-1)}\beta; \quad n \geq 1. \tag{7.5.2}$$

Then the entropy of the bath can be written as Taylor series in the form

$$S_2(E - E_1) = S_2(E - U_r) + \sum_{n=1}^{\infty} \frac{1}{n!} r^{(n-1)}\beta\Big|_{E-U_r} (U_r - E_1)^n, \tag{7.5.3}$$

where we have denoted the equilibrium of energy of the system in the present case as U_r.

Finally, the probability of the microstate with energy E_1 can be written as

$$p_r(E_1) = \frac{1}{Z_r} \exp(\beta\{\Delta E_1\}_r), \tag{7.5.4}$$

where $\Delta E_1 = U_r - E_1$, $\{x\}_r = (\exp(rx) - 1)/r$ and Z_r is the normalising partition function.

The number $\{x\}_r$ has a nice nonadditive property. Thus $\{x_1 + x_2\}_r = \{x_1\}_r + \{x_2\}_r + r\{x_1\}_r\{x_2\}_r$. More of this will be mentioned in the section below on discussion.

The mean value U_r is to be obtained in a self-consistent way from the equation

$$U_r = \sum_{E_1} E_1 p_r(E_1), \tag{7.5.5}$$

noting that p_r is also a function of U_r.

Now it is also easy to see that the distribution (7.5.4) can be obtained from maximisation of the Gibbs-Shannon entropy, subject to two constraints: (i) normalisation of the probability, and (ii) fixed mean value of the quantity, $\langle\{\Delta E_1\}_r\rangle = A_r$. The second constraint is equivalent to the constraint on mean value of energy, imposed within the maximum entropy derivation of canonical exponential distribution. This is due to the fact that, $p_r \to p_0 \propto \exp(-\beta E_1)$, as $r \to 0$ and $U_r \to U_0$, the canonical ensemble mean value. Therefore, we must have $A_0 = 0$ and the constraint (ii) reduces to the canonical constraint.

It is interesting to study the above distribution for real physical systems and compare the results with those obtained from q-exponential distributions or from the predictions of the gaussian ensemble. The details of these studies will be given elsewhere.

7.6 Discussion

After the investigations into finite bath models, we highlight in this section, the connection between q-analysis and nonadditive thermodynamics. The theory of q-analysis was formulated in the beginnings of 20th century and it provides a framework to study basic hypergeometric series [12]. q-series were studied even during the time of Euler and is significant in the theory of partitions [13]. In the previous decade, it was intensively studied in the context of quantum group theory [14, 15]. Quantum groups are mathematically a very rich subject and they constitute generalisations of the Lie symmetries. Further, q-derivative defined as [16]

$$\mathcal{D}_{q,x} f(x) = \frac{f(qx) - f(x)}{(q-1)x}, \tag{7.6.1}$$

plays a central role in the noncommutative q-calculus. Such calculus arises as a set of algebraic relations which are covariant under quantum group transformations.

First note that q-exponential distributions may also be obtained from maximisation of Tsallis entropy subject to the constriaints of normalisation of probability and mean value of energy [10]. Tsallis entropy is given by

$$S_q[p] = -\sum_i \frac{(p_i)^{q-1} - 1}{q-1} p_i. \tag{7.6.2}$$

Now, although nonextensive thermodynamics does not as such embody a quantum group structure, the two subjects can be linked via q-analysis. The connection mainly stems from the observation that a q-number defined as:

$$[x]_q = (q^x - 1)/(q-1), \tag{7.6.3}$$

satisfies a similar nonadditive property as Tsallis entropy. Thus

$$[x_1 + x_2]_q = [x_1]_q + [x_2]_q + (q-1)[x_1]_q[x_2]_q, \tag{7.6.4}$$

whereas for Tsallis entropy, we have

$$S_q[p^{(A+B)}] = S_q[p^{(A)}] + S_q[p^{(B)}] + (1-q)S_q[p^{(A)}]S_q[p^{(B)}]. \tag{7.6.5}$$

It is being assumed, for the purpose of illustration, that the two systems A and B are statistically independent, so that the joint probability distribution is factorisable : $p^{(A+B)} = p^{(A)}p^{(B)}$. Now let us define [17] Shannon entropy for a given distribution $\{p_i\}$ as follows:

$$S = -\frac{d}{ds} \sum_i (p_i)^s \bigg|_{s=1}. \tag{7.6.6}$$

Then Tsallis entropy is obtained by simply replacing the ordinary derivative by a q-derivative. Thus

$$S_q = -\mathcal{D}_{q,s} \sum_i (p_i)^s \bigg|_{s=1}. \tag{7.6.7}$$

For $q \to 1$, one obtains the ordinary derivative as well as the definition of Shannon entropy. Further connection between q-analysis and generalised entropies was explored in [18, 19].

Now within the new model for finite bath, the constraint on the energy is modified. The effective energies $\{\Delta E_1\}_r$ satisfy very similar nonadditive properties as the q-numbers or the deformed bit number in the definition of the Tsallis entropy. On the other hand, to derive Eq. (7.5.4) from maximum entropy principle, we need to optimise only the Gibbs-Shannon entropy. Thus this procedure seems as complementary to the maximisation of Tsallis entropy under the standard mean value constraints.

Concluding, we have considered the approach of a sample system in contact with a heat bath to arrive at equilibrium distributions for the sample system. We have indicated different models for the finite bath. The real baths or environments are expected to have a finite size. In canonical ensemble, we assume the size of the bath to be infinite. In fact, that constitutes the very definition of a reservoir. Different conditions on bath properties yield different equilibirum distributions of sample system energies. Explicitly, the gaussian ensemble and q-exponential distributions have been discussed. The various thermodynamic quantities like entropy and free energy are nonadditive in these formalisms. We have also presented a new model which can be seen as an intermediate case of the above two scenarios. Lastly, we have indicated the possible connection between noadditivity in these models with the deformed numbers in the context of q-analysis.

Acknowledgements

The author is sincerely grateful to Prof. A. Sengupta for giving the opportunity to participate in the conference MPCNS-2004 and contribute to its proceedings. I also acknowledge fruitful discussions and warm company of Professors S. Abe and A.G. Bashkirov during this conference.

References

[1] R.K. Pathria, *Statistical Mechanics*, 2nd. ed. (Butterworth Heinemann, Oxford, 1996).
[2] J.H. Hetherington, J. Low. Temp. Phys. **66**, 145 (1987).
[3] M.S.S. Challa and J.H. Hetherington, Phys. Rev. Lett. **60**, 77 (1988).
[4] M.S.S. Challa and J.H. Hetherington, Phys. Rev. A **38**, 6324 (1988).
[5] R.S. Johal, A. Planes and E. Vives, Phys. Rev. E **68**, 056113 (2003).
[6] M. Costeniuc, R.S. Ellis, H. Touchette, B. Turkington, J. Stat. Phys. **119**, 1283 (2005). See also cond-mat/0505218.
[7] C. Tsallis, J. Stat. Phys. **52**, 479 (1988).

[8] *Nonextensive Statistical Mechanics and Its Applications*, edited by S. Abe and Y. Okamoto, Lecture Notes in Physics Vol. 560 (Springer-Verlag, Heidelberg, 2001).
[9] *Nonextensive Statistical Mechanics and Physical Applications*, edited by G. Kaniadakis, M. Lissia and A. Rapisarda [Physica A **305**, 1 (2002)].
[10] *Nonextensive Entropy-Interdisciplinary Applications*, edited by M. Gell-Mann and C. Tsallis (Oxford University Press, Oxford, 2003).
[11] M.P. Almeida, Physica A **300**, 424 (2001).
[12] H. Exton, *q-Hypergeometric Functions and Applications* (Harwood, Chichester, 1983).
[13] G.E. Andrews, *Theory of Partitions* (Addison Wesley, Reading MA, 1976).
[14] V. Chari and A. Pressley, *A Guide to Quantum Groups* (Cambridge University Press, Cambridge, England, 1994).
[15] S. Majid, *Foundations of Quantum Group Theory* (Cambridge University Press, Cambridge, England, 1996).
[16] F.H. Jackson, Q. J. Pure Appl. Math. **41**, 193 (1910).
[17] S. Abe, Phys. Lett. A **224**, 326 (1997).
[18] R.S. Johal, Phys. Rev. E **58**, 4147 (1998).
[19] R.S. Johal, Phys. Lett. A **294**, 292 (2002).

8

Quantum Black Hole Thermodynamics

Parthasarathi Majumdar

Theory Group, Saha Institute of Nuclear Physics, Kolkata 700 064, INDIA
E-Mail: parthasarathi.majumdar@saha.ac.in

Summary. Black hole thermodynamics is reviewed for non-experts, underlining the need to go beyond classical general relativity. The origin of the microcanonical entropy of isolated, non-radiant, non-rotating black holes is traced, within an approach to quantum spacetime geometry known as Loop Quantum Gravity, to the degeneracy of boundary states of an $SU(2)$ Cherns Simons theory. Not only does one retrieve the area law for black hole entropy, an infinite series of finite and unambiguous corrections to the area law are derived in the limit of large horizon area. The inclusion of black hole radiance is shown to lead to additional effects related to the thermal stability of black holes. A universal criterion for such stability is derived, in terms of the mass and the microcanonical entropy discussed earlier. As a byproduct, a universal form for the *canonical* entropy of black holes is obtained, in terms of the better-understood microcanonical entropy.

8.1 Introduction

Black holes are very special objects, different from anything in the universe:

- They are regions of spacetime (almost) entirely out-of-bounds to external observers. Everything, including electromagnetic waves, are trapped inside these regions by means of gravitational fields. However, these trapped objects, be they matter or radiation, inevitably lose their identity as they fall through the 'hole'. A black hole then cannot be said to 'contain' these objects; they have been devoured and digested, producing the purest source of gravitation ever.
- They are exact solutions of a fundamental equation of physics, viz., the Einstein equation of general relativity. Yet, they are of *macroscopic* size, with 'gravitational radii' ranging from a kilometre to a million kilometres. This prompted S. Chandrasekhar to say that 'black holes ... are the most perfect macroscopic objects there are in the universe ... are the simplest as well.' [1]

P. Majumdar: *Quantum Black Hole Thermodynamics*, StudFuzz **206**, 218–246 (2006)
www.springerlink.com

- However, as we shall see, black holes also delineate the realm of validity of general relativity, and perhaps of the notion of the spacetime continuum itself. In other words, quantization of spacetime geometry, of general relativity in particular, becomes a necessity for understanding black holes, without the need at all for espousing aesthetic motivations like 'since the strong, weak and electromagnetic interactions have quantum theories, so must gravitation, being a fundamental interaction like the other three'; or indeed 'quantum gravity must exist because without that we shall not be able to formulate a unified theory of all fundamental forces.'
- Classically, black holes are quintessentially general relativistic. This means that they cannot occur within Newton's law of gravitation. Why not? What about the Mitchell-Laplace considerations of the 18th century, involving the escape velocity? According to these ideas, the escape velocity from a sphere of radius R and mass M, $v^2 = 2GM/R$ increases with decreasing R, reaching the velocity of light c for $R = R_S \equiv 2GM/c^2$; thus, if the radius decreases slightly below R_S, we have $v > c$, which is impossible since nothing travels faster than light. Hence, it was concluded in the 1780s, that dense gravitating objects with such sizes would become invisible, trapping everything including light. For an object having the earth's mass, if its contents were to be squeezed into a glass marble of a centimetre in radius, this would render the marble invisible. There are a couple of flaws with this reasoning. First of all, the Newtonian gravitation law is founded on the Galilean principle of relativity which provides us with the notion of relative velocity for *all* velocities, *including* that of light. If a source of light moves relative to a stationary observer with a velocity v, light from the source moves, according to Galilean relativity, with a velocity $c + v$ relative to the stationary observer, and *not* c! *There is nothing absolute at all about* c *in Galilean relativity.* Further, in the Newtonian framework, only mass produces and is affected by gravitation. Electromagnetic waves carry no mass, and hence cannot be trapped by gravitation within this framwork.
- How about special relativity? Since Newton's law of gravitation strictly conforms to Galilean relativity, it could not possibly be consistent with special relativity where the velocity of light is an absolute constant. Could not a generalization of the Newtonian law be consistent with special relativity, though? No, because gravitation appears to require **non-inertial** frames of reference! The Principle of Equivalence, first hypothesized by Einstein, asserts that we can locally remove a gravitational field by a suitable acceleration, as has been experienced by anyone in a lift at the start of a descent when the lift momentarily falls freely. A stronger version of this principle states that *physical laws are the same in all reference frames.* Different non-inertial frames are related by coordinate *diffeomorphisms*

$$x^i \to x'^i = x'^i(\{x^j\}) \ , \ i,j = 0,1,2,3$$

Invariance under such diffeomorphisms lies at the heart of general relativity.

In the last part of this introduction, an extremely brief recap of general relativity is given, mainly to set notation and to agree on conventions. Readers uninitiated in general relativity will have to consult any of the excellent textbooks that are now available [2].

- The spacetime manifold is a metrically connected four dimensional pseudo-Riemannian manifold, with the metric $g_{ij}(x)$ having signature $-+++$. Geodesics, affinely parametrised in terms of the *proper time* τ on such a manifold are given by

$$\frac{d^2x^i}{d\tau^2} + \Gamma^i_{jk}\frac{dx^j}{d\tau}\frac{dx^k}{d\tau} = 0, \tag{8.1.1}$$

where the affine connection $\Gamma^i_{jk} = \Gamma^i_{kj}$ is required to be metric compatible, leading to the Christoffel connection

$$\Gamma^i_{jk} = \frac{1}{2}g^{il}[\partial_{(j}g_{k)l} - \partial_l g_{jk}]. \tag{8.1.2}$$

- The Riemann curvature tensor is given by

$$R^l_{ijk} = \partial_{[i}\Gamma^l_{j]k} + \Gamma^l_{m[i}\Gamma^m_{j]k},$$

from which the Ricci tensor and Ricci scalar are obtained by contraction

$$R_{ik} \equiv \delta^j_l R^l_{ijk}, \qquad R \equiv g^{ik}R_{ik}.$$

- The Einstein equation is

$$\mathcal{G}_{ij} \equiv R_{ij} - \frac{1}{2}g_{ij}R = 8\pi G T_{ij}, \tag{8.1.3}$$

where, T_{ij} is the energy-momentum tensor of all matter and radiation. This remarkable equation has two momentous implications
- The geometry of spacetime is a *dynamical* field, determined locally by the energy-momentum tensor of matter and radiation.
- Spacetime geometry and matter/radiation have a symbiotic relationship : 'matter tells space how to curve, space tells matter how to move'.

The plan for the rest of this review is as follows: in the next section, we begin with a description of the spacetime view of black holes together with a very brief description of how they may form out of gravitational collapse of massive stars. We focus on the main attributes of black hole spacetimes : a one-way null 3-surface which functions as an inner boundary of spacetime known as the event horizon, and a *curvature singularity* where continuum spacetime geometry loses its meaning. We then go on to discuss theorems related to the event horizon derived from general relativity, and point out the similarity the results have with the laws of thermodynamics. We discuss the work of Bekenstein, which heralded the entire area of black hole thermodynamics, especially the notion of black hole entropy, underlining the need to formulate a quantum version of spacetime geometry. The next section is devoted to a review of a particular formulation of quantum spacetime geometry often called Loop Quantum Gravity (LQG). The notion of an isolated horizon which corresponds to a non-stationary (in the sense of the absence of a time-like isometry globally) but non-radiant, null inner boundary of spacetime is also briefly described. These two formulations are then employed to analyze the boundary quantum field theory describing the quantum isolated horizon. From a statistical mechanical standpoint, the isolated horizon corresponds to a *microcanonical* ensemble, whose entropy can be computed unambiguously. The results include not only the area law anticipated by Bekenstein (and Hawking) but a whole slew of quantum corrections, for asymptotically large (macroscopic) horizon areas. We then turn our attention to the phenomenon of Hawking radiation. We address the crucial issue of negative heat capacity and the consequent thermal instability. Employing the canonical ensemble of equilibrium statistical mechanics, we derive a universal criterion for thermal stability involving the mass of the isolated black hole and the microcanonical entropy discussed earlier. As a bonus, additional logarithmic corrections to the canonical entropy are obtained. We conclude with our outlook on the present status of the field and what needs to be done in the near future.

8.2 Black holes in general relativity

We adopt here the 'spacetime' view of black holes in which one deals with exact solutions of Einstein's equation with certain isometries. In other words, we avoid the complications associated with the more realistic 'astrophysical' view in which black holes accrete matter whose analysis is crucial for observational detection of these objects. The spacetime viewpoint is described in terms of a region $\mathcal{B}$ defined as $\mathcal{B} \equiv \mathcal{M} - I^-(\mathcal{I}^+)$ where, $\mathcal{M}$ is the entire spacetime manifold, containing all events, and $I^-(\mathcal{I}^+)$ is the set of all events to the past of the asymptotic null infinity $\mathcal{I}^+$. In other words, a black hole spacetime is the set of all events which cannot communicate to (null) observers in the infinite future. The boundary $\partial\mathcal{B}$ of such events is called the *event horizon*, and is a null, one-way hypersurface. $\mathcal{B}$ contains a spacetime *singularity* which is hidden from all distant observers by the trapping boundary $\partial\mathcal{B}$.

As already mentioned, black holes are exact solutions of Eq. (8.1.3) with a vanishing energy-momentum tensor (*vacuum*) or the energy momentum tensor of the Maxwell field. Other more exotic forms of matter/radiation appear to lead to solutions which are unstable under perturbations, and shall therefore not be considered here. The vacuum solutions form a two-parameter family, with the two parameters being identified with the mass and the angular momentum of spacetime. The *electrovac* solutions on the other hand are characterized by one additional parameter identified as the electric charge of the black hole.

The simplest of the vacuum solutions is the spherically symmetric Schwarzschild black hole, whose metric, in spherical polar coordinates, is given by

$$ds^2 = -\left(1 - \frac{2GM}{c^2 r}\right) c^2 dt^2 + \left(1 - \frac{2GM}{c^2 r}\right)^{-1} dr^2 + r^2 d\Omega^2 \qquad (8.2.1)$$

where, M is the mass of the spacetime, and $d\Omega^2$ is the usual Euclidean metric on the 2-sphere. One of the singularities of the metric, on the null surface $r = 2GM/c^2$, is associated with the event horizon. It is not a spacetime singularity, but one that occurs only due to a bad choice of coordinates. Thus, the spacetime metric can be made regular at the event horizon, if we choose as coordinates the affine parameters decribing null geodesics in this geometry. Of course, the induced 3-metric on the event horizon is degenerate, as it must be on any null hypersurface.

The other singularity at $r = 0$ is infinitely more severe, since here, the spacetime curvature diverges. This is geometrically meaningless, and a radical departure from the basic premise of a *smooth* spacetime continuum, inherent in general relativity. For the Schwarzschild geometry, the singularity corresponds to a *spacelike* surface, i.e., to a particular 'moment of time' at which the notion of time itself *ends*. This singularity is unavoidable for all incoming future-directed geodesics, implying that all matter and fields crossing the event horizon must inevitably lose their identity when reaching this surface. Another way of saying this is that classical general relativity is no longer valid as one approaches this region of singularity, just as Maxwell's equations of classical electrodynamics break down very close to sources of electric and magnetic fields. And just as in the case of electrodynamics, the only logical route around the conundrum might lie in a quantum version of spacetime geometry.

But is all this realistic? To answer this, we need to have a brief description of the formation of a black hole from a dying star. As the nuclear fuel which powers stars gets depleted, the outward thermal pressures generated by nuclear fusion is no longer sufficient to balance the inward gravitational attraction. Stars then begin to shrink gravitationally. This continues until the matter inside the star is so compressed (a cubic centimetre weighing a tonne), that quantum mechanics takes over, and the Pauli exclusion principle comes into play. According to this principle, fermions (which constitute most of matter as electrons, protons and neutrons) cannot be compressed arbitrarily, because that will require many of them to occupy the same state, which is not possible. Thus, the 'Pauli degeneracy pressure' becomes a counterforce to gravity, and provided the mass of the star is below a certain limiting value, about 1.5 $M_\odot$ (as deduced by S. Chandrasekhar in 1931), the star may find peace as a white dwarf. White dwarfs are aplenty in the universe. But not all stars follow this usual pattern of evolution. If the star mass exceeds the Chandrasekhar limit, the white dwarf is unstable under gravitational contraction. As the core of the star contracts, becoming denser and accreting more matter, the outer layers become larger and hotter, until the star explodes dramatically as a supernova. A lot of energy-momentum is ejected from the star in this explosion. Meanwhile, the core has collapsed gravitationally to a compact, extremely dense object, usually made up exclusively of neutrons. The Pauli degeneracy pressure of the neutrons is often sufficient to balance their gravitational attraction inwards, producing a *neutron* star. However, if the core mass exceeds that of the sun, this balancing does not work, and another phase of rapid collapse under gravity ensues. Such a collapse of a spherical supernova core is depicted in Fig. 8.1

With further shrinkage of the core, even this becomes impossible. When the local light cones align with the dotted line marked 'event horizon' in Fig. 8.1, light is confined to propagate along the three dimensional lightlike surface known as event horizon. This means that, instantaneously, light will appear to

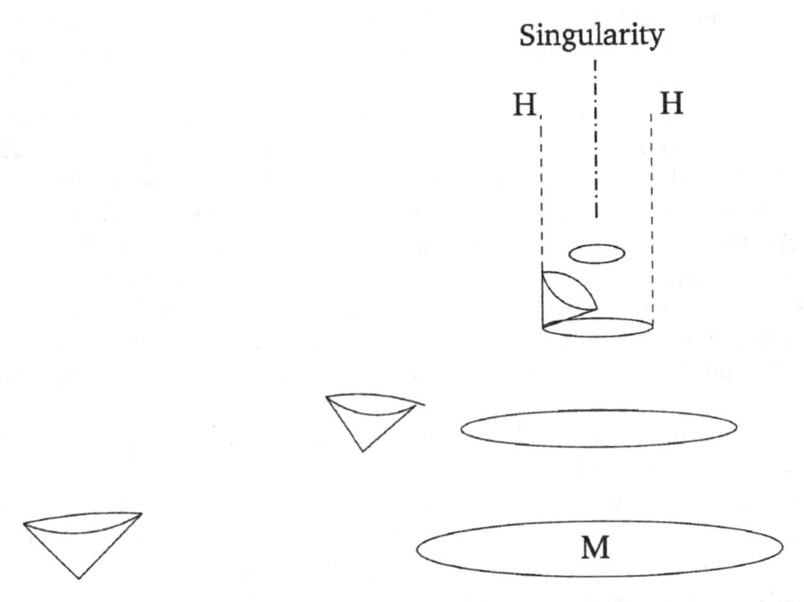

Fig. 8.1. Collapse of a spherical core. The ellipses represent instantaneous sizes of the spherical star core at successive instants of time. The light cones exhibit the nature of the local geometry. Distant observers do not feel the effect of curvature due to the black hole, so that light cones stand erect as in Minkowski space. As one approaches the collapsing star, light cones begin to tilt in response to the curvature of the increasingly dense core. This means that particles and light starting from the bottom of a tilted light cone find it increasingly harder to propagate as in flat spacetime and escape to infinity. Light can still make it out to infinity, but only if it is appropriately aimed.

travel along the outer boundary of the spherical core, and will no longer travel out to infinity. All other objects will be confined to the 'inside' of the event horizon, remaining hidden from external observers. For a core of mass M, the gravitational (Schwarzschild) radius of the core is $R_S = 2GM/c^2$, exactly the same as the limiting radius found earlier from Newtonian considerations. Since the core now traps light and everything else, it is *black*. The event horizon is thus called an *outer trapping surface*. The event horizon also demarkates between the set of events accessible to a distant external observer, and the set that is not. In this sense, from the standpoint of an external observer, the event horizon functions as an *inner* boundary of spacetime, the *outer* boundary being that at null (or spatial) infinity. Finally, the event horizon allows light and other matter to enter it but not to escape from it; in this sense, it is a *one-way membrane*. We shall come back to other very important properties of the event horizon which will require going beyond classical general relativity for a complete understanding.

Meanwhile, the shrinking of the core beyond the event horizon continues unabated, as shown by the small ellipses occurring at time slices beyond that corresponding to the formation of the event horizon. Light cones inside the event horizon tilt even further, such that all matter and radiation emanating from the bottom of these cones stay firmly inside the event horizon. The shrinking continues until the core has turned into a singularity, as disscussed earlier. At this event, a spacetime continuum of events cannot be defined locally at all. If we consider the cone which has been tilted almost to horizontal in Fig. 8.1, it is clear that any light ray or material particle starting at the bottom of that light cone has a very tentative future, in that timelike and null geodesics corresponding to their world lines *end abruptly*, at a finite proper time. Such a pathology is quite generic in general relativity, as has been established with great rigour using the Raychaudhuri equation by Geroch, Hawking and Penrose [3].

8.3 Black hole thermodynamics

The Raychaudhuri equation can also be used to derive certain results known as Laws of Black Hole Mechanics [4], [20], which can be stated as follows :

- *The area of the event horizon of a black hole can never decrease*:

$$\delta \mathcal{A}_h \geq 0 \tag{8.3.1}$$

  In other words, if two black holes of horizon area $\mathcal{A}_1$ and $\mathcal{A}_2$ were to coalesce adiabatically, the area of the horizon of the resultant black hole $\mathcal{A}_{12} \geq \mathcal{A}_1 + \mathcal{A}_2$.
- *The surface gravity of a black hole is a constant on the event horizon.* The surface gravity κ is a geometrical quantity relatd to the proper acceleration at any point of a timelike geodesic. For stationary black holes, $\kappa_h =$ const

- If M is the black hole mass undergoing a change by an amount δM, the area of the horizon must change according to

$$\delta M = \kappa \delta \mathcal{A}_h.$$

For charged rotating black holes, this law states that

$$dM \equiv \Theta d\mathcal{A}_h + \Phi dQ + \mathbf{\Omega} \cdot d\mathbf{L}, \tag{8.3.2}$$

where, $\kappa \equiv (r_+ - r_-)/4\mathcal{A}_h$ is the surface gravity, $\Phi \equiv 4\pi Q r_+/\mathcal{A}_h$ is the electrostatic potential at the horizon and $\mathbf{\Omega} \equiv 4\pi \mathbf{L}/M\mathcal{A}_h$ is the angular velocity at the horizon.

There is an obvious analogy of these laws of black hole mechanics with the standard laws of thermodynamics: the first implies that the horizon area of a stationary black hole behaves like an entropy, since it never decreases. In fact, this becomes clearer if one thinks of an isolated system separated into two parts by a partition; on removing the partition, the total entropy is never less than the sum of the entropies in the two partitioned regions. The second leads to an analogy of the surface gravity with the equilibrium temperature. The third is analogous to a combination of the first and second laws of thermodynamics, with the black hole mass playing the role of the internal energy. However, recall that a black holes contain nothing but pure spacetime curvature; unlike ordinary matter, there are no microstates which could be held responsible for thermodynamic behaviour.

One can adopt two standpoints with regard to these general relativistic (and hence geometrical) theorems:

- These are mere analogies between these theorems and the laws of thermodynamics, since black holes contain no matter or radiation which can have any thermodynamic behaviour at all. There are no atoms or molecules in black hole spacetimes whose microstates will reflect the randomness characteristic of ordinary thermodynamic systems. Hence the analogies are intriguing, but nothing beyond that. There is no fundamental physics underlying these analogies. This is the viewpoint initially adopted by the authors of the theorems.
- These analogies reflect that black holes have a certain thermodynamic behaviour associated with them, whose origin cannot be the proofs of the theorems alone, but a larger underlying framework of which general relativity is only a part. For consistency, this framework must endow black hole spacetimes with an entropy $S_{\rm bh} \propto \mathcal{A}_h$. This is the viewpoint adopted by Bekenstein [9].

For obvious reasons, we focus on the second viewpoint. Since in thermodynamics entropy has the dimensions of the Boltzmann constant k_B, the proportionality stated above should be written as

$$S_{\text{bh}} = \zeta k_B \frac{\mathcal{A}_h}{l^2}, \tag{8.3.3}$$

where, ζ is a dimensionless constant of $O(1)$, and l is a length scale which is independent of parameters characterizing the black hole. Of necessity, then, l must be a fundamental constant, and the only one involving G and c is the Planck length $l_P \equiv (G\hbar/c^3)^{1/2}$. The length that this scale corresponds to is on the order of 10^{-33} centimetres, smaller than any scale ever probed. The Planck length is often associated with the scale of the distance from a point mass at which the *quantum* effects of general relativity can no longer be neglected. In other words, while it is true that classically, black hole spacetimes can have no entropy or thermodynamic behaviour of any sort, this may no longer be the case when quantum gravitational effects are taken into account.

We now come to the issue as to when can the surface gravity κ_h on the event horizon be identified with the black hole temperature T_{bh}? The temperature T_{bh} is not the temperature of the horizon of the black hole in the sense that a freely falling observer will sense it as he crosses the horizon. At the horizon, the redshift factor vanishes, so the observer detects no temperature at all. The notion of black hole temperature is made unambiguous if we imagine placing the black hole in a background of black body radiation in equilibrium at a temperature $T < T_{\text{bh}}$. In his celebrated work, Hawking [31] showed that, for an observer located at infinity with a vanishing ambient temperature, a black hole actually radiates in a Planckian spectrum like a black body at a temperature T_{bh}! The mean number of particles emitted at a frequency ω is given by the Planckian formula

$$n_\omega = \frac{|t_\omega|^2}{\exp(2\pi\omega/\kappa_h) - 1}, \tag{8.3.4}$$

where, t_ω is the absorption coefficient and we have set the Boltzmann constant $k_B = 1$. The larger the mass of the black hole, the smaller is the horizon surface gravity and hence the Hawking temperature T_{bh}. It follows that for most stellar black holes, this temperature $T_{\text{bh}} \ll 2.7°$K, so that Hawking radiation from such black holes is swamped by the cosmic microwave background. In fact, these black holes absorb rather than emit, radiation. One other outcome of Hawking's work is fixing the constant $\zeta = 1/4$ in (8.3.3); this law, $S_{\text{BH}} = \mathcal{A}/4\ell_P^2$ is called the Bekenstein-Hawking Area Law (BHAL) .

As for the status of the second law of thermodynamics in presence of black holes, it is true that when matter falls across the event horizon into a black hole, the entropy of the matter is lost. One might think of this as a decrease in the entropy of the universe. However, as Bekenstein [9] pointed out, when matter falls into a black hole, the mass of the black hole, and hence its area, increases. The hypothesis of black hole entropy then says that the entropy of the black hole must increase as a consequence, by at least the same amount as the entropy lost by the the part of the universe outside the event horizon. Thus, in presence of black holes, the Second law of thermodynamics is modified into the statement that *the total entropy, defined as the sum of the entropy outside the event horizon and the black hole entropy, can never decrease in any physical process,*

$$\delta(S_{\text{ext}} + S_{\text{bh}}) \geq 0. \tag{8.3.5}$$

This relation is called the Generalized Second Law of thermodynamics and its enunciation marks the beginning of the subject known as Black Hole Thermodynamics.

The essence of this section is then that the two prime aspects of black holes, namely the existence of an event horizon as well as the existence of a spacetime singularity, both require abandoning the notion of a spacetime continuum. Black hole physics is perhaps one of two important reasons for seeking a quantum description of spacetime geometry, the other one being the Big Bang singularity. While no complete Quantum General Relativity exists at this point, a formulation that perhaps best retains the basic spirit of general relativity is the so-called Loop Quantum Gravity. In the following section, the aim would be to derive the entropy of an *isolated* black hole within this particular framework.

8.4 Microcanonical entropy

8.4.1 Classical Aspects

The standard formulation of general relativity is based on the notion of a spacetime metric $g_{\mu\nu}$ with indefinite (Lorentzian) signature, as already mentioned. Now, for quantization, one needs to construct the Hamiltonian formulation of the theory, towards which one must at first be able to identify a suitable canonical pair of variables which span the classical phase space of the theory. The so-called ADM (Arnowitt-Deser-Misner) [6] formulation does precisely that. One chooses a spacelike hypersurface M of spacetime, on which Cauchy data, consisting of the 3-metric q_{ab} induced by the embedding of M in spacetime, and its canonical conjugate π^{ab} are chosen as the canonical pair. ADM derive, as a function of these variables, four first class constraints : three of these correspond to the invariance of the theory under spatial diffeomorphisms, while the fourth arises due to invariance under temporal diffeomorphisms. Unfortunately, the constraints are highly intractable within this approach. A key step in the right direction appears to be using the projection of the affine connection to M, rather than the 3-metric, as the basic configuration space variable, with the extrinsic curvature of M being the conjugate momentum. A first order formulation is also used to write down the action.

Further improvements occur with the introduction of the *self-dual* $SU(2)$ connection [5] and finally, by means of canonical transformations, the Barbero-Immirzi class of real $SU(2)$ connections [7] as the configuration space variables, with the densitized triad on M (in 'time gauge') being the canonically conjugate momenta. This connection is defined as ${}^\gamma A^i_a \equiv \Gamma^i_a + \gamma K^i_a$; the rescaled densitized triad $\gamma^{-1} E^{ai}$, with $E^{ai} \equiv \det e e^{ai}$, where e^{ai} is the triad on a chosen spatial slice M. Here, $\Gamma^i_a \equiv q_{ab} \epsilon^i_{jk} \Gamma^{bjk}/2$, with Γ^{bjk} being the pullback of the Levi-Civita spin connection to the spatial slice under consideration, and q_{ab} is the 3-metric on the slice; the extrinsic curvature K^i_a is defined as $K^i_a \equiv q_{ab} \Gamma^{b0i}$; γ, the Barbero-Immirzi parameter [7] is a real positive parameter. Four dimensional local Lorentz invariance has been partially gauge fixed to the 'time gauge' $e^0_a = -n_a$ where n_a is the normal to the spatial slice. This choice leaves the residual gauge group to be $SU(2)$. It is convenient to introduce the quantity ${}^\gamma \Sigma^{ij}_{ab} \equiv \gamma^{-1} e^i_{[a} e^i_{b]}$, in terms of which the symplectic two-form of general relativity can be expressed as

$$\Omega = \frac{1}{8\pi G} \int_M \mathrm{Tr}[\delta^\gamma \Sigma \wedge \delta^\gamma A' - \delta^\gamma \Sigma' \wedge \delta^\gamma A]. \tag{8.4.1}$$

The expression (8.4.1) of course is subject to modification by boundary terms arising from the presence of boundaries of spacetime. The black hole horizon, assumed to have the topology $S^2 \otimes R$, is intersected by M in a two-sphere which thus plays the role of an inner boundary.

Rather than using the notion of event horizon appropriate to stationary situations studied in earlier literature [20], we adopt here the concept of 'isolated' horizon [21]. This has the advantage of being characterized completely locally, without requiring a global timelike Killing vector field. The characterization, for non-rotating situations, involves a null surface H with topology as assumed above, with preferred foliation by two- spheres and ruling by lines transverse to the spheres. l^a and n_a are null vector fields satisfying $l^a n_a = -1$ on the isolated horizon. l^a is a tangent vector to the horizon, which is assumed to be geodesic, twist-free, divergenceless and most importantly, *non-expanding*. The Raychaudhuri equation is then used to prove that it is also free of shear. Similarly, the null normal one-form field n_a is assumed to be shear- and twist-free, and have negative spherical expansion. Finally, while stationarity is not a part of the characterization of an isolated horizon, the vector direction field l^a can be shown [21] to behave like a Killing vector field *on* the horizon, satisfying

$$l^a \nabla_a l^b = \kappa l^b. \tag{8.4.2}$$

Here, κ is the acceleration of l^a on the isolated horizon. Unlike standard surface gravity whose normalization is fixed by the requirement that the global timelike Killing vector generate time translations at spatial infinity, the normalization of κ here varies with rescaling of l^a.

These features imply that while gravitational or other radiation may exist arbitrarily close to the horizon, nothing actually crosses the horizon, thereby emulating an 'equilibrium' situation. This, in turn, means that the area A_H of the isolated horizon must be a constant. Lifting of this restriction leads to dynamical variants (the so-called 'dynamical' horizons) which have also been studied [22]; we shall however not consider these here.

The actual implementation of these properties of the isolated horizon require boundary conditions on the phase space variables on the 2-sphere foliate of the horizon. Recalling that the horizon is an inner boundary of spacetime, it is obvious that one needs to add boundary terms to the classical Einstein action, in order that the variational principle can be used to derive equations of motion. It turns out [8], [21] that the 'boundary action' S_H that one must add to the Einstein action (in the purely gravitational case, using the Ashtekar self-dual connection)

$$S_E = \frac{-i}{8\pi G} \int_{\mathcal{M}} \mathrm{Tr} \Sigma \wedge F, \tag{8.4.3}$$

is an $SU(2)$ Chern Simons (CS) action[1]

$$S_H = \frac{-i}{8\pi G} \frac{A_H}{4\pi} \int_H \mathrm{Tr} \left[A \wedge dA + \frac{2}{3} A \wedge A \wedge A \right], \tag{8.4.4}$$

where, now, A is the CS connection, and F the corresponding curvature. The resultant modification to the symplectic structure (8.4.1) is given by the CS symplectic two-form

$$\Omega_H = -\frac{k}{2\pi} \oint_{\mathrm{S}} \mathrm{Tr}[\delta^\gamma A \wedge \delta^\gamma A'], \tag{8.4.5}$$

where, $k \equiv A_H / 8\pi\gamma G$. In writing the boundary action (8.4.4), we have suppressed other terms like the boundary term at infinity.

The sympletic structure Ω_H remains the same under canonical transformation of the variables to the Barbero-Immirzi connection and the densitized triad defined earlier. In terms of these phase space variables, it is easy to see that the variational principle for the full action is valid, provided we have, on the two-sphere foliation of H, the restriction,

$$\frac{k}{2\pi} F^i_{ab} + \Sigma^i_{ab} = 0. \tag{8.4.6}$$

Eq. (8.4.6) has the physical interpretation of Gauss law for the CS theory, with the two-form Σ playing the role of source current. We shall see shortly that this has crucial implications for the quantum version of the theory.

[1] Strictly speaking, the boundary conditions considered by [8] involve a partial gauge fixing whereby the only independent connection on the horizon is actually an internal radial $U(1)$ projection of the $SU(2)$ CS connection. However, we ignore this subtlety at this point and continue to work with the $SU(2)$ CS theory. The modification to the final answer, had we chosen not to ignore this subtlety, will be discussed later.

8.4.2 Quantum Aspects: General

The classical configuration space consists of the space of smooth, real $SU(2)$ Lie Algebra-valued connections modulo gauge transformations [23]. Alternatively, the space can be described in terms of three dimensional oriented, piecewise analytic networks or graphs embedded in the spatial slice M [23] as shown in Fig. 8.2

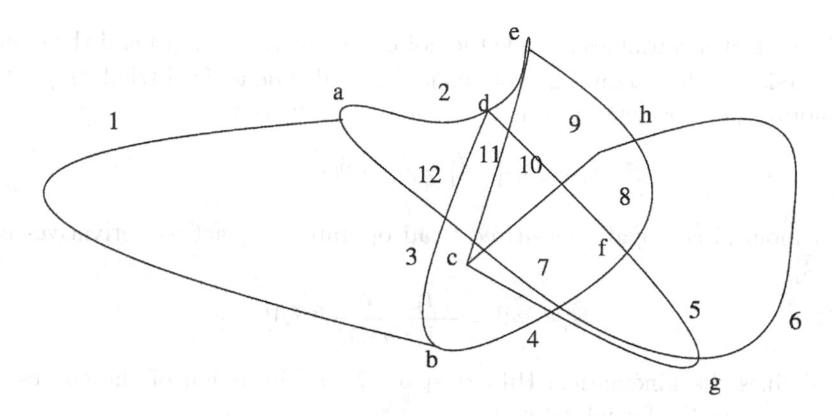

Fig. 8.2. Spin network graph C

Consider a particular graph C with n links (or edges) $e_1, \ldots, e_n$ (shown in Fig. 8.2 as links numbered $1, \ldots, 12$); consider also the pullback of the connection A to C. Consider the holonomies defined as

$$h_C(e_i) \equiv \mathcal{P} \exp \oint_{e_i \in C} {}^{\gamma}A_C, \quad i = 1, 2 \ldots, n, \tag{8.4.7}$$

where ${}^{\gamma}A_C$ represents the restriction of the connection to the graph C; these span the configuration space $\mathcal{A}_C$ of connections on the graph C. This space consists of $[SU(2)]^n$ group elements obtained as n-fold compositions of $SU(2)$ group elements characterised by the spin j_i of the edge e_i for $i = 1, 2, \ldots, n$. The edges of C terminate at vertices $v_1, \ldots, v_n$ (represented in Fig. 8.2 by vertices marked $a, b, c, \ldots, h$) which, in their turn, are characterised by group elements $g(v_1), \ldots, g(v_m)$, which together constitute a set of $[SU(2)]^m$ group elements for a given graph C. The union of spaces $\mathcal{A}_C$ for all networks is then an equally good description of the classical configuration space.

The transition to the *quantum* configuration space is made, first by enhancing the space of connections to include connections ${}^{\gamma}\bar{A}$ which are not smooth but distributional, and then considering the space $\mathcal{H}_C$ of square-integrable functions $\Psi_C[{}^{\gamma}\bar{A}]$ of connections. For the integration measure, one uses n-copies of the $SU(2)$-invariant Haar measure. For a given network C, the wave function $\Psi_C[{}^\gamma \bar{A}]$ can be expressed in terms of a smooth function ψ of the holonomies $\bar{h}_C(e_1), \ldots, \bar{h}_C(e_n)$ of distributional connections,

$$\Psi_C[{}^\gamma \bar{A}] = \psi(\bar{h}_C(e_1), \ldots, \bar{h}_C(e_n)). \tag{8.4.8}$$

The inner product of these wave functions can be defined as

$$\langle \Psi_{1C}, \Psi_{2C} \rangle = \int d\mu \bar{\psi}_{1C} \psi_{2C}. \tag{8.4.9}$$

Basic dynamical variables include the holonomy operator $\widehat{h}_C(e)$ and the operator version of the canonically conjugate γ-rescaled densitized triad ${}^\gamma \widehat{E}^i_a$. The holonomy operator acts diagonally on the wave functions,

$$[\widehat{h}_C(e) \Psi_C][{}^\gamma \bar{A}] = \bar{h}_C(e) \Psi_C[{}^\gamma \bar{A}].$$

The canonical conjugate densitized triad operators $\widehat{E}^i_a$ act as derivatives on $\Psi_C[{}^\gamma \bar{A}]$:

$${}^\gamma \widehat{E}^i_a \Psi[{}^\gamma A] = \frac{\gamma\, l_P^2}{i} \frac{\delta}{\delta {}^\gamma A^a_i} \Psi[{}^\gamma A].$$

One defines the kinematical Hilbert space $\mathcal{H}$ as the union of the spaces of wave functions Ψ_C for all networks.[2]

Particularly convenient bases for the wave functions are the spin network bases. Typically, the spin network (spinet) states can be schematically exhibited as

$$\psi_C(\{h_C\}; \{v\}) = \sum_{\{m\}} \prod_{v \in C} I_v \prod_i D^i_{\ldots}, \tag{8.4.10}$$

where, D^i is the $SU(2)$ representation matrix corresponding to the ith edge of the network C, carrying spin j_i, and I_v is the invariant $SU(2)$ tensor inserted at the vertex v. If one considers all possible spin networks, the set of spinet states corresponding to these is dense in the kinematical Hilbert space $\mathcal{H}$. Spinet states diagonalize the densitized triad (momentum) operators and hence operators corresponding to geometrical observables like area, volume, etc. constructed out of the the triad operators. The spectra of these observables turn out to be *discrete*; e.g., for the area operator corresponding to the area of a two dimensional spacelike physical surface s (like the intersection of a spatial slice with a black hole horizon), one considers the spins $j_1, j_2, \ldots, j_p$ on s at the p punctures made by the p edges of the spinet assumed to intersect the surface. The area operator is defined as [24], [25]

[2] Unfortunately, $\mathcal{H}$ is *not* the physical Hilbert space of the theory; that space is the algebraic dual of $\mathcal{H}$ with no natural scalar product defined on it. However, for the purpose of calculation of the microcanonical entropy, it will turn out to be adequate to use $\mathcal{H}$.

$$\widehat{A}_s \equiv \left\{ \sqrt{n_a n_b \widehat{E}_i^a \widehat{E}_i^b} \right\}_{\mathrm{reg}} , \tag{8.4.11}$$

where, n_a is the normal to the surface, and *reg* indicates that the operator expression within the braces is suitably regularized. The eigenspectrum turns out to be [24, 25]

$$a_s(p; \{j_i\}) = 8\pi\gamma\, l_P^2 \sum_{i=1}^{p} \sqrt{j_i(j_i+1)}. \tag{8.4.12}$$

Spinet basis states correspond to networks without any 'hanging' edge, so that they transform as gauge singlets under the gauge group $SU(2)$. Furthermore, invariance under spatial diffeomorphisms is implemented by the stipulation that the length of any edge of any graph is without physical significance.

8.4.3 Quantum Aspects: entropy calculation

As discussed in the previous subsection, the sphere S_H formed by the intersection of the isolated horizon and a spatial slice M can be thought of as an inner boundary of M. The dynamics of the isolated horizon is described by an $SU(2)$ Chern Simons theory with the bulk gravitational degrees of freedom playing the role of source current. This picture can be implemented at the quantum level in a straightforward manner. Because of the isolation implied by the boundary conditions, the kinematical Hilbert space $\mathcal{H}$ can be decomposed as

$$\mathcal{H} = \mathcal{H}_V \otimes \mathcal{H}_{\mathrm{S}},$$

where, $\mathcal{H}_V$ ($\mathcal{H}_{\mathrm{S}}$) corresponds to quantum states with support on the spatial slice M (on the inner boundary, i.e., these are the Chern Simons states). The boundary conditions also imply the Chern Simons Gauss law, Eq. (8.4.6); the quantum operator version of this equation may be expressed as

$$\frac{k}{2\pi} \mathbf{1} \otimes \widehat{F}^i_{ab} + \widehat{\Sigma}^i_{ab} \otimes \mathbf{1} = 0 \qquad \text{on } S_H. \tag{8.4.13}$$

Now, the bulk spinet states diagonalize the operator $\widehat{\Sigma}$ with distributional eigenvalues,

$$\widehat{\Sigma}(\mathbf{x})|\psi\rangle_V \otimes |\psi\rangle_{\mathrm{S}} = \gamma\, l_P^2 \sum_{i=1}^{p} \lambda(j_i)\delta^{(2)}(\mathbf{x}, \mathbf{x_i})|\psi\rangle_V \otimes |\psi\rangle_{\mathrm{S}}. \tag{8.4.14}$$

Eq. (8.4.13) then requires that the boundary Chern Simons states also diagonalize the Chern Simons curvature operator $\widehat{F}$. In other words, edges of the bulk spin network punctures the horizon foliate S_H, endowing the ith puncture with a *deficit angle* [8] $\theta_i \equiv \theta(j_i)$ for $i = 1, 2, \ldots, p$, such that

$$\sum_{i=1}^{p} \theta(j_i) = 4\pi. \tag{8.4.15}$$

The curvature on S_H is thus vanishingly small everywhere else except at the location of the punctures. This manner of building up the curvature of the two-sphere S_H out of a large but finite number of deficit angles requires that the number of such angles must be as large as possible. This is achieved for the smallest possible value of all spins j_i, namely $j_i = 1/2$ for all i. This is illustrated in Fig. 8.3 below

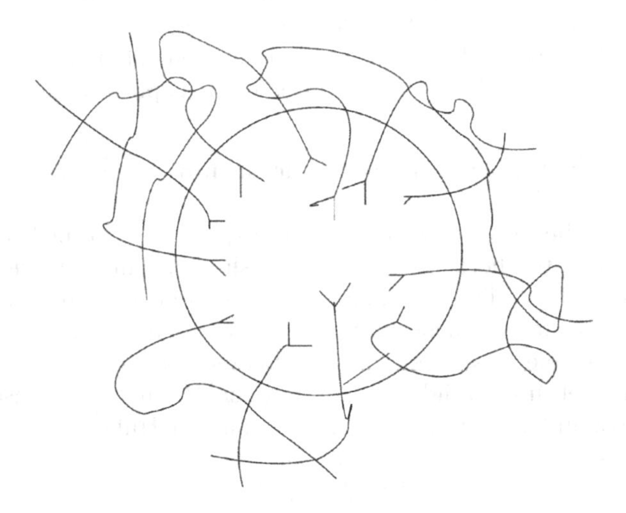

Fig. 8.3. Spin network links puncturing the horizon

The calculation of the entropy now proceeds by treating the isolated horizon as a microcanonical ensemble with fixed area. Recalling the semiclassical relationship between horizon area and mass of the isolated horizon, this is equivalent to considering a standard equilibrium microcanonical ensemble where the (average) energy of the ensemble does not fluctuate thermally. The number of configurations of such a system is equal to the exponential of the microcanonical entropy S_{MC}. Likewise, in this case, the number of boundary Chern Simons states dim $\mathcal{H}_{\mathrm{S}}$ with pointlike sources, as depicted in Eq. (8.4.13) (keeping (8.4.14) in view) yields $\exp S_{\mathrm{MC}}$. This number has been calculated for all four dimensional non-rotating isolated horizons [8], [11] of large macroscopic fixed horizon area $A_H \gg l_{\mathrm{P}}^2$. In ref. [11], the computation makes use of the well-known relation between the dimensionality of the boundary Chern Simons Hilbert space and the number of *conformal blocks* of the corresponding two dimensional $SU(2)_k$ Wess-Zumino-Witten model that 'lives' on the punctured two-sphere S_H. This number is given by

$$\dim \mathcal{H}_{\rm S} = \sum_{p} \prod_{i=1}^{p} \sum_{j_i} \mathcal{N}(p, \{j_i\}), \tag{8.4.16}$$

subject to the constraint that the area eigenvalues are fixed (to within a fator of the Planck area) to the constant macroscopic area A_H,

$$A_H = 8\pi\gamma\, l_P^2 \sum_{i=1}^{p} \sqrt{j_i(j_i+1)}, \tag{8.4.17}$$

where,

$$\mathcal{N}(p, \{j_i\}) = \sum_{m_1=-j_1}^{j_1} \cdots \sum_{m_p=-j_p}^{j_p} \left[\delta_{(\sum_{n=1}^{p} m_n),0} - \frac{1}{2}\delta_{(\sum_{n=1}^{p} m_n),1} - \frac{1}{2}\delta_{(\sum_{n=1}^{p} m_n),-1} \right] \tag{8.4.18}$$

Instead of the area constraint, one may now recall Eq. (8.4.15) which also is a constraint on the spins and number of punctures. Using this result in the area formula (8.4.17) yields the maximal number of punctures

$$p_0 = \frac{A_H}{4\pi\sqrt{3}\gamma l_P^2}. \tag{8.4.19}$$

The corresponding number of Chern Simons states for this assignment of spins is given via (8.4.18) by

$$\mathcal{N}(p_0) \simeq \frac{2^{p_0}}{p_0^{3/2}} \left[1 + \mathrm{const} + O(p_0^{-1})\right]. \tag{8.4.20}$$

Now, the (microcanonical) entropy of the isolated horizon is given by

$$S_{IH} \equiv \log \dim \mathcal{H}_{\rm S},$$

as remarked earlier. For isolated horizons with large macroscopic area , the largest contribution to the *rhs* of Eq.(8.4.16) is given by the contribution of the single term of the multiple sum, corresponding to $j_i = 1/2 \forall i$ and $p = p_0$. This contribution dominates all others in the multiple sum, so that, one has, using Eq.(8.4.20), the microcanonical entropy formula [12]-[16]

$$S_{IH} = S_{\rm MC} = S_{\rm BH} + \delta_Q S_{\rm MC}, \tag{8.4.21}$$

where,

$$S_{\rm BH} \equiv \frac{A_H}{4 l_P^2}$$

is the Bekenstein-Hawking Area Law (BHAL), and we have set the Barbero-Immirzi parameter $\gamma = \log 2/\pi\sqrt{3}$ [8] in order to reproduce the BHAL with the correct normalization. $\delta_Q S_{\rm MC}$, given by

$$\delta_Q S_{\mathrm{MC}} = -\frac{3}{2} \log S_{\mathrm{BH}} + \mathrm{const} + O(S_{\mathrm{BH}}^{-1}), \tag{8.4.22}$$

constitutes an infinite series (in decreasing powers of S_{BH}) of corrections to the BHAL due to quantum fluctuations of spacetime, and can be thought of as 'finite size' corrections. One important aspect of the formula (8.4.21) is that the coefficient of each correction term is finite and unambiguously calculable, after γ has been fixed as mentioned.

8.4.4 It from Bit

Now consider the same object within the very loose lattice structure considered above. We can think of a two dimensional 'floating' (as opposed to a rigid) lattice basically covering the sphere, as shown in Fig. 8.4.

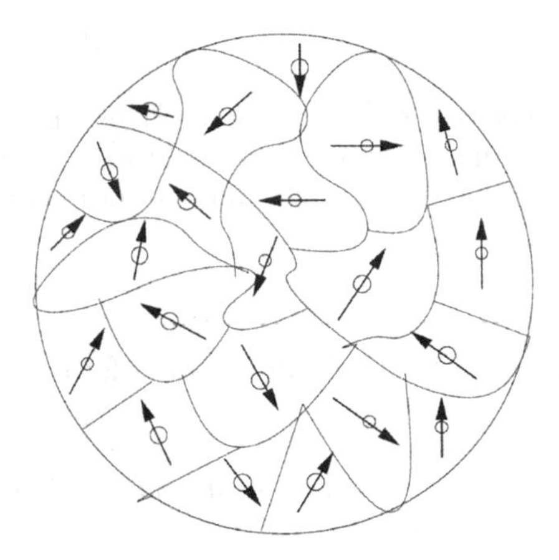

Fig. 8.4. It from Bit

Since the area of a tiny 'plaquette' of this lattice may be taken to be Planck area ℓ_P^2, it stands to reason that the ratio of the macroscopic area of the event horizon to the area of an elementary plaquette $\mathcal{A}/\ell_P^2 \gg 1$. This latter inequality defines our notion of a macroscopic black hole; our treatment in this review will focus on such black holes, and not to those very interesting but difficult cases where this number is $O(1)$.

Let us now place, following [28], binary variables ('bits') on the lattice sites (or equivalently, at the centre of the plaquettes). Then the number of such variables $p \equiv \xi \mathcal{A}/\ell_P^2 \gg 1$, where $\xi = O(1)$. Without any loss of generality, p can be taken to be an even integer. Now, we can think of the two values of each binary variable as characterising two quantum states, so that the size of the 'Hilbert space' of states on the (latticized) horizon is $\mathcal{N}_{\mathrm{bh}} = 2^p$. The entropy corresponding to this system is defined as $S_{\mathrm{bh}} \equiv \log \mathcal{N}_{\mathrm{bh}}$; choosing the constant $\xi = (4 \log 2)^{-1}$, we obtain

$$S_{\rm bh} = \frac{\mathcal{A}}{\ell_P^2} = S_{\rm BH}. \tag{8.4.23}$$

The generality of the above scenario makes it appealing vis-a-vis a quantum theory of black holes in particular and of quantum gravity in general. There is however one crucial aspect of any quantum approach to black hole physics which seems to have been missed in the above, – the aspect of symmetry. Indeed, the mere random distribution of spin 1/2 (binary) variables on the lattice which approximates the black hole horizon, without regard to possible symmetries, possibly leads to a far bigger space of states than the *physical* Hilbert space, and hence to an overcounting of the number of the degrees of freedom, i.e., a larger entropy.

8.4.5 The physical Hilbert (sub)space

But what is the most plausible symmetry that one can impose on states so as to identify the physical subspace? The elementary variables are binary or spin 1/2 variables which can be considered to be standard spin 1/2 variables under spatial rotations (more precisely $SU(2)$ doublets). Recall now that at every point on a curved spacetime one can erect a local Lorentz frame where the basic variables can be subjected to a local Lorentz transformation. Here, of course, we are interested in a spatial slice of the curved spacetime, and so the transformation of interest is a 'local spatial rotation'. Thus, the symmetry that one would like to impose on the degrees of freedom obtained so far would be invariance under these local spatial rotations in three dimensions. However, since one is dealing with black holes of very large area, this amounts to considering 'global' or 'rigid' spatial rotations or $SU(2)$ transformations. On very general grounds then, the most natural symmetry of the physical subspace must be this $SU(2)$ [16].

One is thus led to a symmetry criterion which defines the physical Hilbert space $\mathcal{H}_S$ of horizon states contributing to black hole entropy: *$\mathcal{H}_S$ consists of states that are compositions of elementary $SU(2)$ doublet states with vanishing total spin ($SU(2)$ singlets).* Observe that this criterion has no allusions whatsoever to any specific proposal for a quantum theory of gravitation. Nor does it involve any gauge redundancies (or any other infinite dimensional symmetry like conformal invariance) at this point. It is the most natural choice for the symmetry of physical horizon states simply because in the 'It from bit' picture, the basic variables are spin 1/2 variables. Later on we shall show however that this symmetry arises very naturally in the Non-perturbative Quantum GR approach known also as Quantum Geometry. It will emerge from that approach that horizon states of large macroscopic black holes are best described in terms of spin 1/2 variables at the punctures of a punctured two-sphere which represents (a spatial slice of) the event horizon.

The criterion of $SU(2)$ invariance leads to a simple way of counting the dimensionality of the physical Hilbert space [16]. For p variables, this number is given by

$$\dim \mathcal{H}_{\mathrm{S}} \equiv \mathcal{N}(p) = \binom{p}{p/2} - \binom{p}{(p/2-1)} \tag{8.4.24}$$

There is a simple intuitive way to understand the result embodied in (8.4.24). This formula counts the number of ways of making $SU(2)$ singlets from p spin 1/2 representations. The first term corresponds to the number of states with net J_3 quantum number $m = 0$ constructed by placing $m = \pm 1/2$ on the punctures. However, this term by itself *overcounts* the number of $SU(2)$ singlet states, because even non-singlet states (with net integral spin, for p is an even integer) have a net $m = 0$ sector. Beside having a sector with total $m = 0$, states with net integer spin have, of course, a sector with overall $m = \pm 1$ as well. The second term basically eliminates these non-singlet states with $m = 0$, by counting the number of states with net $m = \pm 1$ constructed from $m = \pm 1/2$ on the p sites. The difference then is the net number of $SU(2)$ singlet states that represents the dimensionality of $\mathcal{H}_{\mathrm{S}}$.

It may be pointed out that the first term in (8.4.24) also has another interpretation. It counts the number of ways binary variables corresponding to spin-up and spin-down can be placed on the sites to yield a vanishing total spin. Alternatively, one can think of the binary variables as unit positive and negative $U(1)$ charges; the first term in (8.4.24) then corresponds to the dimensionality of the Hilbert space of $U(1)$ *invariant* states. As already shown in [15], this corresponds to a *binomial* rather than a random distribution of binary variables.

In the limit of very large p, one can evaluate the factorials in (8.4.24) using the Stirling approximation. One obtains

$$\mathcal{N}(p) \approx \frac{2^p}{p^{3/2}}. \tag{8.4.25}$$

Clearly, the dimensionality of the physical Hilbert space is smaller than what one had earlier, as would be an obvious consequence of imposing $SU(2)$ symmetry. Using the relation between p and the classical horizon area A_{S} discussed in the last section, with the constant ξ chosen to take the same value as in that section, (8.4.25) can be shown [15] to lead to the following formula for black hole entropy,

$$S_{\mathrm{bh}} \equiv \log \mathcal{N}(p) \approx \frac{\mathcal{A}}{4\ell_P^2} - \frac{3}{2}\log\left(\frac{\mathcal{A}}{4\ell_P^2}\right) + \text{const} + O(\mathcal{A}^{-1}). \tag{8.4.26}$$

The logarithmic correction to the BHAL is not unexpected if we think of $S_{\mathrm{bh}}(\mathcal{A})$ as a power series for large $\mathcal{A}$ with $\mathcal{A}/4$ as the leading term; indeed, various approaches to computation of black hole entropy (like the Euclidean path integral [32], Non-perturbative Quantum GR [5], [11], boundary conformal field theory [13], and so on) have been used, and a general result like

$$S_{\mathrm{bh}}(\mathcal{A}) = \left(\frac{\mathcal{A}}{4\ell_P^2}\right) + C\log\left(\frac{\mathcal{A}}{4\ell_P^2}\right), \tag{8.4.27}$$

has been found, with various values of C, both positive and negative. In some of the perturbative approaches, there is an added constant $C' \sim \log(\Lambda)$ where Λ is a length cut-off needed to yield a finite result for S_{bh} [14]. This is quite in contrast to our result (8.4.26) where S_{bh} is *intrinsically finite.* Note also that according to the Second Law of Black Hole Mechanics, if two black holes coalesce, the minimum area of the resultant black hole is the sum of the two horizon areas. For such a coalescence, it is easy to see that, for $C > 0$,

$$S_{\mathrm{bh}}(\mathcal{A}_1 + \mathcal{A}_2) < S_{\mathrm{bh}}(\mathcal{A}_1) + S_{\mathrm{bh}}(\mathcal{A}_2). \tag{8.4.28}$$

Assuming isolated eternal black holes which coalesce adiabatically with no emission of gravitational waves, this property is perhaps not too desirable from the point of view of the Second Law of Thermodynamics. From this point of view also our result $C = -3/2$ appears more preferable. This is precisely the result that was obtained earlier from Non-perturbative Quantum GR (also called Quantum Geometry) [11] on the basis of incipient contributions in ref. [5].

Even when the resultant black hole has a horizon area larger than the sum of the areas, this preference for $C < 0$ seems to hold, although in a slightly weaker form. Also, from the theory of isolated horizons which incorporates radiation present in the vicinity of the horizon without crossing it, this result seems to have a greater appeal than those others with $C > 0$.

8.5 Canonical Entropy

It is obvious that the foregoing analysis does not deal with even a semirealistic situation, since all black holes (but for a set of measure zero of *extremal* black holes) undergo Hawking radiation which creates a thermal bath with which the black hole is in thermal equilibrium. One must therefore address the issue of a *canonical* entropy of the black hole, presumably based upon a canonical, rather than a microcanonical ensemble. Thus, in addition to the effects of quantum spacetime fluctuations that preserve the area of an isolated horizon and yield a microcanonincal entropy for it, we expect additional effects arising out of thermal fluctuations of the area itself to arise in the canonical entropy.

8.5.1 Canonical partition function: holography?

Following [16], we start with the canonical partition function in the quantum case

$$Z_C(\beta) = \mathrm{Tr}(\exp -\beta \widehat{H}). \tag{8.5.1}$$

Recall that in classical general relativity in the Hamiltonian formulation, the bulk Hamiltonian is a first class constraint, so that the entire Hamiltonian consists of the boundary contribution H_S on the constraint surface. In the quantum domain, the Hamiltonian operator can be written as

$$\widehat{H} = \widehat{H}_V + \widehat{H}_S, \tag{8.5.2}$$

with the subscripts V and S signifying bulk and boundary terms respectively. The Hamiltonian constraint is then implemented by requiring

$$\widehat{H}_V|\psi\rangle_V = 0 \tag{8.5.3}$$

for every physical state $|\psi\rangle_V$ in the bulk. Choose as basis for the Hamiltonian in (8.5.2) the states $|\psi\rangle_V \otimes |\chi\rangle_S$. This implies that the partition function may be factorized as

$$\begin{aligned} Z_C &\equiv \mathrm{Tr}\exp -\beta\widehat{H} \\ &= \underbrace{\dim \mathcal{H}_V}_{\text{indep of } \beta} \underbrace{\mathrm{Tr}_S \exp -\beta\widehat{H}_S}_{\text{boundary}} \end{aligned} \tag{8.5.4}$$

Thus, the relevance of the bulk physics seems rather limited due to the constraint (8.5.3). The partition function further reduces to

$$Z_C(\beta) = \dim\, \mathcal{H}_V Z_S(\beta), \tag{8.5.5}$$

where $\mathcal{H}_V$ is the space of bulk states $|\psi\rangle$ and Z_S is the 'boundary' partition function given by

$$Z_S(\beta) = \mathrm{Tr}_S \exp -\beta\widehat{H}_S. \tag{8.5.6}$$

Since we are considering situations where, in addition to the boundary at asymptopia, there is also an inner boundary at the black hole horizon, quantum fluctuations of this boundary lead to black hole thermodynamics. The factorization in Eq.(8.5.5) manifests in the canonical entropy as the appearance of an additive constant proportional to dim $\mathcal{H}_V$. Since thermodynamic entropy is defined only upto an additive constant, we may argue that the bulk states do not play any role in black hole thermodynamics. This may be thought of as the origin of a weaker version of the holographic hypothesis [29].

For our purpose, it is more convenient to rewrite (8.5.6) as

$$Z_C(\beta) = \sum_{n\in\mathcal{Z}} \underbrace{g\left(E_S(\mathcal{A}(n))\right)}_{\text{degeneracy}} \exp -\beta E_S(\mathcal{A}(n)), \tag{8.5.7}$$

where, we have made the assumptions that (a) the energy is a function of the area of the horizon $\mathcal{A}$ and used the result proved earlier that this area is quantized.

Going back to Eq. (8.5.7), we can now rewrite the partition function as an integral, using the Poisson resummation formula

$$\sum_{n=-\infty}^{\infty} f(n) = \sum_{m=-\infty}^{\infty} \int_{-\infty}^{\infty} dx \exp(-2\pi i m x) f(x). \tag{8.5.8}$$

For macroscopically large horizon areas $\mathcal{A}(p)$, $x >> 1$, so that the summation on the *rhs* of (8.5.8) is dominated by the contribution of the $m = 0$ term. In this approximation, we have

$$\begin{aligned} Z_C &\simeq \int_{-\infty}^{\infty} dx\, g(E(A(x))) \exp -\beta E(A(x)) \\ &= \int dE \exp\left[S_{\mathrm{MC}}(E) - \log\left|\frac{dE}{dx}\right| - \beta E\right] \end{aligned} \tag{8.5.9}$$

where $S_{\mathrm{MC}} \equiv \log g(E)$ is the microcanonical entropy.

Now, in equilibrium statistical mechanics, there is an inherent ambiguity in the definition of the microcanonical entropy, since it may also be defined as $\tilde{S}_{\mathrm{MC}} \equiv \log \rho(E)$ where $\rho(E)$ is the density of states. The relation between these two definitions involves the 'Jacobian' factor $|dE/dx|^{-1}$

$$\tilde{S}_{\mathrm{MC}} = S_{\mathrm{MC}} - \log\left|\frac{dE}{dx}\right|. \tag{8.5.10}$$

Clearly, this ambiguity is irrelevant if all one is interested in is the leading order BHAL. However, if one is interested in logarithmic corrections to BHAL as we are, this difference is crucial and must be taken into account.

We next proceed to evaluate the partition function in Eq. (8.5.9) using the saddle point approximation around the point $E = M$ where M is to be identified with the (classical) mass of the boundary (horizon). Integrating over the Gaussian fluctuations around the saddle point, and dropping higher order terms, we get,

$$Z_C \simeq \exp\left\{S_{\mathrm{MC}}(M) - \beta M - \log\left|\frac{dE}{dx}\right|_{E=M}\right\}\left[\frac{\pi}{-S''_{\mathrm{MC}}(M)}\right]^{1/2} \tag{8.5.11}$$

Using $S_C = \log Z_C + \beta M$, we obtain for the canonical entropy S_C

$$S_C = S_{\mathrm{MC}}(M) \underbrace{- \frac{1}{2}\log(\Delta)}_{\delta_{th} S_C}, \tag{8.5.12}$$

where,

$$\Delta \equiv \frac{d^2 S_{\mathrm{MC}}}{dE^2}\left(\frac{dE}{dx}\right)^2\Bigg|_{E=M}. \tag{8.5.13}$$

Eq. (8.5.12) exhibits the equivalence of the microcanonical and canonical entropies, exactly as one expects when thermal fluctuation corrections are ignored. It is interesting that this now leads to the following canonical entropy for non-rotating black holes [34]

$$S_{can} = S_{\mathrm{MC}}(\mathcal{A}) - \frac{1}{2} \log \Delta, \tag{8.5.14}$$

where

$$\Delta \equiv [\mathcal{A}'(x)]^2 \left[S'_{\mathrm{MC}}(\mathcal{A}) \frac{M''(\mathcal{A})}{M'(\mathcal{A})} - S''_{\mathrm{MC}}(\mathcal{A}) \right] \tag{8.5.15}$$

Thus, the canonical entropy is expressed in terms of the microcanonical entropy for an average large horizon area, and the mass which is also a function of the area. Clearly, stable equilibrium ensues so long as $\Delta > 0$ [38].

Additional support for this condition can be gleaned by considering the thermal capacity of the system, using the standard relation

$$C(\mathcal{A}) \equiv \frac{dM}{dT} = \frac{M'(\mathcal{A})}{T'(\mathcal{A})}, \tag{8.5.16}$$

with T being derived from the microcanonical entropy $S_{\mathrm{MC}}(\mathcal{A})$, and hence a function of $\mathcal{A}$. One obtains for the heat capacity the relation

$$C(\mathcal{A}) = \left[\frac{M'(\mathcal{A})}{T(\mathcal{A})\mathcal{A}'(x)} \right]^2 \Delta^{-1}, \tag{8.5.17}$$

so that $C > 0$ if only if $\Delta > 0$. Since the positivity of the heat capacity is certainly a necessary condition for stable thermal equilibrium, it is gratifying that an identical criterion emerges for Δ as found from the canonical entropy (8.5.14).

Using now eq, (8.5.13) for the expression for Δ, the criterion for thermal stability of non-rotating macroscopic black holes is then easily seen to be

$$M(\mathcal{A}) > S_{\mathrm{MC}}(\mathcal{A}) \tag{8.5.18}$$

as already mentioned in the summary. We have been using units in which $G = \hbar = c = k_B = 1$. If we revert back to units where these constants are not set to unity, the lower bound Eq. (8.5.18) can be re-expressed as

$$M(\mathcal{A}) > \left(\frac{\hbar c}{G k_B^2} \right)^{1/2} S_{\mathrm{MC}}(\mathcal{A}). \tag{8.5.19}$$

We remind the reader that in contrast to semiclassical approaches based on specific properties of classical metrics, our approach incorporates crucially the microcanonical entropy generated by quantum spacetime fluctuations that leave the horizon area constant. Apart from the plausible assumption of the black hole mass being dependent only on the horizon area, no other assumption has been made to arrive at the result. Even so, it subsumes most results based on the semiclassical approach. It also supercedes our earlier assay [34] based on an assumption of a power law functional dependence of the mass on the area.

As a byproduct of the above analysis, the canonical entropy for stable black holes can be expressed in terms of the Bekenstein-Hawking entropy S_{BH} as

$$S_{\mathrm{can}} = S_{\mathrm{BH}} - \frac{1}{2}(\xi - 1)\log S_{\mathrm{BH}} - \frac{1}{2}\log\left[\frac{S'_{\mathrm{MC}}(\mathcal{A})\, M''(\mathcal{A})}{S''_{\mathrm{MC}}(\mathcal{A})M'(\mathcal{A})}\right]. \quad (8.5.20)$$

For any smooth $M(\mathcal{A})$, one can truncate its power series expansion in $\mathcal{A}$ at some large order and show that the quantity in square brackets in Eq. (8.5.20) does not contribute to the log(area) term, so that

$$S_{\mathrm{can}} = S_{\mathrm{BH}} - \frac{1}{2}(\xi - 1)\log S_{\mathrm{BH}} + \mathrm{const} + O(S_{\mathrm{BH}}^{-1}). \quad (8.5.21)$$

The interplay between constant area quantum spacetime fluctuations and thermal fluctuations is obvious in the coefficient of the *log(area)* term where the contribution due to each appears with a specific sign. It is not surprising that the thermal fluctuation contribution increases the canonical entropy. The cancellation occurring for horizons on which a residual $U(1)$ subgroup of $SU(2)$ survives, because of additional gauge fixing by the boundary conditions describing an isolated horizon [7], may indicate a possible non-renormalization theorem, although no special symmetry like supersymmetry has been employed anywhere above. It is thus generic for all non-rotating black holes, including those with electric or dilatonic charge. One would expect the result to hold also for rotating black holes, as well, although the details of the microcanonical entropy for such black holes have not yet been worked out.

8.5.2 Canonical entropy of anti de Sitter black holes

The corrections to the canonical entropy due to thermal fluctuations can be calculated in principle for all isolated horizons which includes all stationary black holes. We shall first deal with the case of adS black holes where the calculation makes sense for a certain range of parameters of the black hole solution. Computation of such corrections has been performed in ref. [17]. Here, we recount the computation in a slightly different form, and compare the result with the corrections to the BHAL due to quantum spacetime fluctuations.

BTZ

The non-rotating BTZ metric is given by [26]

$$ds^2 = -\left(\frac{r^2}{\ell^2} - 8G_3M\right)dt^2 + \left(\frac{r^2}{\ell^2} - 8G_3M\right)^{-1} dr^2 + r^2 d\phi^2, \quad (8.5.22)$$

where, $\ell^2 \equiv -1/\Lambda^2$ and Λ is the cosmological constant. The BH entropy is

$$S_{\mathrm{BH}} = \frac{\pi r_H}{2G_3},$$

where, the horizon radius $r_H = \sqrt{8G_3M\ell}$. Quantum spacetime fluctuations produce corrections to the microcanonical BHAL, given for $r_H \gg \ell$ by [27]

$$\delta_Q S_{\mathrm{MC}} = -\frac{3}{2}\log S_{\mathrm{BH}}. \tag{8.5.23}$$

Using (8.5.23), and identifying the mass M of the black hole with the equilibrium energy E_0, the microcanonical entropy S_{MC} has the properties

$$\begin{aligned} S''_{\mathrm{MC}}(M) < 0 \qquad & \text{for } r_H > \ell \\ S''_{\mathrm{MC}}(M) > 0 \qquad & \text{for } r_H < \ell. \end{aligned} \tag{8.5.24}$$

Alternatively, the specific heat of the BTZ black hole is positive, so long as $r_H \geq \ell$. The system can therefore be thought of as being in equilibrium for parameters in this range. It follows that the calculation of $\delta_F S$ yields a sensible result in this range,

$$\delta_F S = \frac{3}{2}\log S_{\mathrm{BH}} = -\delta_Q S_{\mathrm{MC}}. \tag{8.5.25}$$

The import of this for the canonical entropy is rather intriguing, using Eq.(8.5.21)

$$S_C = S_{\mathrm{BH}}. \tag{8.5.26}$$

The quantum corrections to the BHAL in this case are cancelled by corrections due to thermal fluctuations of the area (mass) of the black hole horizon. We do not know the complete significance of this result yet.[3]

4 dimensional adS Schwarzschild

Such black holes have the metric

$$ds^2 = -V(r)\,dt^2 + V(r)^{-1}dr^2 + r^2 d\Omega^2, \tag{8.5.27}$$

where,

$$V(r) = 1 - \frac{2GM}{r} + \frac{r^2}{\ell^2}, \tag{8.5.28}$$

with $\ell^2 \equiv -3/\Lambda$. The horizon area $A_H = 4\pi r_H^2$, where the Schwarzschild radius obeys the cubic $V(r_H) = 0$. It is easy to see that the cubic yields the mass-area relation

[3] We should mention that this result ensues only if one takes recourse to the classical relation between the horizon area and the mass. The validity of that relation in the domain in which the QGR calculation has been performed, is not obvious at this point.

$$M = \frac{1}{2G} \left(\frac{A_H}{4\pi} \right)^{1/2} \left(1 + \frac{A_H}{4\pi \ell^2} \right) \tag{8.5.29}$$

It is clear from Eq.(8.5.29) that

$$S''_{\rm MC}(M) < 0 \qquad \text{for } A_H > \frac{4}{3}\pi \ell^2, \tag{8.5.30}$$

so that, once again the specific heat is positive in this range. The thermal fluctuation contribution for this parameter range is

$$\delta_F S = \log \left(\frac{A_H}{l_P^2} \right). \tag{8.5.31}$$

The net effect on the canonical entropy is a partial cancellation of the effects due to quantum spacetime fluctuations and thermal fluctuations,

$$S_C = S_{\rm BH} - \frac{1}{2} \log S_{\rm BH}. \tag{8.5.32}$$

Note that the thermal and quantum fluctuation effects compete with each other in both cases considered above, with the net result that the canonical entropy is still superadditive

$$S_C(A_1 + A_2) \geq S_C(A_1) + S_C(A_2).$$

The point $r_H \sim \ell$ in parameter space signifies the breakdown of thermal equilibrium; this point has been identified with the so-called Hawking-Page phase transition [18] from the black hole phase to a phase which has been called an 'adS gas'. In this latter phase, the black hole is supposed to have 'evaporated away', leaving behind a gas of massless particles in an asymptotically adS spacetime.

8.6 Conclusions

The exact cancellation of the logarithmic corrections to the BHAL for the BTZ black hole canonical entropy, due to quantum spacetime fluctuations and thermal fluctuations, may hold a deeper significance which warrants further analysis. In ref. [27], the microcanonical entropy is calculated from the exact Euclidian partition function of the $SU(2) \times SU(2)$ Chern Simons theory which describes the BTZ black hole. Perhaps this calculation can be extended to a finite temperature canonical quantum treatment of the problem, to include the effect of thermal fluctuations. Such an analysis is necessary to allay suspicions about using (semi)classical relations between the mass and the area of the black hole. Likewise, for four dimensional black holes, the precise relation admitted within QGR between the boundary Hamiltonian and the area operator needs to be ascertained. Presumably, this relation will not qualitatively change the results for adS black holes, although it might lead to a better understanding of the Hawking-Page phase transition. More importantly, this issue needs to be addressed in order to determine if the thermodynamic instability found for generic asymptotically flat black holes is an artifact of a semiclassical approach.

It is conceiveable that inclusion of charge and/or angular momentum for adS black holes in dimensions ≥ 4 will present no conceptual subtleties, so long as the (outer) horizon area exceeds in magnitude the inverse cosmological constant. However, to the best of our knowledge, the formulation of a higher dimensional (i.e., > 4) QGR has not been completed; this will have to be done before comparison of quantum and thermal fluctuation effects can be made in higher dimensions.

The thermodynamic instability discerned for asymptotically flat black holes also appears to emerge for de Sitter Schwarzschild black holes [17]. Since current observations appear to point enticingly to an asymptotically de Sitter universe, this instability must be better understood. In its present incarnation, it would imply that massive black holes will continue to get heavier *without limit.* On the other end of the scale, the instability can be interpreted in terms of disappearance of primordial black holes due to Hawking radiation, except for the possible existence of Planck scale remnants. It should be possible to estimate the density of such remnants and check with existing bounds from cosmological data.

Acknowledgement

We thank A. Ashtekar, R. Bhaduri, A. Chatterjee, S. Das, G. Date, A. Ghosh, R. Kaul, J. Samuel, S. SenGupta, S. Sinha and G.'t Hooft for many useful discussions.

References

[1] S. Chandrasekhar, *The Mathematical Theory of Black Holes*, Cambridge (1984).
[2] R. Wald, *General Relativity*, Chicago (1984).
[3] S. W. Hawking and G. F. R. Ellis, *The Large Scale Structure of Spacetime*, Cambridge (1973).
[4] R. Wald, *Quantum Field Theory in Curved Spacetime and Black Hole Thermodynamics*, Chicago (1992).
[5] A. Ashtekar, *Lectures on Non-perturbative Canonical Gravity.* World Scientific, Singapore (1991).
[6] R. Arnowitt, S. Deser and C. W. Misner, in *Gravitation: an introduction to current research*, ed. L. Witten (Wiley, New York, 1962).

[7] A. Ashtekar and J. Lewandowski, *Class. Quant. Grav.* 21, R53 (2004).
[8] A. Ashtekar, J. Baez, A. Corichi and K. Krasnov, *Adv. Theor. Math. Phys.* 4 (2000) 1 and references therein.
[9] J. D. Bekenstein, *Phys. Rev.* D7 (1973) 2333.
[10] S. W. Hawking, *Phys. Rev.* D13 (1976) 191.
[11] R. Kaul and P. Majumdar, *Phys. Lett.* B439 (1998) 267.
[12] R. Kaul and P. Majumdar, *Phys. Rev. Lett.* 84 (2000) 5255 (gr-qc/0002040).
[13] S. Carlip, *Class. Quant. Grav.* 7 (2000) 4175.
[14] R. B. Mann and S. Solodukhin, *Phys. Rev.* D55, 3622 (1997).
[15] S. Das, R. K. Kaul and P. Majumdar, *Phys. Rev.* D 63 (2001) 044019 (hep-th/0006211).
[16] P. Majumdar, *Pramana* 55 (2000) 511; hep-th/0011284, hep-th/0110198.
[17] S. Das, P. Majumdar and R. Bhaduri, *Class. Quant. Grav.* 19 (2002) 2355.
[18] S. W. Hawking and D. N. Page, Commun. Math. Phys. 87 (1983) 577.
[19] G. Immirzi, *Quantum Gravity and Regge Calculus*, gr-qc/9701052 and references therein.
[20] J. Bardeen, B. Carter and S. Hawking, *Comm. Math. Phys.* 31 (1973) 161.
[21] A. Ashtekar, C. Beetle and S. Fairhurst, *Class. Quant. Grav.* 17 (2000) 253; A. Ashtekar, C. Beetle and J. Lewandowski, *Class. Quant. Grav.* 19 (2000) 1195.
[22] A. Ashtekar, B. Krishnan, *Phys. Rev. Lett.* 89 (2002) 261101.
[23] A. Ashtekar, J. Lewandowski, Donald Marolf, J. Mourao, T. Thiemann, *J. Math. Phys.* 36 (1995) 6456.
[24] C. Rovelli and L. Smolin, *Nucl. Phys.* B442 (1995) 593.
[25] A. Ashtekar and J. Lewandowski, *Class. Quant. Grav.* 14 (1997) 55.
[26] M. Banados, C. Teitelboim, J. Zanelli, *Phys. Rev. Lett.* 69 (1992) 1849.
[27] T. R. Govindarajan, R. Kaul and S. Varadarajan, *Class. Quant. Grav.* 18 (2001) 2877
[28] J. A. Wheeler, *It from Bit*, in Sakharov Memorial Lectures, Vol. II, Nova (1992).
[29] G. 't Hooft, *Dimensional reduction in quantum gravity*, gr-qc/9310026.
[30] S. W. Hawking, *Nature* 248, 30-31 (1974).
[31] S. W. Hawking, *Commu. Math. Phys.*, 43, 199-222 (1975).
[32] S. W. Hawking and D. N. Page, *Commu. Math. Phys.* 87, 577 (1983).
[33] A. Chatterjee and P. Majumdar, ArXiv:hep-th/0303030.
[34] A. Chatterjee and P. Majumdar, *Phys. Rev. Lett.* 92, 141301 (2004).
[35] A. Chatterjee, and P. Majumdar, *Pramana*, 63, 851-858 (2004).
[36] T. Thiemann, *Phys. Lett.* B380 257-264 (1996).
[37] A. Chatterjee and P. Majumdar, *Phys. Rev.* D 71, 024003 (2005).
[38] A. Chatterjee and P. Majumdar, *Phys. Rev.* D 72, 044005 (2005).

9

Complexity in Organizations: A Paradigm Shift

Russ Marion

Department of Leadership, Clemson University, Clemson, South Carolina, USA
E-mail: Marion2@clemson.edu

Summary. Complexity theory is applied to organizational sciences in this paper. The implications of this application are significant, so much so that they signal a paradigm shift in the way we understand organization and leadership. Complexity theory alters core perceptions about the logic of organizational behavior and, consequently, "discovers" the significant importance of firms' informal social dynamics (informal behaviors have long been treated as something that should be suppressed or channeled). This altered perspective has implications for how we coordinate, motivate, and lead in firms. A complexity view of organizations is particularly useful and germane in light of recent movements among industrialized nations toward knowledge-based, rather than production-based, economies.

9.1 Complexity in Organizations: A Paradigm Shift

Complexity theory has emerged in the past twenty years as a dramatically new way to understand nature. It envisions adaptive systems (species, animals, plants, viruses, etc.) as neural-like interactive networks of agents and seeks to understand the dynamics of network behaviors, Langston [41], Marion and Uhl-Bien [49], Marion and Uhl-Bien [50], Miles et al. [55]. Complexity dynamics interact with natural selection processes to produce order (Kauffman [37]), but they otherwise differ from natural selection (the currently dominant theory of behaviors in nature) in rather significant ways. Complexity's causal logic, for example, is based on recursion (multi-way chains of cause and effect; Marion and Uhl-Bien [49]) rather than the linear, functional relationships of Darwinian logic (one-way temporal chains of causation, called process logic; MacIver [44]). New behaviors emerge seemingly unbidden and cannot typically be traced to simple input events such as mutations. Complex systems behave in quite complicated ways because of the nature of interdependent interactions, Reynolds [65], and they thrive from cooperation more than competition Margulis and Sagan [46]. Complex systems are probably best described as information processing systems because of their dual ability to store, yet dynamically process and change, knowledge.

R. Marion: *Complexity in Organizations: A Paradigm Shift*, StudFuzz **206**, 247–269 (2006)
www.springerlink.com

Complexity theory is becoming an important tool for describing formal social organizations such as businesses and government agencies (e.g., Brown and Eisenhardt [14], Carley and Hill [19], Dooley and de Ven [24], Gronn [28], Lichtenstein. and McKelvey [42], Marion and Uhl-Bien [48], McKelvey [52]). This paper seeks to extend these applications of complexity and to underscore how radically a complexity perspective of organization departs from traditional perspectives.

A secondary intent of this paper is to develop the science of complex organizations to a new level. Early works in this area tended to assume that organizations are uniformly complex. A few scholars, however, are beginning to argue that complexity is unevenly distributed across an organization and across time (C.Thomas et al. [20], Uhl-Bien et al. [73]). That is, certain organizational functions, strategic goals, or production processes may demand more complex structuring than others. Thus while the prime focus of this paper is on developing a complexity theory perspective of organizational behavior, it will do so in a manner consistent with the reality of uneven distribution and will further provide a framework within which unevenness can be understood.

9.2 A Paradigm Shift

Complexity theory first emerged in the biological and physical sciences and has migrated to the social sciences. It is not uncommon for theories from the sciences, particularly from biology, to be adopted to explain social behavior. The earliest known systematic theory of social behavior in the modern era appeared in the early to mid-1700s; called social physics, it was directly derived from Newtonian physics. In the late 19th and early 20th centuries, the prevailing understanding of social behavior was based on Spencer's work with Social Darwinism. Spencer's theory was discredited by the mid-20th century largely because of its misuse by German Nazis to justify genocide, but it reemerged in a more benign form in the 1980s and still enjoys currency in organizational theory. Open systems theory from the 1950s and '60s was grounded in biology; Contingency theories of the 1960s and 1970s were grounded in the same rational epistemology that underlay physics.

The explanation for this science-to-sociology phenomenon is quite straightforward: Since the period of the Enlightenment, science has defined the worldview of Western societies (and of many Eastern societies). By worldviews (or paradigms), I refer to the manner in which society perceives reality in all realms of its existence; thus scientific logic influences religion, politics, philosophy, perceptions of causality, our understanding of day-to-day events, and, topically, organizational behavior (Kuhn [40]). That is, the rational sciences tend to define reality for the rest of society.

According to Kuhn, science undergoes periodic "paradigm shifts," or sudden changes in perspectives regarding natural behaviors. Complexity theory represents just such a shift. Paradigm shifts produce dramatically new ways of understanding present reality; they allow us to "see" new realities that, in retrospect, were there all along but were ignored or were unseen. Paradigm shifts generate whole new sets of propositions about the world and often require new analytical tools to conduct the studies of those propositions.

The complexity paradigm is grounded in chaos theory thus it shares the uncertainty and nonlinearity of chaos — although in more muted form. A major difference between chaos and complexity (aside from complexity being less dynamic) is the fact that complex systems (and the agents that comprise them) are adaptive: They "intelligently" change their behavior and structure to adapt to environmental contingencies. Complex systems naturally (without volition) "seek" sufficient chaos to enable dynamicism (hence their ability to change, to process information, and to create unpredictable outcomes) and sufficient stability to enable change and information to be used and developed. This state, called the "edge of chaos" Langston [41], provides optimal fitness to a system thus is favored by natural selection, Kauffman [37]. At the edge, complex systems can process, alter, and store information with amazing effectiveness. They are capable of engaging in change yet the stable-chaotic balance of complexity minimizes the possibility of catastrophic, destructive change.

Adaptive systems accomplish this by aggregating into neural-like networks, or networks of adaptive agents. Aggregated neural networks are characterized by interactions among agents, by moderately coupled interdependency, and by tension (e.g., from predators or scarce resources). The nature of moderate coupling allows them to dynamically process information while absorbing perturbations that would disrupt less robust systems.

The paradigmatic implications of complexity theory for organizational theories are dramatic. Table 1 contrasts differences between the current worldview of organizational theory (beliefs, perceptions, accepted values and definitions, etc.) and the worldview offered by complexity theory. The differences are not simply matters of style, they get at the very heart of how we think about organization and leadership.

Complexity theory permits us to see organizational behavior in new ways, to restructure the role of leadership, and to envision new ways to organize, coordinate, and motivate workflows. This is possible because complexity changes the way we understand organizations: it alters our core paradigmatic focus.

9.3 Core Paradigmatic Focus

Complexity theory alters the basic logics that underlie perceptions of reality and in doing so it challenges the way we perceive organizational structures. The current paradigm of organization bears imprimatur of Newtonian and Humean arguments that science is defined by functional relationships among variables. Although this logical empiricist perspective was discredited by philosophers by 1977 (Suppe [70]), it is still evident in social research

	Currently dominant paradigm of organizational behaviour	Complexity theory perspective of organizational behaviour
Core paradigmatic focus	Top-down, convergent on leadership	Bottom-up, convergent on interactive dynamics
The function of organization	Organizations enable humans to efficiently produce useful outcomes on a large scale	Organizations enable humans to effectively create knowledge that can produce useful outcomes on a large scale
Structural requirements	Bureaucracy or commitment based unity	Bottom-up, complex organizations
Causation	(a) Linear, process theory	Nonlinear, recursive theory
	(b) Epistemology based on variables	Epistemology based on mechanisms and variables
	(c) Temporal flow worldview	Interaction worldview
Causal implication	(a) Outcomes are planned	Outcomes are emergent surprises
	(b) Leaders are causal stimulants	Leadership is an outcome
Motivation	Motivation by central structures (CEO's, bureaucratic rules etc)	Motivation by interactive dynamics
Vision	Unity of vision	Heterogeneous and indeterminate visions
Definition of leadership	Leaders are individuals who create organizational energy through charisma, intelligence, interpersonal consideration, inspiration etc.	Leadership is energy that emerges across the organization under given enabling conditions

Table 9.1. Old versus new paradigms of organizational behavior.

(Maxwell [51]) and in everyday behaviors — individuals intrinsically assume that events in their lives can be logically explained.

If this logic of simple causes is accepted as given, then prediction and control are likewise uncontestable. Knowledge of causal structures allows one to predictably control future events by adjusting current input variables. Accordingly, even complex events should be controllable by rather simple coordinating and planning structures. This is evident in assumptions about leadership. Meindl [54] argued that people glorify leaders as causal agents and that they invoke them as causal explanations across situations. That is, leadership is perceived as simple causal structures that can accurately and effectively plan and control organizational futures. The core paradigmatic focus derived from this is an assumption of top-down control and the controlling efficacy of leadership.

Complexity theory is premised on diametrically different assumptions. Outcomes derive from recursive interactions among numerous events (variables) and mechanisms (defined as ··· not a variable but an account of the makeup, behaviour and interrelationship of ··· [causal] processes, Pawson and Tilley [61], pp 67-68). Complexity theory does not envision causality in terms of stable, persistent relationships among variables, nor does it see outcome as a sequence of consecutive events (the logic that underlies natural selection). Rather it describes causality as an emergent, often nonlinear function of complex, neural-like interactions. Marion and Uhl-Bien [49] have labeled this, *recursion theory.*

When agents interact within organizations, they adapt their structures and behaviors to accommodate one another. When a recursively interacting network of agents is properly tuned (moderately coupled; Kauffman [37], Kauffman [38]), agents and ideas combine and recombine unpredictably and at times collapse together such that new structures or knowledge emerge from the dynamic process. Such recursive dynamics deny prediction and at times defy explanation; for this reason, complexity theorists seek to describe the mechanisms by which change and emergence occurs rather than seeking to predict outcomes. By understanding these mechanisms, social complexity theorists can devise strategies to enable complex behaviors that optimizing the capacity of organizations for innovation and knowledge production.

Recursive causality bears some similarity to process logic. Process logic argues that events follow events in a causal chain, and that the change from one event to the next is caused by random occurrences. Natural selection, for example, is based on process logic. Recursive causality differs in that its dynamics are not conceived as temporal-linear (one event leading to another). All events occur within the flow of time, of course, but complexity theory focuses on interactive, rather than on sequential, causality. Further, patterns of behavior in both process and recursive systems are non-replicable, but the dynamics in recursive system are also nonlinear (process systems are regular and progressive) — the scope of emergent events are not always proportionate to their inputs (to use a uninventive example, nonlinear emergence occurs when someone calmly says "fire" in a crowded room and produces explosive panic).

One important implication of the recursive logic for organizations is that futures are difficult to predict and plan, as in traditional models of organizational control; rather, they occur as emergent surprises. Popper [63] argued that outcomes are products of numerous variables; traditional planners simply choose from the set of possible predictors of a future outcome but can never identify all causal agents for that given outcome. Complexity theory argues that future outcomes are emergent products of nonlinear, highly complex interactions among variables and mechanisms, and that prediction of anything but the immediate future or the future of highly stable systems is likely unreliable. Complex organizations, then, do not engage in planning in the traditional sense, rather they engage in non-determinate visioning and mission-setting in order to foster emergent innovations.

The core paradigmatic focus of a complex perspective, then, envisions recursive, interactive dynamics as a key driving force in an organization. Interactive dynamics can occur at any given hierarchical level (and across levels); they are informal, unplanned and uncoordinated by positional authority. This perspective de-emphasizes the centrality of authority and emphasizes instead the core importance of effective network dynamics. It shifts the traditional perspective of organizations from top-down or centralized coordination to informal interaction among organizational agents at all hierarchical levels. Leadership's role becomes less to plan and coordinate and more to foster conditions that enable emergence and embedded coordination and motivation. This perspective does not deny authority, rather it adds a focus on informal networked interactions within organizations. Complexity represents, then, a radical shift from 20th century modes of thinking.

9.4 The Functions of Organization

The paradigmatic differences between traditional and complexity perspectives of organization parallel collateral and equally significant changes in the economic systems of industrialized nations. Indeed, the paradigm shifts represented by complexity theory are driven by these economic changes. These changes affects the very definition of organizational function.

Until the last couple of decades of the 20th century, the role of organizations in Western societies was rather clearly one of enabling humans to efficiently manufacture salable products on a large scale. Tangible goods were the essential organizational product of that era — goods such as textiles, TVs, and appliances (Boisot [12]). In that era, a core paradigmatic focus on top-down management functioned well, for commodity production is typically standardized and predictable.

The last decades of the 20th century witnessed dramatic economic changes, however (Drucker [25], Drucker [26], M.A.Hitt et al. [45], Hitt et al. [31]). Globalization intensified competition and made it far less necessary to produce commodities near major market outlets. Trade imbalances and consumer pressures for cheaper goods favored imports from 3rd world economies. Consequently, commodities production shifted from the major industrialized countries of the world to nations where labor was cheaper.

Concurrently, knowledge of electronics, nanotechnology, genetic engineering, communications, and pharmaceuticals (etc.) grew (and continues to grow) at exponential rates and created new demands from consumers. This movement has been labeled the knowledge economy by numerous organizational theorists, including Boisot [12], Burton-Jones [17], Drucker [26], Hitt et al. [31], Nonaka [57], and Nonaka and Nishiguchi [58], Cusumano [21]. A knowledge economy is one for which the principal resource is knowledge. Unlike commodity-based organizations, activities in knowledge organizations are unpredictable and non-standardized. The challenge in such organizations is to structure in ways that maximize individual and organizational learning rather than to structure for efficient production. Knowledge businesses have to rethink structural relationships among people and among organizations, work flows, and even the locus of work (knowledge organizations can benefit from a geographically dispersed work force).

Consequently, the purpose of current knowledge organizations is defined as one of enabling humans and networks of humans to effectively create knowledge that can produce useful outcomes. Knowledge and the applications of knowledge are the primary outcome of these organizations.

9.5 Organizational Control

One important implication of the shift from a commodity-based to knowledge based economy involves the nature of organizational control. Bureaucracy has been the dominant organizational control structure of the 20th century. Max Weber (1947) proposed an ideal bureaucratic model coordinated by impersonal rules, meritocracy, specialization, and role specification (see also Udy 1959). Bureaucracy enabled the coordination of large numbers of people, complex operations, and large scale planning; it fueled the significant industrial growth of the last century. Weber was concerned, however, that bureaucracy would enslave society in a rigid, dehumanizing structure. In his book, *Protestant Ethic and the Spirit of Capitalism* (Weber [75]) he wrote that once humanity embarked on the path of bureaucracy it would be imprisoned "perhaps until the last ton of fossilized coal is burnt" (181-182). He could envision no alternative to bureaucracy's rigid hierarchical structures for enabling the wealth and productivity that derives from the ability to organize on large scales.

In the late 20th century it became evident that organizations might be coordinated by internal (commitment- or trust-based management) rather than external (bureaucratic rules) restraints. "This new set of governance strategies ... is directed at access to and leverage of intangible resources like employee commitment, tacit knowledge and learning behaviours" (Bijlsma-Frankema and Koopman [10], Zaccaro and Klimoski [77]: 204). Smircich and G.Morgan [69] refer to this as management of meaning, defined as "'sense-making' on behalf of others and develop[ing] a social consensus around the resulting meaning" (Bryman [15]). It is this social consensus, rather than rules, that coordinate in this organizational strategy. Weber was incorrect then, and there appear to be alternatives to the "iron cage" of bureaucracy.

Complexity theory suggests yet another control mechanism, one in which coordination is built into network dynamics rather than implemented by managers via bureaucratic rules or by co-opting meaning. This represents a dramatic departure from the traditional assumption (in both bureaucracy and management of meaning) that coordination is the responsibility of leaders within a context of top-down authority. Rather, coordination is embedded into the structure and activities of the complex organization. Such coordination strategies enable maximum flexibility and the capacity to respond effectively to highly volatile environments — a strategy that is ideal in knowledge-based economies.

Coordination derives from several generalized sources in complex organizations. First it derives from a key characteristic of such systems: *moderately coupled interdependency*. Interdependency exists when one agent's fitness is influenced by the fitness-seeking actions of another agent (see Uhl-Bien et al. [73]) for discussion of fitness). When the actions of one agent conflict with the fitness preferences of another, a conflicting constraint exists. Such constraints pressure the agents to act to resolve the issues involved.

Networks of agents are characterized by networks of conflicting constraints, thus the problems of constraints are more complex than those experienced by a pair of agents. Notably, the constraint-accommodating actions of an actor affect constraints in other loci of the network. If a network is tightly coupled, changes by one agent affects a large number of other agents and the resulting network of constraints will be too complex to be resolved. Such systems tend to freeze, for change is too destructive (Kauffman [37]). If the network is loosely coupled, then the network of constraints is too weak to stimulate change in the system. A moderately coupled system is sufficiently tight to stimulate network elaboration but sufficiently loose to enable reasonable resolution of the pattern of constraints (Kauffman [37]). Returning to the issue of coordination, the network of constraints in a moderately coupled system impose sufficient pressure on agents to coordinate their actions but not so much pressure that creativity is constrained.

Interdependency and coordination are further enabled by *enabling rules*, or rules that define action boundaries without limiting creativity. They differ from more traditional bureaucratic rules in that bureaucracy tends to limit behavior to carefully rationalized procedures. Enabling rules, by contrast, expand more than restrict behavioral opportunities. Microsoft, for example, performs its highly complex tasks with independent work groups responsible for programming separate elements of a given software program. Such loosely coupled structuring enables innovation and rapid response to new technologies, but presents significant problems of coordination. Microsoft overcomes this with a key enabling rule that creates interdependency and shifts the structure from loose to moderately coupled: At regular intervals, each programming group must compile its code and run it against every other group's code. If a group's code does not function properly, then that group stays at their desks until the problem is fixed (Cusumano [21]). Such rules enable cross-group coordination, create cascading adjustments to the overall project (problems/solutions by one group may create problems/solutions for others), and expands the scope of responsibility for each work group — key attributes of a complex network.

Coordination derives from *vision and mission.* Vision is a projection of the future; mission is "product, or solution, oriented to be executed within certain defined parameters" (Mumford et al. [56]). Inappropriate vision and mission can inhibit innovation in complex organizations, thus these definitions must be understood in the context of complexity theory. Complex vision envisions an *indeterminate* future, or a future that is unconstrained by current beliefs or understanding. Indeterminate visions rally action and create perceptual boundaries without limiting the future. Such visions typically anticipate future activities or behaviors rather than projecting predetermined outcomes: For example, a vision that anticipates future creativity (an activity) is indeterminate because it does not anticipate a definable outcome; a vision that projects the future state of an existing technology is a determinate vision. The former enables innovation (which by definition cannot be preordained) and creative knowledge growth while the latter simple unfolds what is already known. (Uhl-Bien et al. [73])

Mumford's definition of mission is likewise limiting. They argue, correctly, that mission encourages innovation but they offer contradictory definitions that limit that ability. Mission, they propose, contributes to creativity and innovation in four ways. Missions have clear objectives, they provide a structure for addressing problems, they provide a framework for idea development that does not unduly restrict the autonomy and potential unique contributions of team members, and they provide a framework for sense-making (Mumford et al. [56]). In complexity theory, the strictures regarding visional limitations of the future apply equally to mission, thus Mumford's requirement for clear objectives is problematic. Objectives seek to pre-define innovative outcomes and as such are inconsistent with the fact that innovation is an event that diverges from current understanding (as Popper [63], stated, we cannot anticipate today what we shall know only tomorrow). Complex objectives are sensitive to such constraint in the same manner that complexity theory is sensitive about the nature of vision.

9.5.1 Top-Down Versus Bottom-Up Coordination

The discussion thus far has not addressed the unequal distribution of complexity that was alluded to in the opening paragraphs of this paper. Unequal distribution translates into varying coordination strategies across an organization and across time. In its fullest reality, coordination in knowledge-oriented organizations is complicated — there is no simple, single way to understand it. Like everything else about complex organizations, coordination structures are dynamic, meshed networks of formal and informal hierarchies, formal and informal rules, and personal influence relationships. Coordination can emanate from numerous sources, and top leadership is only one source of coordination (and may not be the most important source).

I propose, however, that patterns of coordination, from wherever it emanates, are heavily influenced by three interrelated forces: technological, managerial, and interactive. *Technology* refers to what needs to be done; it is the core production imperative of the system. The systems we describe here produce knowledge; other systems produce more tangible commodities. Profits are subsumed within technology since the two are closely linked. Without them, the organization does not typically survive.

The *managerial* function is to be understood from two perspectives: its responsibilities and its personalities. Its responsibilities are broadly to foster effective functioning by employees and to manage routine tasks (what J.A.Kelso [35], calls — referencing the brain — motor functions). Routine tasks include such things as budgeting and purchasing; fostering effectiveness involves maximizing the capacity of employees in their performance of technological functions (this is the traditional focus of leadership studies). Secondly, the managerial function is to be understood in terms of personalities and individual preferences, particularly at upper echelons. Much has been written about power preferences (Jermier [36], Kinchloe and McLaren [39], Oakes et al. [59]) and legitimacy preferences (DiMaggio and Powell [22], Dirsmith and MacIver [23], Scott [66], Selznick [68], Westphal et al. [76]) of leaders, for example; such preferences shape structures and behaviors of organizations.

The *interactive* dimension of organization is the unique contribution of complexity theory. Interaction occurs at all hierarchical levels and refers to the bottom-up, unpredictable activities of agents within and without the organization. It is related to what Philip Selznick [67] called irrational forces, George Homans [32], the informal group, and Charles Perrow [62], the unstable environment. Interactive dynamics influence all social-based functions of an organization; they are a source of attitudes, movements, knowledge and creativity. Complexity describes how these dynamics generate negentropic energy, hence emergent structure and innovation. Interactive dynamics self-produce their own coordinating protocols (e.g., Homan's informal group rules) and require no external intervention to do so. Interactive dynamics have absorbed organizational practitioners and scholars alike for over a century, and much effort has been expended attempting to subdue this "creature." A complexity theory of organization seeks ways to take advantage of the dynamicism and creativity of this force rather than fighting it.

Each of these forces — technology, managerial, and interactive — influence the nature of coordination in any given organization. Importantly, however, the forces are intermeshed: Each is shaped to some degree by the other two. For example, legitimacy preferences of management shape the organizational structure within which production occurs and from which interactive dynamics emerge, but both production and interaction likewise limit the scope of managerial discretion regarding legitimacy (see, for example, Westphal et al. [76]).

Any one of the forces can exert itself over the others; management, for example, can seek to dominate interaction (indeed this was the "golden ring" of leadership theory in the twentieth century), or interaction can dominate management (as in mob behavior). However, complexity theory suggests that organizational effectiveness is optimized when the three functions are appropriately harmonized, and that the nature of effective balance changes over time. Management forces function best when they work with interactive dynamics in ways consistent with the demands of technology, and the nature of this balance influences the particular admixture of coordinating structures (top-down versus bottom-up versus technology driven). For example, Dell Computer® was driven by market considerations (technology) to take profits in 2005; consequently it de-emphasized bottom-up creativity (interactive dynamics) and emphasized efficiency (top-down controls). Dell may very well find that, after a few years, it will need to swing the balance back in favor of interaction dynamic and consequent innovative behaviors.

The interactive force in this organizational triptych produces optimal innovation and creativity. The central question, then, for knowledge organizations (or the knowledge components of organizations) that desire such traits is, simply, how is it accomplished and how is coordination achieved without compromising creativity? In the previous section, we presented behaviors and actions that enable innovation and knowledge growth (interaction, interdependency, enabling rules, indeterminate vision, etc.). The answer to these questions then is straight forward: leaders foster these characteristics. The following paragraphs explain by examining the role of leadership in enabling and coordinating complex organizations.

9.5.2 The Role of Leadership in Coordinating Complex Organizations

To address the role of leadership in fostering effective application of interactive forces, we first ask: Why do complex social systems even need top-down leadership? Ant colonies (the mascot of the American school of complexity) are amazingly complex without board rooms and long-range plans. Birds manage to flock without Gantt charts. Among humans, mass movements can emerge without benefit of planning committees. Complex systems have the capacity to coordinate themselves without input from a coordinating authority.

There are several possible answers. First, the activities of firms are far more complex than are the activities of ants and birds, and such complexity may be dependent on intelligent interventions, Lichtenstein. and McKelvey [43]. Second, organizational behaviors often depend on not just adaptability, but *creative* adaptability, and leadership may help enable this. Third, humans are free-will agents; their work behaviors are not controlled by genetic dispositions, thus humans require the organizing and coordinating actions of managers to accomplish the sort of motivation that ants and birds accomplish instinctively. Thus complexity in human systems is more complex that is that for instinctual systems, and that added complexity may contribute to an enhanced capacity for creativity and productivity.

So leadership is important in human endeavors. The predilection of leaders, however, is to impose their wills and to coordinate according to their personal mental schema — to subdue complexity. For the last 100 years (starting with Frederick Taylor), our textbooks have taught that direction and leadership are synonymous. Complexity science redirects this perspective, and asks instead: How can leaders coordinate complex organizations in a manner that does not impede interactive dynamics?

At the most basic level, leaders of complexity work with followers to build organizations that permit emergence of complex dynamics and internal coordinating structures. They structure the organization to encourage interaction, and they structure relationships and rules to foster interdependency. The illustration at Microsoft has been discussed. Some organizations have structured workspaces without office to encourage interaction, or have implemented a virtual workspace to encourage interaction that is not bound by geography. Leaders allocate budgets to encourage entrepreneurialship within the firm. They find ways to permit organizational dynamics to solve coordination problems (such as scheduling, task allocation, and organizing) rather than micromanaging their solution, see Bonabeau and Meyer [13].

Leaders of complexity act as advocates and symbols that define the functions and perceptual boundaries of a movement; Mahatma Gandhi, for example, defined the boundaries of the Indian movement against colonialism relative to peaceful resistance. They coordinate by serving as rallying focal points. They spark growth and innovation. They are agents that bond groups and enable interactive and interdependent dynamics. Stuart Kauffman [37] refers to such functions as "catalysts;" he proposes that catalysts are created by a movement and are not themselves creators of movements.

Leaders of complexity coordinate by performing what J.A.Kelso [35] and Uhl-Bien et al. [73] refer to metaphorically as the motor functions of an organization. Just as the brain functions to keep the heart and lungs operating, leaders (in their managerial roles) serve to control underlying organizational functions (budgeting and accounting, purchasing, hiring, etc.) required to keep an organization functioning.

Positional leaders are the agents by which organizational strategy — the balance between technological, managerial, and interactive forces — is changed. Thus, leaders may signal a need to shift organizational fitness strategies if they perceive current strategies are moving out of sync with the environment. They may, for example, perceive a need to shift into an efficiency, profit-taking mode thus consolidating the advances of a previous, knowledge-producing (complexity) mode (what C.Thomas et al. [20], and Uhl-Bien et al. [73], call oscillation). Boal and Hooijberg [11] argue that this requires that leaders develop wisdom, or sound judgment about when change is needed.

Finally, drawing from the European perspective of complexity, top-down leaders manipulate certain variables — what Haken [29] calls control parameters — in order to force organizations into a complex state (Prigogine [64]'s, far-from-equilibrium). Complex states possess such levels of pent up energy that phase transition, or new order, is inevitable. Thus leaders from the European perspective are agents, progenitors, in fostering complexity.

9.6 Emergence

Emergence is a function of the nature of recursive causation in complex systems (recursive logic was introduced earlier as a core paradigmatic feature of complexity). According to the recursion perspective, events emerge from complex interactive dynamics involving neural-like networks of adaptive agents. That is, emergent events are products of unpredictable combinations and recombinations among interdependent agents. Networked, interdependent interactions pressure agents to adapt to shifting constraint landscapes (see earlier discussion of conflicting constraints) by elaborating their structures (Kauffman [37]). Emergence occurs when sets of agents, pressured and coordinated by constraints and visions and rules, begin to "resonate in sync" with one another. Emergence is the appearance of mob behavior after a local sports team defeats a rival, or the "implosion" of various technologies in 1975 from which the microcomputer emerged (Anderson [1], Marion [47]). Emergence is unpredictable and nonlinearly related to its input causes. The logic of recursion focuses on the mechanisms of emergence rather than relationships among variables, thus research in complexity examine such things as the changing nature of complex systems, Parunak et al. [60], or the effect of perturbations on a complex network, Carley and Hill [19].

The European perspective of emergence is a bit different. As noted earlier and only briefly revisited here, tension pushes a system away from equilibrium; at a point that Prigogine [64] has called far-from-equilibrium, the tension is dissipated by phase transition, or emergence.

There is a more subtle, and important difference between the American and European schools, however. From the American perspective, emergences is largely a function of internal dynamics and no external work is required to make it happen (Darwin's environmental selection is an example of external work — selection by an environment — performed on internal dynamics — mutations — to produce new structure). The European perspective proposes that external agents create tension in a system via control parameters; thus this perspective requires some degree of work to produce emergent events. Applied to organizations, the European perspective is an argument for direct input by formal authority while the American perspective calls for more indirect input. That is, the American school perspective proposes that leaders create conditions in which emergence can occur (interdependency, rules, etc); the European perspective proposes that leaders exert work to create tension and thus enable emergence.

I propose a blended model in which leaders both create tension and create enabling conditions. Both hierarchical authority and interactive dynamics are realities of social organization, and to limit social action to one or the other would be to deny the full capacity of human organizations for creativity and productivity. Human networks can indeed produce order for free, but they have the advantage of intelligent actions by formal leaders. The trick, as stated in several ways in this paper, is to make the two dynamics, interaction and hierarchy, work together to create useful complex dynamics.

9.7 Motivation

Coordination, productivity, and motivation are three major challenges for businesses. From the complexity perspective, all three assume a vastly different complexion than one normally finds in the leadership literature. This is attributable primarily to the interactive nature of complexity. In traditional perspectives, the top-down leader is responsible for these functions; from the complexity perspective, they are built in to the interactive structure.

Motivation is a function of some of the same dynamics that contribute to coordination. Motivation, for example, is a function of networks of interdependent relationships in which individuals are responsible to each other for productivity. That is, complex systems are structured such that agents pressure one another to produce. Performance pressure is enhanced by enabling rules that require interdependent actions (e.g., Microsoft's rule that programmers must run their code together periodically and correct any problems; Cusumano [21].

Motivation is enhanced by adaptive tension, or pressure to act (Haken [29], McKelvey [52]). Tension, discussed earlier as a role of leaders of complexity, pushes an organization to a far-from-equilibrium state where phase transitions occur to release pressures. One of the more famous examples of adaptive tension was perpetuated by Jack Welch, who told his employees at GE that they had to be number one or number two in their respective fields or be canceled, Tichy and Sherman [71]. Adaptive tension in complex organizations specifies an indeterminate rather than determinate future state so that phase transitions can produce creative outcomes.

Individuals or groups can be motivated by allocating resources to follow entrepreneurial behavior rather than just allocating resources to fund given functions (Bonabeau and Meyer [13]). Any given organization must fund core functions, of course; complexity theory suggests that extra resources be allocated to functions/departments which demonstrate exceptional activity. The principle is simple; the organization supports behaviors that exhibit motivated initiative.

There is a bit of a trap in this strategy that must be finessed, however. Arthur [2], see also Marion [47], argues that success begets further success: Growth is product of increasing returns, and growth shuts out competition (contrary to free market philosophy). Dominating ideas offer stability, effectiveness, efficiency, and profits, thus growth dynamics should be encouraged by any given organization. At the same time, complex organizations must also foster competitive ideas, and a policy of allocating resources only to successful programs inhibits development of new ideas. If funding follows exceptional *activity* rather than merely following success, this problem is ameliorated. That is, complex organizations support entrepreneurial behaviors in addition to supporting successful strategies.

Motivation is fostered by transformational leadership strategies (Avolio and Bass [3], Avolio and Bass [4]. Bass and Avolio [7], J.A.Conger [34], Hunt [33]). Transformational leadership theory proposes that the leader looks for "potential motives in followers, seeks to satisfy higher needs, and engages the full person of the follower" (Burns [16]: 4), thus transforming followers into self-motivated "leaders" and creating a culture of organizational effectiveness (Bass [6]). The goal is to transform the worker's sense of meaning to one that is productive for the organization (Bryman [15]) — to lead followers in such a way that they willingly embrace the goals of the organization. Transformational leadership is "predicated upon the inner dynamics of a freely embraced change of heart in the realm of core values and motivation" (Bass and Steidlmeier [8]: 192). Transformational leaders accomplish this with charisma, individualized consideration, inspiration, and intellectual stimulation, Bass and Avolio [7].

Complexity theory adds an important caveat, however: Complexity leadership accomplishes transformation with heterogeneous, rather than unitary, vision. Transformational leadership focuses on centralized visions of the future. Bass [6], a central figure in the transformational leadership movement, argued that "The transforming leader provides followers with a cause around which they can rally" (p. 467). Avolio et al. [5] add that "one of the authentic leader's [a recent mutation of transformational leadership] core challenges is to identify followers' strengths ··· while linking them to a common purpose or mission" (p. 806). Berson and Avolio (2004) are even more explicit:

"A core responsibility for organizational leaders is to direct followers towards achieving organizational purposes by articulating the organization's mission, vision, strategy, and goals, Zaccaro and Klimoski [77]. Leaders at all levels are responsible for the dissemination of strategic organizational goals, as well as for convincing their constituents to effectively implement those goals. Canella and Monroe [18] indicated that transformational leaders form relationships with followers that may make it easier for them to disseminate and implement strategic goals" (p. 626).

Complexity theory seeks to enable and enhance creativity, and the centralized visioning that is at the core of transformational leadership inhibits this important function. Core visions limit the expression of creativity to the capacity of the leaders that generate the vision. As McKelvey, paraphrasing Yaneer Bar Yam, argued when a central individual (or vision) creates direction for an organization, that organization can be no more creative, no more intelligent, than that one individual/vision is capable of being (McKelvey et al. [53]).

Complex organizations instead embrace heterogeneous vision. Heterogeneous vision refers to diverse visions and aspirations of multiple agents. Heterogeneous visions interact interdependently across a neural network of sometime conflicting, sometime congruous visions. Such interactions pressure agents to elaborate their visions and to form vision alliances. Networks of interacting visions adapt and change as organizational knowledge and environmental contexts change and develop. Unlike centralized, leader-determined visions, networked, heterogeneous visions are dynamic, organically growing, thus such networks can adjust quickly to hyper-turbulent environments. Importantly, these dynamics enable and produce knowledge and creativity much more robustly than do centralized, top-down structures.

9.8 Definition of Leadership

All this begs the question, "What is leadership?" The definitive or final definition of leadership doesn't exist, for definitions are dependent on one's perspective of what leaders do and how organizations function — that is, it is dependent on one's paradigm of organization. Bryman [15] identifies two general categories of leadership definitions: The first defines leadership in terms of influencing workers to achieve organizational goals (management of influence); the second defines leadership relative to focusing worker sense of meaning (e.g., transformational leadership). Both influence and meaning models tend to orient toward top-down influence over interpersonal relationships, and involve creating a "will" or "energy' to perform in productive ways. Consequently, traditional leadership definitions can be summarized as the creation of organizational energy (productive activity) through charisma, intelligence, interpersonal consideration, inspiration, force, authority, etc. The process of creating such energy is a top-down process that is vested heavily in hierarchical authorities.

The complexity paradigm suggests a dramatically different definitional category that can be labeled, *management of emergence*. Positional leadership in complex organizations focuses more on creating conditions that enable emergence of distributed leadership than it does on directing worker behaviors (Uhl-Bien et al. [73]). Positional leaders accomplish this by maneuvering structures, organizational patterns, enabling rules, tension, and motivation in order to foster interactive sources of *energy*. By fostering energy, a second level of leadership can emerge, one that is dispersed across authority levels. "Energy" is defined as inter-influence behaviors that contribute to or foster knowledge and creativity. For example, when two individuals interact over divergent ideas and create new understanding or knowledge, then an instance of distributed, or interactive, leadership occurs. These "energic bits" interact with other "energic bits" within complex networks of interdependent interaction to create emergent organizational knowledge and innovation.

Leadership, then, occurs at two levels in complex organizations: at an enabling level conducted largely by agents in positional roles, and as a behavior that emerges among individuals and groups across the organization. Consequently, complexity leadership is defined as (a) actions that foster conditions which enable complex dynamics, and (b) the energy-expanding activities that emerge across the organization under given enabling conditions.

9.9 Complexity versus Top-Down?

Complexity would seem to pose an alternative to formal structured organization. Indeed it is tempting to conceptualize elements of social dynamics as top-down versus bottom-up opposites — formal structure versus informal complexity in organizations, Western authority structures versus Eastern Taoism, MS Window's closed structure versus Linux's open structure, standing armies versus guerilla warfare. But such dichotomies over-simplify reality. Both dynamics — top-down and bottom-up — coexist because both are necessities. They are interdependent, often embedded within one another (executive groups are simultaneously authority-oriented and interactive), and inter-supportive (e.g., top-down structures in organizations support complexity by managing routine tasks and organizing spaces and work processes to enable interactive dynamics).

The relative dominance of top-down and bottom-up dynamics in organizations is dependent on technological and managerial forces. Managerial forces include personal preferences of authorities (managers who are driven to be in control will tend toward top-down). Technological forces are related to what needs to be accomplished. Generally, knowledge oriented processes demand a more bottom-up structure while commodity-based processes require a more top-down structure.

More general questions of social complexity (Windows versus Linux, democracy versus totalitarianism, structured militaries versus guerilla forces) are evaluated in a similar manner. The nature of a given phenomenon is heavily dependent on contextual conditions — the beliefs and dispositions of the people involved, the available resources, and the nature of tasks to be performed. The strongest strategy is a context-appropriate mix of complexity and top-down coordination. Based on this, we can tentatively address "versus" questions. Will democracy be able to defeat terrorism? The answer will depend on which side is the more robust (adaptive, complex) and has the top-down capacity to enable and support that robust dynamic. Will Windows or Linux dominate the future of computing, or will democracy defeat totalitarianism? Again the same answer applies.

Complexity helps explain the processes by which a system optimizes its strategic fitness, but the very nature of dynamic systems precludes predictions about its future. One can say, however, that the more robust a system (characterized by both complexity and top-down coordination), the more likely it will succeed. As Ross Ashby put it in the late 1950s. it takes variety to defeat variety.

9.10 Conclusions

The core differences between traditional perspectives of organization and a complexity perspective can be summarized in terms of perspective. That is, both traditional and complexity theorists are looking at the same entity (the organization) but seeing different things. This is the essence of a paradigm shift: seeing phenomena from a dramatic new angle.

Complexity theorists metaphorically turn the organization over and observe dynamics occurring below the radar of perspectives that define organizations in terms of CEOs, strategic planning, and power. Thus traditional theorists perceive unstructured social elements that need to be structured by top-down, intelligent leadership; complexity theorists perceive the value of those unstructured social elements for producing innovation and knowledge. Traditional theorists devise strategies for upper echelon authorities (Hambrick and Mason [30]); complexity theorists devise strategies to capitalize on interactive dynamics. Traditional theorists seek to plan the future; complexity theorists seek unplanned, creative futures.

Organizations are neural networks that can produce knowledge and innovations if effectively led and organized. Complexity offers tools for managing the globalized, hyper-turbulent economy of the 21st century. It offers viable alternatives to bureaucracies. It offers new ways of understanding leadership. Complexity is not the do-all of organizational science and theorists in the field must carefully avoid the implication that this, at last, is the unified field theory of social dynamics. This science does, however, expand understanding of organizational behavior in a rather significant way.

Acknowledgments

I would like to acknowledge the invaluable comments of Mary Uhl-Bien in the "emergence" of this paper.

References

[1] Anderson, P. (1995) Microcomputer manufacturers In *Organizations in Industry*, ed. G. Carroll and M. Hannan, Oxford University Press New York pp. 37–58.

[2] Arthur, W. (1989) The economy and complexity In *Lectures in the sciences of complexity*, ed. D. Stein, Vol. 1 Addison-Wesley Redwood City, CA pp. 713–740.

[3] Avolio, B. and Bass, B. M. (1987) Transformational leadership, charisma, and beyond In *Emerging leadership vistas*, ed. J. G. Hunt, H. Baliga, H. Dachler and C. Schriesheim, Heath Lexington, MA.

[4] Avolio, B. and Bass, B. M. (1998) Individual consideration viewed at multiple levels of analysis: A multi-level framework for examining the diffusion of transformational leadership In *Leadership: Multiple-level approaches: Contemporary and alternative*, ed. F. Dansereau and F. Yammarino, JAI Press Stamford, CT pp. 53–74.

[5] Avolio, B., Gardner, W., Walumbwa, F., Luthans, F. and May, D. (2004) Unlocking the mask: A look at the process by which authentic leaders impact follower attitudes and behaviors. *The Leadership Quarterly* **15**(6), 801–823.

[6] Bass, B. M. (1985) *Leadership and performance beyond expectations* Free Press New York.

[7] Bass, B. M. and Avolio, B. (1993) The implications of transactional and transformational leadership for individual, team, and organizational development. *Research in Organizational Change and Development* **4**, 231–272.

[8] Bass, B. M. and Steidlmeier, P. (1999) Ethics, character, and authenic transformational leadership behavior. *The Leadership Quarterly* **10**(2), 181–217.

[9] Berson, Y. and Avolio, B. (2004) Transformational leadership and the dissemination of organizational goals: A case study of a telecommunication firm. *The Leadership Quarterly* **15**(5), 625–646.

[10] Bijlsma-Frankema, K. and Koopman, P. (2004) The oxymoron of control in an era of globalisation: Vulnerabilities of a mega myth. *Journal of Managerial Psychology* **19**(3), 204–217.

[11] Boal, K. and Hooijberg, R. (2001) Strategic leadership research: Moving on. *The Leadership Quarterly* **11**(4), 515–549.

[12] Boisot, M. H. (1998) *Knowledge assets: Securing competitive advantage in the information economy* Oxford University Press Oxford.

[13] Bonabeau, E. and Meyer, C. (2001) Swarm intelligence: A whole new way to think about business. *Harvard Business Review* **79**(5), 107–114.

[14] Brown, S. L. and Eisenhardt, K. M. (1998) *Competing on the edge* Harvard Business School Press Boston, MA.

[15] Bryman, A. (1996) Leadership in organizations In *Handbook of Organization Studies*, ed. S. Clegg, C. Hardy and W. Nord, Sage Publications London pp. 276–292.

[16] Burns, J. M. (1978) *Leadership* Harper & Row New York.

[17] Burton-Jones, A. (1999) *Knowledge capitalism: Business, work and learning in the new economy* Oxford University Press Oxford.

[18] Canella, A. and Monroe, M. (1997) Contrasting perspectives on strategic leaders: Toward a more realistic view of top managers. *Journal of Management* **23**(3), 213–230.

[19] Carley, K. and Hill, V. (2001) Structural change and learning within organizations In *Dynamics of organizational societies*, ed. A. Lomi and E. Larsen, AAAI/MIT Press Cambridge, MA pp. 63–92.

[20] C.Thomas, Kaminska-Labbé, R. and McKelvey, B. (2003) Managerial problems from coevolving causalities: Unraveling entangled organizational dynamics. In 'INSEAD conference on Expanding Perspectives on Strategy Processes' Fontainebleau, France.

[21] Cusumano, M. (2001) Focusing creativity: Microsoft's "Synch and Stabilize" approach to software product development In *Knowledge emergence: Social, technical, and evolutionary dimensions of knowledge creation*, ed. I. Nonaka and T. Nishiguchi, Oxford University Press Oxford pp. 111–123.

[22] DiMaggio, P. and Powell, W. (1983) The iron cage revisited: Institutional isomorphism and collective rationality in organizational fields. *American Sociological Review* **48**, 147–160.

[23] Dirsmith, M. and MacIver, R. M. (1988) An institutional perspective on the rise, social transformation, and fall of a university budget category. *Administrative Science Quarterly* **33**, 562–587.

[24] Dooley, K. J. and de Ven, A. V. (n.d.) Explaining complex organizational dynamics. *Organization Science*.

[25] Drucker, P. (1993) *Post-capitalist society* Butterworth-Heinemann Oxford.

[26] Drucker, P. (1999) *Management Challenges for the 21st Century* HarperCollins New York.

[27] Drucker, P., Dorothy, L., Straus, S., Brown, J. and Garvin, D., eds (1998) *Harvard Business Review on Knowledge Management* Harvard Business School Publishing Boston.

[28] Gronn, P. (2002) Distributed leadership as a unit of analysis. *The Leadership Quarterly* **13**, 423–451.

[29] Haken, H. (1983) *Synergetics, an introduction* 3rd edn Springer-Verlag Berlin.

[30] Hambrick, D. C. and Mason, P. (1984) Upper echelons: The organization as a reflection of its top managers. *Academy of Management Review* **9**(2), 193–206.

[31] Hitt, M., Keats, B. and DeMarie, S. (1998) Navigating in the new competitive landscape: Building strategic flexibility and competitive advantage in the 21st century. *Academy of Management Executive* **12**(4), 22–42.

[32] Homans, G. C. (1950) *The human group* Harcourt, Brace & World New York.

[33] Hunt, J. G. (1999) Transformational/charismatic leadership's transformation of the field: An historical essay. *The Leadership Quarterly* **10**(2), 129–144.

[34] J.A.Conger (1999) Charismatic and transformational leadership in organizations: An insider's perspective on these developing streams of research. *The Leadership Quarterly* **10**(2), 145–179.

[35] J.A.Kelso (1995) *Dynamic patterns: The self-organization of brain and behavior* MIT Press Cambridge, MA.

[36] Jermier, J. M. (1998) Introduction: Critical perspectives on organizational control. *Administrative Science Quarterly* **43**(2), 235–256.

[37] Kauffman, S. A. (1993) *The origins of order* Oxford University Press New York.

[38] Kauffman, S. A. (1995) *At home in the universe: The search for the laws of self-organization and complexity* Oxford University Press New York.

[39] Kinchloe, J. and McLaren, P. (1994) Rethinking critical theory and qualitative research In *Handbook of Qualitative Reseach*, ed. N. Denzin and Y. Lincoln, Sage Publications Thousand Oaks, CA. pp. 279–314.

[40] Kuhn, T. (1970) *The structure of scientific revolutions* 2nd edn The University of Chicago Press Chicago. Loaner.

[41] Langston, C. (n.d.) Studying artificial life with cellular automata. *Physica.*

[42] Lichtenstein., B. and McKelvey, B. (2003) Four degrees of emergence: A typology of complexity and its implications for management. Working paper, University of Syracuse.

[43] Lichtenstein., B. and McKelvey, B. (2004) Toward a theory of emergence by stages: Complexity dynamics, self-organization, and power laws in firms.

[44] MacIver, R. (1964) *Social causation* Harper & Row New York.

[45] M.A.Hitt, Ireland, R. and Hoskisson, R. (1995) *Strategic management: Competitiveness and globalization* West St. Paul, MN.

[46] Margulis, L. and Sagan, D. (1986) *Microcosmos* Oxford University Press Oxford.

[47] Marion, R. (1999) *The edge of organization: Chaos and complexity theories of formal social organization* Sage Newbury Park, CA.

[48] Marion, R. and Uhl-Bien, M. (2001) Leadership in complex organizations. *The Leadership Quarterly* **12**, 389–418.

[49] Marion, R. and Uhl-Bien, M. (2003) Complexity theory and al-Qaeda: Examining complex leadership. *Emergence: A Journal of Complexity Issues in Organizations and Management* **5**, 56–78.

[50] Marion, R. and Uhl-Bien, M. (2005) A model of complex leadership. *Working Paper.*

[51] Maxwell, J. A. (2004) Causal explanation, qualitative research, and scientific inquiry in education. *Educational Researcher* **33**(2), 3–11.

[52] McKelvey, B. (2003) Emergent order in firms: Complexity science vs. the entanglement trap In *Complex systems and evolutionary perspectives on organizations*, ed. E.Mitleton-Kelly, Elsevier Science Amsterdam, NL pp. 99–125.

[53] McKelvey, B., Mintzberg, H., Petzinger, T., Prusak, L., Senge, P. and Shultz, R. (1999) The gurus speak: Complexity and organizations. *Emergence: A journal of complexity issues in organizations and management* **1**(1), 73–91.
[54] Meindl, J. R. (1990) On leader ship: An alternative to the conventional wisdom In *Research in organizational behavior*, ed. B. M. Staw and L. L.Cummings, Vol. 12 JAI Greenwich, CT pp. 159–204.
[55] Miles, R., Snow, C. C., Matthews, J. A. and Miles, G. (1999) Cellular-network organizations In *Twenty-first century economics*, ed. W. E. Halal and K. B. Taylor, Macmillan New York pp. 155–173.
[56] Mumford, M. D., Eubanks, D. L. and Murphy, S. T. (2005) Creating the conditions for success: Best practices in leading for innovation.
[57] Nonaka, I. (1994) A dynamic theory of organizational knowledge creation. *Organization Science* **5**(1), 14–37.
[58] Nonaka, I. and Nishiguchi, T., eds (2001) *Knowledge emergence: Social, technical, and evolutionary dimensions of knowledge creation* Oxford University Press Oxford.
[59] Oakes, L., Townley, B. and Cooper, D. J. (1998) Business planning as pedagogy: Language and control in a changing institutional field. *Administrative Science Quarterly* **43**(2), 257–292.
[60] Parunak, H. V. D., Savit, R. and L.Riolo, R. (1998) Agent-based modeling vs. equation-based modeling: A case study and users' guide. In *Workshop on Modeling Agent Based Systems*, ed. J. S. Sichman, G. Gilbert and R. Conte, Springer Paris.
[61] Pawson, R. and Tilley, N. (1997) *Realistic evaluation* Sage London.
[62] Perrow, C. (1970) *Organizational analysis: A sociological view* Wadsworth Publishing Company, Inc. Belmont, CA.
[63] Popper, K. (1986) *The poverty of historicism* Routledge London.
[64] Prigogine, I. (1997) *The end of certainty* The Free Press New York.
[65] Reynolds, C. (1987) Flocks, herds, and schools: A distributed behavioral model. *Computer Graphics* **21**, 25–32.
[66] Scott, W. R. (1987) The adolescence of institutional theory. *Administrative Science Quarterly* **32**(4), 493–511.
[67] Selznick, P. (1949) *TVA and the grass roots* University of California Press Berkeley, CA.
[68] Selznick, P. (1996) Institutionalism "Old" and "New". *Administrative Science Quarterly* **41**, 270–277.
[69] Smircich, L. and G.Morgan (1982) Leadership: The management of meaning. *Journal of Applied Behavioral Science* **18**, 257–273.
[70] Suppe, F., ed. (1977) *The Structure of Scientific Theories* 2nd edn University of Illinois Press Urbana.
[71] Tichy, N. M. and Sherman, S. (1994) *Control your destiny or someone else will* HarperCollins New York.
[72] Udy, S. (1959) "Bureaucracy" and "rationality" in Weber's organization theory: An empirical study. *American Sociological Review* **24**, 791–795.

[73] Uhl-Bien, M., Marion, R. and McKelvey, B. (2004) Complex Leadrship: Shifting Leadership from the Industrial Age to the Knowledge Era.

[74] Weber, M. (1947) *The theory of social and economic organization* Free Press Glencoe, IL.

[75] Weber, M. (1952) *Protestant ethic and the spirit of capitalism* Scribner New York.

[76] Westphal, J., Ranjay, G. and Shortell, S. (1997) Customization or conformity: An institutional and netork perspective on the content and consequences of TQM adoption. *Administrative Science Quarterly* **42**, 366–394.

[77] Zaccaro, S. and Klimoski, R. (2001) The nature of organizational leadership: An introduction In *The nature of organizational leadership*, ed. S. Zaccaro and R. Klimoski, Jossey-Bass San Francisco pp. 3–41.

10

Chaos, Nonlinearity, Complexity: A Unified Perspective

A. Sengupta

Department of Mechancal Engineering, Indian Institute of Technology Kanpur, Kanpur 208016, INDIA
E-Mail: `osegu@iitk.ac.in`

Summary. In this paper we employ the topological-multifunctional mathematical language and techniques of non-injective illposedness developed in [30] to formulate a notion of *ChaNoXity* — Chaos, Nonlinearity, Complexity — in describing the specifically nonlinear dynamical evolutionary processes of Nature. Non-bijective ill-posedness is the natural mode of expression for chanoxity that aims to focus on the nonlinear interactions generating dynamical evolution of real irreversible processes. The basic dynamics is considered to take place in a matter-negmatter (regulating matter, defined below) *kitchen space* $X \times \mathfrak{X}$ of Nature that is inaccessible to both the matter (X) and negmatter ($\mathfrak{X}$) components. These component spaces are distinguished by opposing evolutionary directional arrows and satisfy the defining property

$$(\forall A \subseteq X, \exists \mathfrak{A} \subseteq \mathfrak{X}) \, \text{s.t.} \, (A \cup \mathfrak{A} = \emptyset).$$

Dynamical equilibrium is considered to be represented by such *competitively collaborating* homeostatic states of the matter-negmatter constituents of Nature.

The reductionist approach to science today remains largely the dominant model. It fosters the detailed study of limited domains in individual subdisciplines within the vast tree of science. However, over the past 30 years or so, an alternative conceptual picture has emerged for the study of large areas of science which have been found to share many common conceptual features, regardless of the subdiscipline, be it physics, chemistry or biology. Self-organization and complexity are the watchwords for this new way of thinking about the collective behaviour of many basic but interacting units. In colloquial terms, we are talking about systems in which 'the whole is greater that the sum of parts'.

Complexity is the study of the behaviour of large collection of such simple, interacting units, endowed with the potential to evolve with time. The complex phenomena that emerge from the dynamical behaviour of these interacting units are referred to as self-organizing. More technically, self-organization is the spontaneous emergence of non-equilibrium structural reorganizations on a macroscopic level, due to the collective interactions

A. Sengupta: *Chaos, Nonlinearity, Complexity: A Unified Perspective*, StudFuzz **206**, 270–352 (2006)
`www.springerlink.com`

between a large number of (usually simple) microscopic objects. Such structural organizations may be of a spatial, temporal or spatio-temporal nature, and is thus an emergent property.

For self-organization to arise, a system needs to exhibit two properties: it must be both dissipative and nonlinear. Self-organization and complexity are essential scientific concepts for understanding integrated systems whether in physics, biology or engineering ⋯ with a much more 'holistic', yet equally rigorous, scientific perspective compared with the reductionist methods, and so provide new insights into many of the more intellectually challenging concepts, including the large-scale structure of the Universe, the origin and evolution of life on Earth (and more widely in the cosmos), consciousness, intelligence and language.

There is, therefore, a general and conceptual framework for the description of self-organizing phenomena, of a theoretical and essentially mathematical nature. This more or less boils down to the theory of nonlinear dissipative dynamical systems.

Coveney [8]

10.1 Introduction

A dissipative structure is an open, out-of-equilibrium, unstable system that maintains its form and structure by interacting with its environment through the exchange of energy, matter, and entropy, thereby inducing spontaneous evolutionary convergence to a complex, and possibly chaotic, equilibrated state. These systems maintain or increase their organization through exergy destruction in a locally reduced entropy state by increasing the entropy of the "global" environment of which they are a part. This paper applies the mathematical language and techniques of non-bijective, and in particular non-injective, ill-posedness and multifunctions introduced and developed in [30] to formulate an integrated approach to dissipative systems involving chaos, nonlinearity and complexity (ChaNoXity), where a complex system is understood to imply

- an assembly of many *interdependent parts*
- interacting with each other through *competitive nonlinear collaboration*
- leading to *self-organized, emergent* behaviour.[1]

[1] Competitive collaboration — as opposed to reductionism — in the context of this characterization is to be understood as follows: The interdependent parts retain their individual identities, with each contributing to the whole in its own characteristic fashion *within a framework of dynamically emerging global properties of the whole.* A comparison with reductionism as summarized in Fig. 10.10*c*, shows that although the properties of the whole are generated by the parts, the individual units acting independently on their own cannot account for the emergent global behaviour of the total.

We will show how each of these defining characteristics of complexity can be described and structured within the mathematical framework of our multifunctional graphical convergence of a net of functions (f_α). In this programme, convergence in topological spaces continues to be our principal tool, and the particular topologies of significance that emerge are the topology of saturated sets and the exclusion topology. We will demonstrate that a complex system can be described as an association of independent expert groups, each entrusted with a specific specialized task by a top-level coordinating command, that consolidates and regulates the inputs received from its different constituent units each working independently of the others within the global framework of the coordinating authority, by harmonizing and combining them into an emerging whole; thus the complexity of a system, broadly speaking, is the amount of information needed to describe it. In this task, and depending on the evolving complexity of the dynamics, the coordinating unit delegates its authority to subordinate units that report back to it the data collected at its own level of authority.

Recall that

(i) a multifunction — which constitutes one of the foundational notions of our work — and the non-injective function are related by

$$\begin{aligned} f \text{ is a non-injective function} &\Longleftrightarrow f^- \text{ is a multifunction} \\ f \text{ is a multifunction} &\Longleftrightarrow f^- \text{ is a non-injective function.} \end{aligned} \tag{10.1.1}$$

and

(ii) the neighbourhood of a point $x \in (X, \mathcal{U})$ — which is a generalization of the familiar notion of distances of metric spaces — is a nonempty subset N of X containing an open set $U \in \mathcal{U}$; thus $N \subseteq X$ is a neighbourhood of x iff $x \in U \subseteq N$ for some open set U of X. The collection of all neighbourhoods of x

$$\mathcal{N}_x \stackrel{\text{def}}{=} \{N \subseteq X : x \in U \subseteq N \text{ for some } U \in \mathcal{U}\} \tag{10.1.2}$$

is the neighbourhood system at x, and the subcollection U of $\mathcal{U}$ used in this expression constitutes a neighbourhood (local) base or basic neighbourhood system, at x. The properties

(N1) x belongs to every member N of $\mathcal{N}_x$,

(N2) The intersection of any two neighbourhoods of x is another neighbourhood of x: $N, M \in \mathcal{N}_x \Rightarrow N \cap M \in \mathcal{N}_x$,

(N3) Every superset of any neighbourhood of x is a neighbourhood of x: $(M \in \mathcal{N}_x) \wedge (M \subseteq N) \Rightarrow N \in \mathcal{N}_x$

characterize $\mathcal{N}_x$ completely and imply that a subset $G \subseteq (X, \mathcal{U})$ is open iff it is a neighbourhood of each of its points. Accordingly if $\mathcal{N}_x$ is an arbitrary collection of subsets of X associated with each $x \in X$ satisfying (N1) – (N3), then the special class of neighbourhoods G

$$\mathcal{U} = \{G \in \mathcal{N}_x : x \in B \subseteq G \text{ for some } B \in \mathcal{N}_x \text{ and each } x \in G\} \tag{10.1.3}$$

defines a unique topology on X containing a basic neighbourhood B at each of its points x for which the neighbourhood system is the prescribed collection $\mathcal{N}_x$. Among the three properties (N1) – (N3), the first two now re-expressed as

(NB1) x belongs to each member B of $\mathcal{B}_x$.

(NB2) The intersection of any two members of $\mathcal{B}_x$ contains another member of $\mathcal{B}_x$: $B_1, B_2 \in \mathcal{B}_x \Rightarrow (\exists B \in \mathcal{B}_x : B \subseteq B_1 \cap B_2)$.

are fundamental in the sense that the resulting subcollection $\mathcal{B}_x$ of $\mathcal{N}_x$ generates the full system by appealing to (N3). This basic neighbourhood system, or *local base*, at x in $(X, \mathcal{U})$ satisfies

$$\mathcal{B}_x \stackrel{\text{def}}{=} \{B \in \mathcal{N}_x : x \in B \subseteq N \text{ for each } N \in \mathcal{N}_x\} \tag{10.1.4}$$

which reciprocally determines the full neighbourhood system

$$\mathcal{N}_x = \{N \subseteq X : x \in B \subseteq N \text{ for some } B \in \mathcal{B}_x\} \tag{10.1.5}$$

as all the supersets of these basic elements.

The topology of saturated sets is defined in terms of equivalence classes $[x]_\sim = \{y \in X : y \sim x \in X\}$ generated by a relation $\sim$ on a set X; the neighbourhood system $\mathcal{N}_x$ of x in this topology consists of all supersets of the equivalence class $[x]_\sim \in X/\sim$. In the x-exclusion topology of all subsets of X that exclude x (plus X, of course), the neighbourhood system of x is just $\{X\}$. While the first topology provides, as in [30], the motive force for an evolutionary direction in time, the second will define a complementary *negative* space $\mathfrak{X}$ of (associated with, generated by) X, with an oppositely directed evolutionary arrow. With *dynamic equilibrium* representing a state of *homeostasis*[2] between the associated opposing motives of evolution, *equilibrium* will be taken to mark the end of a directional evolutionary process represented by convergence of the associated sequence to an adherence set.

[2] *Homeostasis* (Greek, *homoio-*: same, similar; *stasis*: a condition of balance among various forces, literally means "resistance to change") is the property of an open system to maintain its structure and functions by means of a multiplicity of dynamical equilibria rigorously controlled by interdependent regulation mechanisms. Homeostatic systems by opposing changes to maintain internal balance — with failure to do so eventually leading to its death and destruction — represent the action of negative feedbacks in sustaining a constant state of equilibrium by adjusting its physiological processes.

Examples: (a) Homeostasis is the fundamental defining character of a healthy living organism that allows it to function more efficiently by maintaining its internal environment within acceptable limits in competitive collaboration with its environment: the internal processes are regulated according to need. With respect to a parameter, an organism may maintain it at a constant level regardless of the environment, while others can allow the environment to determine its parameter through behavioral adaptations. It is the second type that is relevant for homeostasy. (b) The gravitational collapse of a cloud of interstellar matter raises its temperature until the nuclear fuel at the center ignites halting the collapse. The

Let $f: X \to Y$ be a function and $f^-: Y \rightrightarrows X$ its multi-inverse: hence $ff^-f = f$ and $f^-ff^- = f^-$ although $f^-f \neq \mathbf{1}_X$ and $ff^- \neq \mathbf{1}_Y$ necessarily. Some useful identities for subsets $A \subseteq X$ and $B \subseteq Y$ are shown in Table 10.1, where the complement of a subset $A \subseteq X$ is denoted by $A^c = \{x : (x \in X - A) \wedge (x \notin A)\}$. Let the *f-saturation* of A and the *f-component* of B on the image of f

$$\mathcal{S}_f(A) = f^- f(A)$$
$$\mathcal{C}_f(B) = ff^-(B) = B \bigcap f(X)$$

define generalizations of injective and surjective mappings in the sense that any f behaves one-one and onto on its saturated and component sets respectively; in particular f is injective iff $\mathcal{S}_f(A) = A$ for all subsets $A \subseteq X$ and surjective iff $B = \mathcal{C}_f(B)$ for all $B \subseteq Y$. It is possible therefore to replace each of the relevant assertions of Table 10.1 with the more direct injectivity and surjectivity conditions on f. Indeed

$$\begin{aligned} f(x) = y &\Longrightarrow f(f^- f(x)) = y = ff^-(y) \\ &\Longrightarrow f(\mathcal{S}_f(x)) = \mathcal{C}_f(y) \qquad (10.1.6a) \\ x = f^-(y) &\Longrightarrow f^- f(x) = [x] = f^-(ff^-(y)) \\ &\Longrightarrow \mathcal{S}_f(x) = f^-(\mathcal{C}_f(y)) \qquad (10.1.6b) \end{aligned}$$

demonstrate the bijectivity of $f : \mathcal{S}_f(x) \to \mathcal{C}_f(y)$; hence in the bijective inverse notation the corresponding functional equation takes the form

$$f(\mathcal{S}_f(A)) = \mathcal{C}_f(B) \Longleftrightarrow \mathcal{S}_f(A) = f^{-1}(\mathcal{C}_f(B)). \qquad (10.1.7)$$

This significant generalization of bijectivity of functions is noteworthy because our notion of chaos and complexity is based on ill-posedness of non-bijective functional equations, and one of the principal objectives of this work is to demonstrate that *the natural law of entropy increase is caused by the urge of the system $f(x) = y$ to impose an effective state of uniformity throughout X by the generation of saturated and component open sets.*

All statements of the first column of the table for saturated sets $A = \mathcal{S}_f(A)$ apply to the quotient map q; observe that $q(A^c) = (q(A))^c$. Moreover combining the respective entries of both the columns, it is easy to verify the following results for the saturation map $\mathcal{S}_f$ on saturated sets $A = \mathcal{S}_f(A)$.

(a) $\mathcal{S}_f(\cup A_i) = \cup \mathcal{S}_f(A_i)$: The union of saturated sets is saturated.

---

consequent thermal pressure gradient of expansion inhibits the dominant gravitational force of compression resulting in the birth of a star that is a state of dynamical equilibrium between these opposing forces.

Homeostasis, as the ability or tendency of an organism or cell to maintain internal equilibrium by adjusting its physiological processes, will be used in this work to denote a state of dynamical equilibrium among various forces acting on the system.

	$f: X \to Y$	$f^-: Y \rightarrow X$
1	$A_1 \subseteq A_2 \Rightarrow f(A_1) \subseteq f(A_2)$ $\Leftarrow$ iff $A = \mathcal{S}_f(A)$	$B_1 \subseteq B_2 \Rightarrow f^-(B_1) \subseteq f^-(B_2)$ $\Leftarrow$ iff $B = \mathcal{C}_f(B)$
2	$f(A) \subseteq B \Leftrightarrow A \subseteq f^-(B)$ $B \subseteq f(A) \Rightarrow f^-(B) \subseteq A$ iff $A = \mathcal{S}_f(A)$	$f(A) \subseteq B \Leftrightarrow A \subseteq f^-(B)$ $B \subseteq f(A) \Leftarrow f^-(B) \subseteq A$ iff $B = \mathcal{C}_f(B)$
3	$A = \emptyset \Leftrightarrow f(A) = \emptyset$	$f^-(\emptyset) = \emptyset$ $f^-(B) = \emptyset \Rightarrow B = \emptyset$ iff $B = \mathcal{C}_f(B)$
4	$f(A_1) \cap f(A_2) = \emptyset \Rightarrow$ $A_1 \cap A_2 = \emptyset \Leftarrow$ iff $A = \mathcal{S}_f(A)$	$f^-(B_1) \cap f^-(B_2) = \emptyset \Leftarrow$ $B_1 \cap B_2 = \emptyset \Rightarrow$ iff $B = \mathcal{C}_f(B)$
5	$f(\cup_\alpha A_\alpha) = \cup_\alpha f(A_\alpha)$	$f^-(\cup_\alpha B_\alpha) = \cup_\alpha f^-(B_\alpha)$
6	$f(\cap_\alpha A_\alpha) \subseteq \cap_\alpha f(A_\alpha)$, "=" iff $A = \mathcal{S}_f(A)$	$f^-(\cap_\alpha B_\alpha) = \cap_\alpha f^-(B_\alpha)$
7	$f(A^c) = (f(A))^c \cap f(X)$ iff $A = \mathcal{S}_f(A)$	$f^-(B^c) = ((f^-(B))^c$

Table 10.1. The role of saturated and component sets in a function and its inverse; here $A = \mathcal{S}_f(A)$ and $B = \mathcal{C}_f(B)$ are to be understood to hold for all subsets $A \subseteq X$ and $B \subseteq Y$, with the conditions ensuring that f is in fact injective and surjective respectively. Unlike f, f^- preserves the basic set operations in the sense of 5, 6, and 7. This makes f^- rather than f the ideal instrument for describing topological and measure theoretic properties like continuity and measurability of functions.

(b) $\mathcal{S}_f(\cap A_i) = \cap \mathcal{S}_f(A_i)$: The intersection of saturated sets is saturated.
(c) $X - \mathcal{S}_f(A) = \mathcal{S}_f(X - A)$: The complement of a saturated set is saturated.
(d) $A_1 \subseteq A_2 \Rightarrow \mathcal{S}_f(A_1) \subseteq \mathcal{S}_f(A_2)$
(e) $\mathcal{S}_f(\cap A_i) = \emptyset \Rightarrow \cap A_i = \emptyset$.

While properties (a) and (b) lead to the topology of saturated sets, the third makes it a complemented topology when the (closed) complement of an open set is also an open set. In this topology there are no boundaries between sets which are isolated in as far as a sequence eventually in one of them converging to points in the other is concerned.

As the guiding incentive for this work is an understanding of the precise role of irreversibility and nonlinearity in the dynamical evolution of (irreversible) real processes, we will propose an index of nonlinear irreversibility in the kitchen space $X \times \mathfrak{X}$ of Nature, wherein all the evolutionary dynamics are postulated to take place. The physical world X is only a projection of this multifaceted kitchen that is distinguished in having a complementary "negative" component $\mathfrak{X}$ interacting with X to generate the dynamical reality perceived in the later. This nonlinearity index, together with the dynamical synthesis of opposites between opposing directional arrows associated with X and its complementing negworld $\mathfrak{X}$, suggests a description of time's arrow that is specifically nonlinear with chaos and complexity being the prime manifestations of strongly nonlinear systems.

The entropy produced within a system due to irreversibilities within it [15] are generated by nonlinear dynamical interactions between the system and its negworld, and the objective of this paper is to clearly define this interaction and focus on its relevance in the dynamical evolution of Nature.

10.2 ChaNoXity: Chaos, Nonlinearity, Complexity

10.2.1 Entropy, Irreversibility, and Nonlinearity

Here we provide a summary of the "modern" approach to entropy — which is a measure of the molecular disorder of the system, generated as it does work: entropy relates the multiplicity associated with a state so that if one state can be achieved in more ways than another then it is more probable with a larger entropy — due to De Donder [15], incorporating explicitly irreversibility into the formalism of the Second Law of Thermodynamics[3] thereby making it unnecessary to consider ideal, non-physical, reversible processes for computing (changes in) entropy. This follows from the original Clausius inequality

$$dS \geq \frac{dQ}{T}$$

written as

$$dS = dS + d\mathbb{S}, \tag{10.2.1}$$

where dS is the change in the entropy of the system due to heat exchanged by it with its exterior and $d\mathbb{S}$, the "uncompensated heat" of Clausius, represents the entropy generated within the system from real irreversible processes occurring in it. Although $dS = dQ/T$ can be either positive or negative, $d\mathbb{S}$ must always be positive due to the irreversibilities produced in the system, implying that although entropy can either increase or decrease through energy transport across its boundary, the system can only generate and never

[3] The Second Law of Thermodynamics for non-equilibrium processes essentially requires the *exergy* (the maximum useful work that can be obtained from a system at a given state in a specified environment, Eq. 10.2.2) of *isolated* systems to be continuously degraded by diabatic *irreversible* processes that drive systems towards equilibrium by generating entropy, eventually leading to a dead equilibrium state of maximum entropy. Statistically, the equilibrium state is interpreted to represent the most probable state. For a closed system, entropy gives a quantitative measure of the amount of thermal energy not available to do work, that is of the amount of unavailable energy.

destroy entropy. In an isolated system since the energy exchange is zero, the entropy will continue to increase due to effective irreversibilities, and reach the maximum possible value leading to a steady state of dynamical equilibrium in which all (irreversible) processes must cease. When the system exchanges entropy with its surroundings, it is driven to an out-of-equilibrium state and entropy producing irreversibilities begin to operate leading to a more probable disordered state. The entropy flowing out of an *adiabatic* system must, by Eq. (10.2.1), be larger than that flowing into it with the difference being equal to the amount generated by the irreversibilities. The basic point, as will be elaborated in the following, is that dissipative systems in communion with its exterior utilize the exergy (or thermodynamic availability) to organize emerging structures within itself: for a system to be in a non-equilibrium steady state, $dS \leq 0$; hence $d\mathcal{S}$ must be negative of magnitude greater than or equal to $d\mathscr{S}$. The exergy

$$\mathcal{E} = (U - U_{\text{eq}}) + P_0(V - V_{\text{eq}}) - T_0(S - S_{\text{eq}}) - \sum_{j=1}^{J} \mu_{j,0}(N_j - N_{j,\text{eq}}) \quad (10.2.2)$$

of a system is a measure of its deviation from thermodynamic equilibrium with the environment, and represents the maximum capacity of energy to perform useful work as the system proceeds to equilibrium, with irreversibilities increasing its entropy at the expense of exergy; here $_{\text{eq}}$ marks the equilibrium state, and $_0$ represents the environment with which the system interacts.

In postulating the existence of an entropy function $S(U, V, N)$ of the extensive parameters U, V, and $\{N\}_{j=1}^{J}$ of the internal energy, volume, and mole numbers of the chemical constituents comprising a composite compound system that is defined for all equilibrium states, we follow Callen [4] in postulating that in the absence of internal constraints the extensive parameters assume such values that maximize S over all the constrained equilibrium states. The entropy of the composite system is additive over the constituent subsystems, and is continuous, differentiable, and increases monotonically with respect to the energy U. This last property implies that $S(U, V, N)$ can be inverted in $U(S, V, N)$; hence

$$\begin{aligned} dU(S, V, \{N_j\}) = \frac{\partial U(S, V, \{N_j\})}{\partial S} dS + \frac{\partial U(S, V, \{N_j\})}{\partial V} dV \\ + \sum_{j=1}^{J} \frac{\partial U(S, V, \{N_j\})}{\partial N_j} dN_j \quad (10.2.3) \end{aligned}$$

defines the *intensive parameters*

$$\frac{\partial U}{\partial S} \stackrel{\text{def}}{=} T(S, V, \{N_j\}_{j=1}^{J}), \qquad V,\ \{N_j\}\,\text{held const} \qquad (10.2.4)$$

$$\frac{\partial U}{\partial V} \stackrel{\text{def}}{=} -P(S, V, \{N_j\}_{j=1}^{J}), \qquad S,\ \{N_j\}\,\text{held const} \qquad (10.2.5)$$

$$\frac{\partial U}{\partial N_j} \stackrel{\text{def}}{=} \mu_j(S, V, \{N_j\}_{j=1}^J), \qquad S, V \text{ held const} \tag{10.2.6}$$

of absolute temperature T, pressure P, and chemical potential μ_j of the jth component, from the macroscopic extensive ones. Inversion of (10.2.3) gives the differential *Gibbs entropy* definition

$$dS(U, V, \{N_j\}) = \frac{1}{T(U, V, \{N_j\})} dU + \frac{P(U, V, \{N_j\})}{T(U, V, \{N_j\})} dV - \sum_{j=1}^{J} \frac{\mu_j(U, V, \{N_j\})}{T(U, V, \{N_j\})} dN_j \tag{10.2.7}$$

that provides an equivalent correspondence of the partial derivatives

$$(\partial S/\partial U)_{V,N_j} = \frac{1}{T(U, V, \{N_j\})} \tag{10.2.8}$$

$$(\partial S/\partial V)_{U,N_j} = \frac{P(U, V, \{N_j\})}{T} \tag{10.2.9}$$

$$(\partial S/\partial N_j)_{U,V} = -\frac{\sum_{j=1}^{J} \mu(U, V, \{N_j\})}{T} \tag{10.2.10}$$

with the intensive variables of the system.

In the spirit of the Pffafian differential form, dependence of the intensive variables of the First Law

$$\begin{aligned} dU(S, V, \{N_j\}) &= dQ(S, V, \{N_j\}) + dW(S, V, \{N_j\}) + dM(S, V, \{N_j\}) \\ &= dQ(S, V, \{N_j\}) - P(S, V, \{N_j\})\, dV \\ &\quad + \sum_{j=1}^{J} \mu_j(S, V, \{N_j\})\, dN_j \end{aligned} \tag{10.2.11}$$

on the respective extensive macroscopic variables U, V, or N_j serves to decouple the (possibly nonlinear) bonds between them; this is necessary and sufficient for the resultant thermodynamics to be classified as *quasi-static* or *reversible.* These ideal states as pointed out in [4] are simply an ordered class of equilibrium states, neutral with respect to time-reversal and without any specific directional property, distinguished from natural physical processes of ordered *temporal successions* of equilibrium and non-equilibrium states: *a quasi-static reversible process is a directionless collection of elements of an ordered set.*[4] From the definition (10.2.4) of the absolute temperature T it follows that *under quasi-static conditions,*

[4] The most comprehensive view of *irreversibility* follows from the notion of a *time-(a)symmetric theory* that requires the (non)existence of a backward process $\mathcal{P}_r :=$

$$dQ \stackrel{\text{def}}{=} T(S)\, dS \tag{10.2.12}$$

reduces the heat transfer dQ to formally behave work-like, permitting (10.2.11) to be expressed in the *combined first and second law* form

$$dU(S, V, \{N_j\}) = T(S)\, dS - P(V)\, dV + \sum_{j=1}^{J} \mu(N_j)\, dN_j \tag{10.2.13a}$$

$$dS(U, V, \{N_j\}) = \frac{1}{T(U)}\, dU + \frac{P(V)}{T(U)}\, dV - \sum_{j=1}^{J} \frac{\mu(N_j)}{T(U)}\, dN_j \tag{10.2.13b}$$

which are just the integrable *quasi-static versions* of Eqs. (10.2.3) and (10.2.7). Note that the total energy input and the corresponding entropy transfer in this quasi-static case reduces to a simple sum of the constituent parts of the change. For non quasi-static real processes, this linear superposition of the solution into its individual components is not justified as the resulting Pfaffian equation solves as the arbitrary $U(S, V, \{N_j\}) = \text{const}$. For any natural non-cyclic real process therefore, the identification

$$dQ(S, V, \{N_j\}) \stackrel{\text{def}}{=} T(S, V, \{N_j\})\, dS \tag{10.2.14}$$

reduces (10.2.3) to the first law form (10.2.11) for real processes that no longer decomposes into individual, non-interacting component parts like its quasi-static counterpart (10.2.13a). Equation (10.2.14) is graphically expressed [15] in the spirit of (10.2.1) as

$$\begin{aligned} dS &= \frac{dQ(S, V, \{N_j\})}{T(S, V, \{N_j\})} \\ &= \frac{d\mathcal{Q}(S, V, \{N_j\})}{T(S, V, \{N_j\})} + \frac{d\mathfrak{Q}(S, V, \{N_j\})}{T(S, V, \{N_j\})} \\ &= d\mathcal{S} + d\mathfrak{S}, \end{aligned} \tag{10.2.15}$$

where the total entropy exchange is expressed as a sum of two parts: the first

---

$\{r(-t) : -t_f \leq t \leq -t_i\}$ for every permissible forward process $\mathcal{P} := \{s(t) : t_i \leq t \leq t_f\}$ of the theory; here $r = Rs$ with $R^2 = \mathbf{1}$, is the time-reversal of state s. Although in contrast with mechanics thermodynamics has no equations of motion, the second law endows it with a time-asymmetric character and a thermodynamic process is irreversible iff its reverse $\mathcal{P}_r$ is not allowed by the theory, iff time-symmetry is broken in the sense that an irreversible process cannot be reversed without introducing some change in the surroundings, typically by work transforming to heat. Reversible processes are useful idealizations used to measure how well we are doing with real irreversible processes. Entropy change in the universe is a direct quantification of irreversibility indicating how far from ideal the system actually is: irreversibility is directly related to the lost opportunity of converting heat to work.

$$dS = \frac{dQ}{T} \gtrless 0 \tag{10.2.16}$$

may be positive, zero or negative depending on the specific nature of energy transfer dQ with the (infinite) exterior reservoir, but the second

$$d\mathcal{S} = \frac{d\mathcal{Q}}{T} \geq 0 \tag{10.2.17}$$

representing the entropy produced by irreversible nonlinear processes within the system is always positive. Equation (10.2.13b) for a composite body $C = A \cup B$ of two parts A and B, each interacting with its own infinite reservoir under the constraint $U = U_A + U_B$, $V = V_A + V_B$ and $N = N_A + N_B$, yields the Gibbs expression

$$dS_C(U,V,N) = \left[\frac{1}{T_A}\,dU_A + \frac{1}{T_B}\,dU_B\right] + \left[\frac{p_A}{T_A}\,dV_A + \frac{p_B}{T_B}\,dV_B\right] - \left[\frac{\mu_A}{T_A}\,dN_A + \frac{\mu_B}{T_B}\,dN_B\right] \tag{10.2.18}$$

for the entropy exchanged by C in reaching a state of *static equilibrium* with its infinite environment; here T, P and μ are the parameters of the reservoirs that completely determine the internal state of C. This exchange of energy with the surroundings perturbs the system from its state of equilibrium and sets up internal irreversible nonlinear processes between the two subsystems, driving C towards a new state of *dynamic equilibrium* that can be represented ([12], [15]) in terms of flows of extensive quantities set up by forces generated by the intensive variables. Thus for a composite *dynamically* interacting *compound system* C consisting of two chambers A and B of volumes V_A and V_B filled with two nonidentical gases at distinct temperatures, pressures, and mole numbers, the entropy generated by nonlinear irreversible processes within the system on removal of the partition between them, can be expressed in Gibbs form as

$$d\mathcal{S}_C(U,V,N) = \left[\frac{1}{T_A} - \frac{1}{T_B}\right] dU_A + \left[\frac{p_A}{T_A} - \frac{p_B}{T_B}\right] dV_A - \left[\frac{\mu_A}{T_A} - \frac{\mu_B}{T_B}\right] dN_A$$
$$U = U_A + U_B,\ V = V_A + V_B, \text{ and } N = N_A + N_B \text{ remain constants.} \tag{10.2.19}$$

Each term on the right — a product of an intensive thermodynamic force and the corresponding extensive thermodynamic flow — contributing to the uncompensated heat [15] generated within the system from the nonlinear irreversible interactions between its subsystems, is responsible for the increase in entropy accompanying all natural processes leading to the eventual degradation of energy in the universe to a state of inert uniformity.

The interaction of two finite subsystems is to be compared with the static interaction between a finite system and an infinite reservoir. Contrasted with the later, for which the time evolution is unidirectional with the system unreservedly acquiring the properties of the reservoir which undergoes no perceptible changes resulting in the *static equilibrium* from *passive interaction* of the system with its reservoir, the system-system interaction is fundamentally different as it evolves bidirectionally such that the properties of the composite are not of either of the systems, but an average of the individual properties that defines an eventual state of *dynamic interactive equilibrium.* This distinction between passive and dynamical equilibria resulting respectively from the uni- and bi-directional interactions is clearly revealed in Eqs. (10.2.18) and (10.2.19), with bi-directionality of the later being displayed by the *difference form of the generalized forces.* Accordingly, subsystems A and B have two directional arrows imposed on them: the first due to the evolution of the system opposed by a reverse arrow arising from its interactive interaction with the other, see Fig. 10.4. Evolution requires all macroscopic extensive variables — and hence all the related microscopic intensive parameters — to be functions of time so that equilibrium, in the case of (10.2.19) for example, demands

$$\frac{d\mathbb{S}_C}{dt} = 0 \Longrightarrow \left(\frac{dU_A}{dt} = 0\right) \bigwedge \left(\frac{dV_A}{dt} = 0\right) \bigwedge \left(\frac{dN_A}{dt} = 0\right)$$
$$\Longleftarrow (T_A(t) = T_B(t)) \bigwedge (p_A(t) = p_B(t)) \bigwedge (\mu_A(t) = \mu_B(t)). \quad (10.2.20)$$

While we return to this topic subsequently using the tools of directed sets and convergence in topological spaces, for the present it suffices to note that for an emerging, self-organizing, complex evolving system far from stable equilibrium, the reductionist linear proportionality between cause and effect[5] that decouples the entropy change into two independent parts, one with the exterior and the other the consequent internal generation as given by (10.2.15), is open to question as these constitute a system of interdependent evolutionary interlinked processes, depending on each other for their sustenance and contribution to the whole. Thus, "life" forms in which $d\mathcal{S}$, arising from the energy exchanged as food and other sustaining modes with the exterior, depends on the capacity $d\mathbb{S}$ of the life to utilize these resources, which in turn is regulated by $d\mathcal{S}$. These interdependent, non-reductionist, contributions of constituent parts to the whole is a direct consequence of nonlinearity that effectively implies $f(\alpha x_1 + \beta x_2) \neq \alpha f(x_1) + \beta f(x_2)$ for the related processes. The other "non-life" example requires the change to be determined by such internal parameters as mass, specific heat and chemical concentration of the constituents parts. Thus, for example, in the adiabatic mixing of a hot and cold body A and B the equilibrium temperature, given in terms of the respective mole

[5] Which, we recall, allows breaking up of the system into its constituent parts, studying their micro-dynamics and putting them back together in a linear sum to generate the macro-dynamics, thereby presuming that the macroscopic behaviour of a system of a large number of interacting parts is directly proportional to the character of its microscopic constituents.

numbers N, specific heat c and temperature T, by

$$N_A c_A (T_A - T) = N_B c_B (T - T_B) \tag{10.2.21}$$

sets up a state of dynamical equilibrium in which the bi-directional evolutionary arrow prevents A from annihilating B with the equilibrium condition $T = T_A$, $P = P_A$, $\mu = \mu_A$. Putting the heat balance equation in the form

$$dQ_A + (-dQ_B) = 0, \quad dQ = N\,c\,dT$$

suggests that the heat transfer out of a body, considered as a negative real number, be treated as the additive inverse to the positive transfers into the system. This sets up a one-to-one correspondence between two opposing directional real process that evolves to a state of dynamic equilibrium.

The basic feature of this evolutionary thermodynamics — based entirely on (linear) differential calculus — is that it reduces the dynamics of Eqs. (10.2.3) and (10.2.7) to a separation of the governing macroscopic extensive variables, thereby raising the question of the validity of such decoupling of the motive forces of evolution in strongly nonlinear, self-organizing, complex dynamical systems of nature[6]. Such a separation of variables tacitly implies, as in the example considered above, that the total energy exchange taking place

[6] The following extracts from the remarkably explicit lecture MIT-CTP-3112 by Michel [2], delivered possibly in 2000/2001, are worth recalling . *Chaos is still not part of the American university's physics curriculum; most students get physics degrees without ever hearing about it. The most popular textbook in classical mechanics does not include chaos. Why is that? The answer is simple. Physicists did not have the time to learn chaos, because they were fascinated by something else. That something else was 20th century physics of relativity, quantum mechanics, and their myriad of consequences. Chaos was not only unfamiliar to them; it was slightly distasteful!*

In offering an explanation for this, Baranger argues that in discovering calculus, Newton and Leibnitz *provided the scientific world with the most powerful new tool since the discovery of numbers themselves. The idea of calculus is simplicity itself. Smoothness (of functions) is the key to the whole thing. There are functions that are not smooth* $\cdots$. The discovery of calculus led to that of analysis and *after many decades of unbroken success with analysis, theorists became imbued with the notion that all problems would eventually yield to it, given enough effort and enough computing power. If you go to the bottom of this belief you find the following. Everything can be reduced to little pieces, therefore everything can be known and understood, if we analyze it to a fine enough scale. The enormous success of calculus is in large part responsible for the decidedly reductionist attitude of most twentieth century science, the belief in absolute control arising from detailed knowledge.*

Nonetheless, *chaos is the anti-calculus revolution, it is the rediscovery that calculus does not have infinite power. Chaos is the collection of those mathematical truths that have nothing to do with calculus. Chaos theory solves a wide variety of scientific and engineering problems which do not respond to calculus.*

when the gases are allowed to mix completely is separable into independent parts arising from changes in temperature, volume, and diffusion mixing of the gases, *with none of them having any effect on the others.* Observing that the defining property of a complex system responsible for its "complexity" is the interdependence of its interacting parts leading to non-reductionism, this contrary implication of independence of the extensive parameters conflicts with the foundational tenets of chaos and complexity.

Nonetheless, it should be clear from the above considerations that a non-isolated, "non-equilibrium" system can maintain a steady state of low entropy not only by discarding its excess entropy to the surroundings, but more importantly by utilizing [15] a part of this generation by the nonlinearities within itself to enhance its own state of organization consistent with the irreversibilities. Thus when the heated earth at a high level of non-equilibrium instability radiates heat to the cooler atmosphere through evaporation, the earth-atmosphere system is not scorched to the earth's temperature but instead stabilizes itself by "attracting, as it were, a stream of negative entropy upon itself" [29], through condensation of the water vapour back to the earth that essentially opposes this attempt to move the earth-atmosphere system away from its stable equilibrium by acting a gradient dissipator of the temperature difference. As the temperature difference increases, so does the opposition making it more and more difficult for the system to be away from equilibrium. The Second Law of Thermodynamics for non-equilibrium systems — recall footnote 3 — can accordingly be reformulated [27] to require that as the system is forced away from thermodynamic equilibrium it utilizes every possible avenue in "sucking orderliness from its environment" [29], to counter the applied gradients, with its ability to oppose continued displacement increasing with the gradient itself. For such systems the Second Law becomes a law of continuity for the entropy transferred in and out of the system.

The objective of this paper is to propose an explicitly nonlinear, topological formulation of dynamical evolution from an integrated chanoxity perspective that focuses on nonlinearly generated self-organization, adaption, and emergence of systems far from thermodynamic equilibrium. In this perspective, the following observations of Bertuglia and Vaio [3] are worth noting.

Linear approximations become increasingly unacceptable the further away we get from a condition of stable equilibrium. The world of classical science has shown a great deal of interest in linear differential equations for a very simple reason: apart from some exceptions, these are the only equations of an order above the first that can be solved analytically. The simplicity of linearization and the success that it has at times enjoyed have imposed the perspective from which scientists observed reality, encouraging scientific investigation to concentrate on linearity in its description of dynamic processes. On one hand this led to the idea that the elements that can be treated with techniques of linear mathematics prevail over nonlinear ones, and on the other hand it ended up giving rise to the idea that linearity is intrinsically "elegant" because it is expressed in simple, concise formulae, and that a linear model is aesthetically more "attractive" than a nonlinear one. The practice of considering linearity as elegant encouraged a sort of self-promotion and gave rise to a real scientific prejudice: mainly linear aspects were studied. The success that was at times undeniably achieved in this ambit increasingly convinced scholars that linearization was the right way forward for other phenomena that adapted badly to linearization.

However, an arbitrary forced aesthetic sense led them to think (and at times still leads us to think) that finding an equation acknowledged as elegant was, in a certain sense, a guarantee that nature itself behaved in a way that adapted well to an abstract vision of such mathematics.

Linear systems cannot generate dynamics that is sensitive to initial conditions with non-repeating orbits that remain confined in a bounded region of space. This defining character of chaos can be generated only by nonlinear interactions leading to increasing unpredictability of the system's future with increasing time: nonlinearity produces unexpected outcomes, linearity does not. Newtonian classical mechanics is reductionist and the solution of the equations of motion are uniquely determined by the initial conditions for all times.

10.2.2 Maximal Noninjectivity is Chaos

Chaos was defined in [30] as representing maximal non-injective ill-posedness in the temporal evolution of a dynamical system and was based on the purely set theoretic arguments of Zorn's Lemma and Hausdorff Maximal Chain Theorem. It was, however, necessary to link this with topologies because evolutionary directions are naturally represented by adherence and convergence of the associated nets and filters, which require topologies for describing their eventual and frequenting behaviour. For this we found the topology of saturated sets generated by the increasingly non-injective evolving maps to provide the motivation for maximally non-injective, degenerate ill-posedness leading to the concept of the *ininality of topologies* generated by a function $f\colon X \to Y$ that is simultaneously image and preimage continuous. In this case, the topologies on the range $\mathcal{R}(f)$ and domain $\mathcal{D}(f)$ of f are locked with respect to each other as far as further temporal evolution of f is concerned by having the respective topologies defined as the *f-images in Y of f^--saturated open sets of X*. Thus Eqs. (10.1.6a, b), and (10.1.7) taken with the definitions[7]

[7] If $(f_\alpha : X \to (Y_\alpha, \mathcal{V}_\alpha))_{\alpha\in\mathbb{D}}$ is a family of functions into topological spaces $(Y_\alpha, \mathcal{V}_\alpha)$, then the topology generated by the subbasis $\{f_\alpha^-(V_\alpha) : V_\alpha \in \mathcal{V}_\alpha\}_{\alpha\in\mathbb{D}}$ is the *initial topology* of X induced by the family $(f_\alpha)_{\alpha\in\mathbb{D}}$. Reciprocally, if $(f_\alpha\colon (X_\alpha, \mathcal{U}_\alpha) \to Y)_{\alpha\in\mathbb{D}}$ is a family of functions from topological spaces $(X_\alpha, \mathcal{U}_\alpha)$, then the collection $\{G \subseteq Y : f_\alpha^-(G) \in \mathcal{U}_\alpha\}_{\alpha\in\mathbb{D}}$ is the *final topology* of Y of the family $(f_\alpha)_{\alpha\in\mathbb{D}}$. A topology that is both initial and final is *ininal*.

$$\mathrm{IT}\{e;\mathcal{V}\} \stackrel{\mathrm{def}}{=} \{U \subseteq X : U = e^{-}(V),\, V \in \mathcal{V}\} \tag{10.2.22a}$$

and

$$\mathrm{FT}\{\mathcal{U};q\} \stackrel{\mathrm{def}}{=} \{V \subseteq Y : q^{-}(V) = U,\, U \in \mathcal{U}\} \tag{10.2.22b}$$

of initial and final topologies — that denote the coarsest (smallest) and finest (largest) topologies in X and Y respectively making f continuous — implies for open sets $V \in \mathcal{V}$ of Y and $G \subseteq U \in \mathcal{U}$ of X satisfying $U = \mathcal{S}_q(G)$ so that q acts only on saturated open sets,

$$f(\mathcal{S}_f(A)) = \mathcal{C}_f(B) \begin{cases} \stackrel{\mathrm{IT}}{\Longrightarrow} (\mathcal{S}_e(U) = U)\,(e(U) = \mathcal{C}_e(V)) \\ \stackrel{\mathrm{FT}}{\Longrightarrow} (q(\mathcal{S}_q(G) = V)\,(V = \mathcal{C}_q(V)); \end{cases} \tag{10.2.23}$$

see also column 2, row 1 of Table 10.1. As these equations show, preimage and image continuous functions need not be open functions: a preimage continuous function is open iff $e(U)$ is an open set in Y and an image continuous function is open iff the q-saturation of every open set of X is also an open set. The generation of new topologies on the domain and range of a function — which will generally be quite different from the original topologies the spaces might have possessed — by the evolving dynamics of increasingly nonlinear maps is a basic property of the evolutionary process that constitutes the motive for such dynamical changes. Putting implications (10.2.23) together yields

$$U,\, V \in \mathrm{IFT}\{\mathcal{U};f;\mathcal{V}\} \Longleftrightarrow (\mathcal{U} = f^{-}(\mathcal{V}))\,(f(\mathcal{U}) = \mathcal{V}) \tag{10.2.24a}$$

that effectively renders both e and q open functions, and reduces to

$$U,\, V \in \mathrm{HOM}\{\mathcal{U};f;\mathcal{V}\} \Longleftrightarrow (\mathcal{U} = f^{-1}(\mathcal{V}))\,(f(\mathcal{U}) = \mathcal{V}) \tag{10.2.24b}$$

for a bijection satisfying both $\mathcal{S}_f(A) = A,\ \forall A \subseteq X$ and $\mathcal{C}_f(B) = B,\ \forall B \subseteq Y$; observe that the only difference between Eqs. (10.2.24a) and (10.2.24b) is in the bijectivity of f.

There are two defining components, temporal and spatial, in any natural evolutionary processes. However, these are equivalent in the sense that both can be represented as pre-ordered sets with the additional directional property of a *directed set* $(\mathbb{D}, \preceq)$ which satisfies

(DS1) $\alpha \in \mathbb{D} \Rightarrow \alpha \preceq \alpha$ ($\preceq$ is reflexive)
(DS2) $\alpha, \beta, \gamma \in \mathbb{D}$ such that $(\alpha \preceq \beta \wedge \beta \preceq \gamma)$ implies $\alpha \preceq \gamma$ ($\preceq$ is transitive)
(DS3) For all $\alpha, \beta \in \mathbb{D}$, there exists a $\gamma \in \mathbb{D}$ such that $\alpha \preceq \gamma$ and $\beta \preceq \gamma$

with respect to the *direction* $\preceq$. While the first two properties are obvious and constitutes the preordering of $\mathbb{D}$, the third replaces antisymmetry of an order with the condition that every pair of elements of $\mathbb{D}$, whether ordered or not, always has a successor. This directional property of $\mathbb{D}$, that imparts to the static pre-order a sequential arrow by allowing it to choose a forward path between possible alternatives when non-comparable elements bifurcate at the arrow, will be used to model evolutionary processes in space and time.

Besides the obvious examples $\mathbb{N}$, $\mathbb{R}$, $\mathbb{Q}$, or $\mathbb{Z}$ of totally ordered sets, more exotic instances of directed sets imparting directions to neighbourhood systems in X tailored to the specific needs of convergence theory are summarized in Table 10.2, where $\beta \in \mathbb{D}$ is the directional index. Although the neighbourhood

Directed set $\mathbb{D}$	Direction $\preceq$ induced by $\mathbb{D}$
${}_{\mathbb{D}}N = \{N : N \in \mathcal{N}_x\}$	$M \preceq N \Leftrightarrow N \subseteq M$
${}_{\mathbb{D}}N_t = \{(N,t) : (N \in \mathcal{N}_x)(t \in N)\}$	$(M,s) \preceq (N,t) \Leftrightarrow N \subseteq M$
${}_{\mathbb{D}}N_\beta = \{(N,\beta) : (N \in \mathcal{N}_x)(x_\beta \in N)\}$	$(M,\alpha) \leq (N,\beta) \Leftrightarrow (\alpha \preceq \beta) \wedge (N \subseteq M)$

Table 10.2. Natural directions of decreasing subsets in $(X,\mathcal{U})$ induced by some useful directed sets of convergence theory. Significant examples of directed sets that are only partially ordered are $(\mathcal{P}(X),\subseteq)$, $(\mathcal{P}(X),\supseteq)$; $(\mathcal{F}(X),\supseteq)$; $(\mathcal{N}_x,\subseteq)$, $(\mathcal{N}_x,\supseteq)$ for a set X, We take $\mathcal{N}_x$, suitably redefined if necessary, to be always a system of nested subsets of X.

system ${}_{\mathbb{D}}N$ at a point $x \in X$ with the *reverse-inclusion* direction $\preceq$ is the basic example of natural direction of the neighbourhood system $\mathcal{N}_x$ of x, the directed sets ${}_{\mathbb{D}}N_t$ and ${}_{\mathbb{D}}N_\beta$ are more useful in convergence theory because unlike the first, these do not require a simultaneous application of the Axiom of Choice to every $N \in \mathcal{N}_x$.

Chaos as manifest in the limiting adhering attractors is a direct consequence of the increasing nonlinearity of the map under increasing iterations and with the right conditions, appears to be the natural outcome of the characteristic difference between a function f and its multi-inverse f^-. Equivalence classes of fixed points stable and unstable, as generated by the saturation operator $\mathcal{S}_f = f^- f$, determine the ultimate behaviour of an evolving dynamical system, and since the eventual (as also frequent) nature of a filter or net is dictated by topology on the set, chaoticity on a set X leads to a reformulation of the open sets of X to equivalence classes generated by the evolving map f. In the limit of infinite iterational evolution in time resulting in the multifunction Φ, the generated open sets constitute a basis for a topology on $\mathcal{D}(f)$ and the basis for the topology of $\mathcal{R}(f)$ are the corresponding Φ-images of these equivalent classes. It follows that the motivation behind evolution leading to chaos is the drive toward a state of the dynamical system that supports ininality of the limit multi Φ[8]. In this case therefore, the open sets of

[8] For the logistic map $f_\lambda(x) = \lambda x(1-x)$ with chaos setting in at $\lambda = \lambda_* = 3.5699456$, this drive in ininality implies an evolution toward values of the spatial parameter $\lambda \geq \lambda_*$; this is taken to be a spatial parameter as it determines the degree of surjectivity of f_λ. Together with the temporal evolution in increasing noninjectivity for any λ, this comprises the full evolutionary dynamics of the

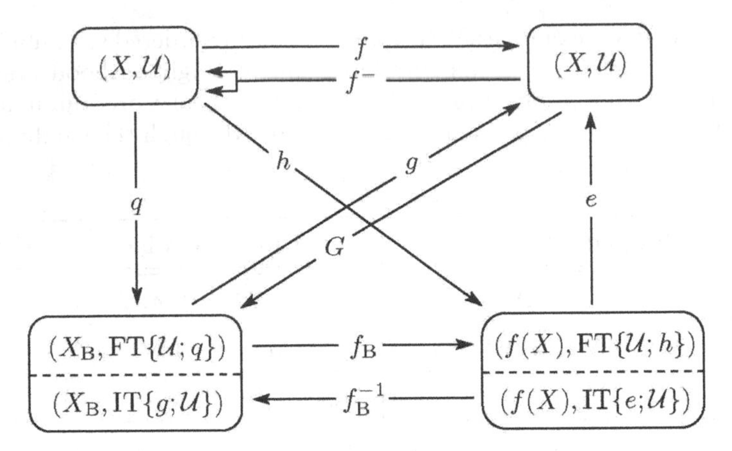

Fig. 10.1. Generation of a multifunctional inverse $x = f^-(y)$ of the functional equation $f(x) = y$ for $f : X \to X$; here $G : Y \to X_B$ is a generalized inverse of f because $fGf = f$ and $GfG = G$ that follows from the commutativity of the diagrams. g and h are the injective and surjective restrictions of f; these will be topologically denoted by their generic notations e and q respectively.

the range $\mathcal{R}(f) \subseteq X$ are the multi images that graphical convergence generates at each of these inverse-stable fixed points. As readily verified from Fig. 10.1, X has two topologies imposed on it by the dynamics of f: the first of equivalence classes generated by the limit multi Φ in the domain of f and the second as Φ-images of these classes in the range of f. Hence while subdiagrams $X - (X_B, \mathrm{FT}\{\mathcal{U};q\}) - (f(X), \mathcal{U}_2)$ and $(X_B, \mathcal{U}_1) - (f(X), \mathrm{IT}\{e;\mathcal{U}\}) - X$ apply to the final and initial topologies of X_B and $f(X)$ respectively, their superposition $X - (X_B, \mathrm{FT}\{\mathcal{U};q\}) - (f(X), \mathrm{IT}\{e;\mathcal{U}\}) - X$ *under the additional requirement of a homeomorphic* f_B leads to the conditions $\mathcal{U}_1 = \mathrm{IT}\{g;\mathcal{U}\}$ and $\mathcal{U}_2 = \mathrm{FT}\{\mathcal{U};h\}$ that X_B and $f(X)$ must possess. For this to be possible,

$$\mathrm{FT}\{\mathcal{U};q\} = \mathrm{IT}\{g;\mathcal{U}\}$$
$$\mathrm{IT}\{e;\mathcal{U}\} = \mathrm{FT}\{\mathcal{U};h\}$$

requires the image continuous q and the preimage continuous e to be also be open functions which translates to the ininality of f on $(X, \mathcal{U})$, and hence for the topology of X to be simultaneously the direct and inverse images of itself under f; compare Eq. (10.2.24a). Since the map f and the topology $\mathcal{U}$ of X are already provided, this is interpreted to mean that the increasing nonlinear

logistic map. These two distinct dynamical mechanisms of increasing surjectivity and decreasing injectivity are not independent, however. Thus λ — which may be taken to represent the energy exchanges of all possible types that the system can have with the surroundings — determines the nature of the internal forward-backward stasis that leads to the eventual equilibrium of the system with its environment.

ill-posedness of the time-iterates of f is driven by ininality of the maximally "degenerate" ill-posed limit relation Φ on X^2. In this case Φ acts as a non-bijective open and continuous relation that forces the sequence of evolving functions (f^n) on X to eventually behave, by (10.1.7), homeomorphically on the saturated open sets of equivalence classes and their f^n-images in X. A homeomorphism, by establishing an equivalence between spaces $(X,\mathcal{U})$ and $(Y,\mathcal{V})$ — algebraically through bijectivity and topologically by setting up a one-to-one correspondence between the respective open sets — renders the spaces as "essentially the same", with the non-bijective ininal function acting as an effective bijection $f\colon \mathcal{S}_f(A) \to \mathcal{C}_f(B)$ for all subsets A and B between X and Y. For a function defined *on* a space X, this means that, under ininality, the domain and range spaces are "effectively the same" thereby precluding any further interaction between them, which corresponds to a condition of equilibrium entropic death. We define the resulting ininal topology on X to be the *chaotic topology on X associated with* f. Neighbourhoods of points in this topology cannot be arbitrarily small as they consist of all members of the equivalence class to which any element belongs; hence a sequence converging to any of these elements necessarily converges to all, and the eventual objective of chaotic dynamics is to generate a topology in X (irrespective of the original $\mathcal{U}$) with respect to which elements of the space are grouped together in large equivalence classes such that if a net converges simultaneously to points $x \neq y \in X$ then $x \sim y$: x is of course equivalent to itself while x, y, z are equivalent to each other iff they are simultaneously in every open set where the net may eventually be in. This signature of chaos eradicates existing separation properties of the space: it makes X uniformly homogeneous and flat, devoid of any interaction inducing inducement among its parts, signifying thereby "death".

The generation of a new topology on X by the dynamics of f on X is a consequence of the topology of pointwise biconvergence $\mathcal{T}$ defined on the set of relations $\mathrm{Multi}((X,\mathcal{U}),(Y,\mathcal{V}))$, [30]. This generalization of the topology of pointwise convergence defines neighbourhoods of f in $\mathrm{Multi}((X,\mathcal{U}),(Y,\mathcal{V}))$ to consist of those functions in $(\mathrm{Multi}((X,\mathcal{U}),(Y,\mathcal{V})),\mathcal{T})$ whose images at any point $x \in X$ lie not only close enough to $f(x) \in Y$ (this gives the usual pointwise convergence) but additionally whose inverse images at $y = f(x)$ contain points arbitrarily close to x. Thus the graph of f apart from being sufficiently close to $f(x)$ at x in $V \in \mathcal{V}$, but must also be constrained such that $f^-(y)$ has at least one branch in the open set $U \in \mathcal{U}$ about x. This requires all members of a neighbourhood $\mathcal{N}_f$ of f to "cling to" f as the number of points on the graph of f increases with the result that unlike for simple pointwise convergence, no gaps in the graph of the limit relation is possible not only on the domain of f but on its range too.

For a given integer $I \geq 1$, the open sets of $(\mathrm{Multi}(X,Y),\mathcal{T})$ are

$$\begin{aligned} B((x_i),(V_i);(y_i),(U_i)) = \{g \in \mathrm{Map}(X,Y)\colon (g(x_i) \in V_i) \\ \bigwedge \left(g^-(y_i) \bigcap U_i \neq \emptyset\right), i = 1,2,\cdots,I\}, \quad (10.2.25) \end{aligned}$$

where $(x_i)_{i=1}^I \in X$, $(y_i)_{i=1}^I \in Y$, $(U_i)_{i=1}^I \in \mathcal{U}$, $(V_i)_{i=1}^I \in \mathcal{V}$ are chosen arbitrarily with reference to $(x_i, f(x_i))$. A local base at f, for $(x_i, y_i) \in \text{Graph}(f)$, is the set of functions of (10.2.25) with $y_i = f(x_i)$, and the collection of all local bases $B_\alpha = B((x_i)_{i=1}^{I_\alpha}, (V_i)_{i=1}^{I_\alpha}; (y_i)_{i=1}^{I_\alpha}, (U_i)_{i=1}^{I_\alpha})$, for every choice of $\alpha \in \mathbb{D}$, is a base ${}_{\text{T}}\mathcal{B}$ of $(\text{Multi}(X,Y), \mathcal{T})$; note that in this topology $(\text{Map}(X,Y), \mathcal{T})$ is a subspace of $(\text{Multi}(X,Y), \mathcal{T})$.

The basic technical tool needed for describing the adhering limit relation in $(\text{Multi}(X,Y), \mathcal{T})$ is a generalization of the topological concept of neighbourhoods to the algebraic concept of a filter which is a collection of subsets of X satisfying

(F1) The empty set $\emptyset$ does not belong to $\mathcal{F}$,

(F2) The intersection of any two members of a filter is another member of the filter: $F_1, F_2 \in \mathcal{F} \Rightarrow F_1 \cap F_2 \in \mathcal{F}$,

(F3) Every superset of a member of a filter belongs to the filter: $(F \in \mathcal{F}) \wedge (F \subseteq G) \Rightarrow G \in \mathcal{F}$; in particular $X \in \mathcal{F}$,

and is generated by a subfamily $(B_\alpha)_{\alpha \in \mathbb{D}} = {}_{\text{F}}\mathcal{B} \subseteq \mathcal{F}$ of itself, known as the filter-base, characterized by

(FB1) There are no empty sets in the collection ${}_{\text{F}}\mathcal{B}$: $(\forall \alpha \in \mathbb{D})(B_\alpha \neq \emptyset)$

(FB2) The intersection of any two members of ${}_{\text{F}}\mathcal{B}$ contains another member of ${}_{\text{F}}\mathcal{B}$: $B_\alpha, B_\beta \in {}_{\text{F}}\mathcal{B} \Rightarrow (\exists B \in {}_{\text{F}}\mathcal{B} \colon B \subseteq B_\alpha \cap B_\beta)$.

Hence any family of subsets of X that does not contain the empty set and is closed under finite intersections is a base for a unique filter on X, and the filter-base

$$ {}_{\text{F}}\mathcal{B} \stackrel{\text{def}}{=} \{B \in \mathcal{F} \colon B \subseteq F \text{ for each } F \in \mathcal{F}\} \tag{10.2.26} $$

determines the filter

$$ \mathcal{F} = \{F \subseteq X \colon B \subseteq F \text{ for some } B \in {}_{\text{F}}\mathcal{B}\} \tag{10.2.27} $$

as all its supersets. Since filters are purely algebraic without any topological content, to use it as a tool of convergence, a comparison of (F1)-(F3) and (FB1)-(FB2) with (N1)-(N3) and (NB1)-(NB2) of Sec. 10.1 show that the neighbourhood system $\mathcal{N}_x$ at x is the *neighbourhood filter at* x and any local base at x is a filter-base for $\mathcal{N}_x$ and generally for any subset A of X, $\{N \subseteq X : A \subseteq \text{Int}(N)\}$ is a filter on X at A. All subsets of X containing a point $p \in X$ is the *principal filter* ${}_{\text{F}}\mathcal{P}(p)$ on X at p, and the collection of all supersets of a nonempty subset A of X is the principal filter ${}_{\text{F}}\mathcal{P}(A)$ at A. The singleton sets $\{\{x\}\}$ and $\{A\}$ are particularly simple examples of filter-bases that generate the principal filters at $\{x\}$ and A; other useful examples that we require subsequently are the set of all residuals

$$ \text{Res}(\mathbb{D}) = \{\mathbb{R}_\alpha \colon \mathbb{R}_\alpha = \{\beta \in \mathbb{D} \colon \alpha \preceq \beta\}\} $$

of a directed set $\mathbb{D}$, and the neighbourhood systems $\mathcal{B}_x$ and $\mathcal{N}_x$. By adjoining the empty set to the principal filters yields the p-inclusion and A-inclusion topologies on X respectively[9].

The utility of filters in describing convergence in topological spaces is because a filter $\mathcal{F}$ on X can always be associated with the net $\chi_{\mathcal{F}} \colon {}_{\mathbb{D}}F_x \to X$ defined by

$$\chi_{\mathcal{F}}(F, x) \stackrel{\text{def}}{=} x \tag{10.2.28}$$

where ${}_{\mathbb{D}}F_x = \{(F, x) : (F \in \mathcal{F})(x \in F)\}$ is the directed set with direction $(F, x) \preceq (G, y) \Rightarrow (G \subseteq F)$; reciprocally a net $\chi \colon \mathbb{D} \to X$ corresponds to the filter-base

$${}_{\mathrm{F}}\mathcal{B}_{\chi} \stackrel{\text{def}}{=} \{\chi(\mathbb{R}_{\alpha}) \colon \mathrm{Res}(\mathbb{D}) \to X \text{ for all } \alpha \in \mathbb{D}\}, \tag{10.2.29}$$

with the corresponding filter $\mathcal{F}_{\chi}$ being obtained by taking all supersets of the elements of ${}_{\mathrm{F}}\mathcal{B}_{\chi}$. Filters and their bases are extremely powerful tools for maximal, non-injective, degenerate ill-posedness in the context of the algebraic Hausdorff Maximal Principle and Zorn's Lemma, that is now summarized below[10].

Let f be a noninjective function in $\mathrm{Multi}(X)$ and $\mathfrak{I}(f)$ be the number of injective branches of f and let

$$F = \{f \in \mathrm{Multi}(X) \colon f \text{ is a noninjective function on } X\} \in \mathcal{P}(\mathrm{Multi}(X))$$

be the collection of all noninjective functions on X satisfying the properties

(a) For every $\alpha \in \mathbb{D}$, F has the extension property

$$(\text{For any } f_{\alpha} \in F)(\exists f_{\beta} \in F) \colon \mathfrak{I}(f_{\alpha}) \leq \mathfrak{I}(f_{\beta}).$$

Define a partial order $\preceq$ on $\mathrm{Multi}(X)$ as

$$\mathfrak{I}(f_{\alpha}) \leq \mathfrak{I}(f_{\beta}) \Longleftrightarrow f_{\alpha} \preceq f_{\beta}, \tag{10.2.30}$$

with $\mathfrak{I}(f) := 1$ for the smallest f. This is actually a preorder on $\mathrm{Multi}(X)$ in which all function with the same number of injective branches are equivalent

[9] A filter is almost a topology: both are closed under finite intersections and arbitrary unions, and both contain the base set X. It is only the empty set that must always be in the topology but never in a filter; adding it to a filter makes it a special type of topology that might be termed a filtered topology. Whereas any arbitrary family of sets can generate a topology as its subbase through finite intersections followed by arbitrary unions, the family must satisfy the finite intersection property before qualifying as a filter subbase; hence, every filter subbase is a topological subbase but not conversely.

[10] **Hausdorff Maximal Principle** (HMP)**:** Every partially ordered set has a maximal chain.

A partially ordered set X is said to be *inductive* if every chain of X has an upper bound in X.

Zorn's Lemma: Every inductive set has at least one maximal element.

to each other. Note that Multi(X) has two orders imposed on it: the first $\preceq$ between its elements f, and the second the usual $\subseteq$ that orders subsets of these functional elements.

(b) Let

$$\mathcal{X} = \{C \in \mathcal{P}(F) : C \text{ is a chain in } (\text{Multi}(X), \preceq)\} \in \mathcal{P}^2(\text{Multi}(X)) \quad (10.2.31)$$

be a collection of chains in Multi(X) with respect to the order (10.2.30) where

$$C_\nu = \{f_\alpha \in \text{Multi}(X) : f_\alpha \preceq f_\nu\} \in \mathcal{P}(\text{Multi}(X)), \qquad \nu \in \mathbb{D}, \qquad (10.2.32)$$

are the chains of non-injective functions where $f_\alpha \in F$ is to be identified with the iterates f^i, the number of injective branches $\mathcal{I}(f)$ depending on i. The chains are to be built from the smallest $C_0 = \mathcal{D}$ the domain of f, by application of a choice function g_c that generates the immediate successor

$$C_j := g(C_i) = C_i \bigcup g_c(\mathcal{G}(C_i) - C_i) \in \mathcal{X}$$

of C_i by picking one from the many

$$\mathcal{G}(C_i) = \{f \in F - C_i : \{f\} \bigcup C_i \in \mathcal{X}\}$$

that C_i may possibly possess. Application of g to C_0 n-times generates the chain $C_n = \{\mathcal{D}, f(\mathcal{D}), \cdots, f^n(\mathcal{D})\}$, and the smallest common chain

$$\begin{aligned} \mathcal{C} &= \{C_j \in \mathcal{P}(\text{Multi}(X)) : C_i \subseteq C_k \text{ for } i \le k\} \subseteq \mathcal{X} \quad (10.2.33) \\ &= \{\mathcal{D}, \{\mathcal{D}, f(\mathcal{D})\}, \{\mathcal{D}, f(\mathcal{D}), f^2(\mathcal{D})\}, \cdots\} \qquad \mathcal{D} := C_0 \end{aligned}$$

of all the possible g-towered chains $\{C_i\}_{i=0,1,2,\cdots}$ of Multi(X) constitutes a principal filter of totally ordered subsets of (Multi$(X), \subseteq$) at C_0. Notice that while $\mathcal{X} \in \mathcal{P}^2(\text{Multi}(X))$ is a set of sets, $C \in \mathcal{P}(\text{Multi}(X))$ is relatively simpler as a set of elements of $f \in \text{Multi}(X)$, which at the base level of the tree of interdependent structures of Multi(X), is canonically the simplest.

To continue further with the application of Hausdorff Maximal Principle to the partially ordered set $(\mathcal{X}, \preceq)$ of sets, it is necessary to postulate that

(i) There exists a smallest element C_0 in $\mathcal{X}$ with no predecessor,

(ii) Every element C of $\mathcal{X}$ has an immediate successor $g(C)$ in $\mathcal{X}$; hence there is no element of $\mathcal{X}$ lying strictly between C and $g(C)$, and

(iii) $\mathcal{X}$ is an *inductive set* so that every chain $\mathcal{C}$ of $(\mathcal{X}, \preceq)$ has a supremum $\sup_{\mathcal{X}}(\mathcal{C}) = \cup_{C \in \mathcal{C}} C$ in $\mathcal{X}$.

Any subset $\mathcal{T}$ of $\mathcal{X}$ satisfying these properties is known as a *tower*; $\mathcal{X}$ is of course a tower by definition. The intersection of all possible towers of $\mathcal{X}$ is the towered chain $\mathcal{C}$ of $\mathcal{X}$, Eq. (10.2.33). Criterion (iii) is especially crucial as it effectively disqualifies $(F, \preceq)$ as a likely candidate for HMP: the supremum of the chains of increasingly non-injective functions need not be a *function*, but is more likely to be a multifunction. Hence $\mathcal{X}$ in the conditions above is the

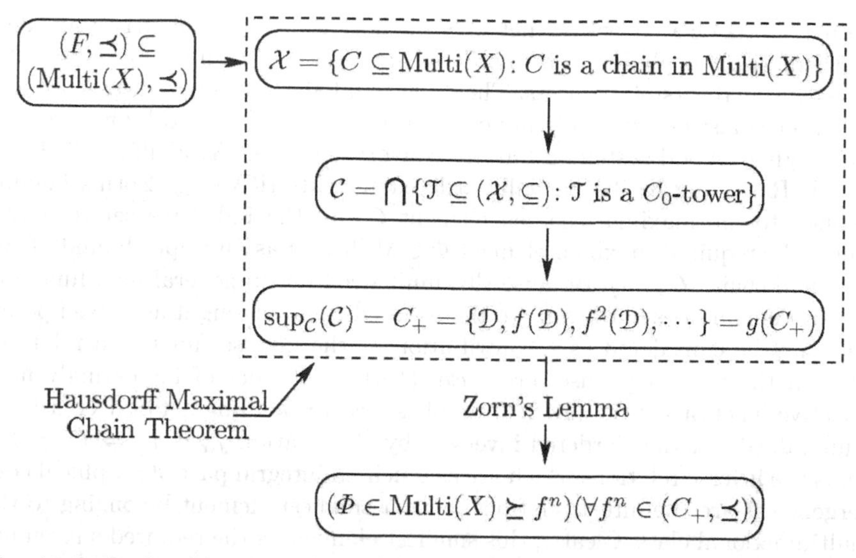

Fig. 10.2. Application of Zorn's Lemma to a partially ordered set F of non-injective functions f in Multi(X). $\mathcal{C} = \{\mathcal{D}, \{\mathcal{D}, f(\mathcal{D})\}, \{\mathcal{D}, f(\mathcal{D}), f^2(\mathcal{D})\}, \cdots\}$ is a chain of towered chains of *functions* in Multi(X) with $C_0 = \mathcal{D}$, the domain of f. Notice that to obtain a maximal Φ at the base level Multi(X), it is necessary to go two levels higher: $\mathcal{X} \in \mathcal{P}^2(\mathrm{Multi}(X)) \to C \in \mathcal{P}(\mathrm{Multi}(X)) \to \Phi \in \mathrm{Multi}(X)$ is a three-tiered structure with the two-tiered HMP feeding to the third of Zorn's Lemma.

space of relations, and it is necessary to consider C of (10.2.32) as a subset of this Multi(X) rather than of F. The careful reader cannot fail to note that the requirement of inductivity of $\mathcal{X}$ effectively leads to an "extension" of Map(X) to Multi(X) where the supremum of the chain of non-injective functions can possibly lie. However since this is purely in an algebraic setting without topologies on the sets, the supremum constitutes only a static cap on the family of equilibrium ordered states: the chains being only ordered and not directed are devoid of any dynamical evolutionary property.

(c) The Hausdorff Maximal Principle applied to $(\mathcal{X}, \subseteq)$ now yields

$$\begin{aligned}\sup_{\mathcal{C}}(\mathcal{C}) &= C_+ = \{f_\alpha, f_\beta, f_\gamma, \cdots\} \\ &= \{\mathcal{D}, f(\mathcal{D}), f^2(\mathcal{D}), \cdots\} = g(C_+) \in \mathcal{C} \qquad (10.2.34)\end{aligned}$$

as the supremum of $\mathcal{C}$ in $\mathcal{C}$, defined as a fixed-point of the tower generator g, without any immediate successor. Identification of this fixed-point supremum as one of the many possible maximal elements of $(\mathcal{X}, \subseteq)$ completes the application of Hausdorff Principle, yielding C_+ as the required maximal chain of $(\mathcal{X}, \subseteq)$.

The technique of HMP is of interest because it presents a graphic step-wise algorithmic rule leading to an equivalent filter description and the algebraic notion of a *chained tower.* Not possessing any of the topological directional properties associated with a net or sequence, the tower comprises an ideal mathematical vocabulary for an ordered succession of equilibrium states of a quasi-static, reversible, process. The directional attributes of convergence and adherence must be externally imposed on towered filters like $\mathcal{C}$ by introducing the neighbourhood system: a filter $\mathcal{F}$ converges to $x \in (X, \mathcal{U})$ iff $\mathcal{N}_x \subseteq \mathcal{F}$.

(d) Returning to the partially ordered set $(\text{Multi}(X), \preceq)$, Zorn's Lemma applied to the maximal chained element C_+ of the inductive set $\mathcal{X}$ finally yields the required maximal element $\Phi \in \text{Multi}(X)$ as an upper bound of the maximal chain $(C_+, \preceq)$. Because this limit need not in general be a function, the supremum does not belong to the towered chain having it as a fixed point, and may be considered as a contribution of the inverse functional relations (f_α^-) in the following sense. From Eq. (10.1.1), the net of increasingly non-injective functions of Eq. (10.2.30) implies a corresponding net of decreasingly multivalued functions ordered inversely by the relation $f_\alpha \preceq f_\beta \Leftrightarrow f_\beta^- \preceq f_\alpha^-$. Thus the inverse relations which are as much an integral part of graphical convergence as are the direct relations, have a smallest element belonging to the multifunctional class. Clearly, this smallest element as the required supremum of the increasingly non-injective tower of functions defined by (10.2.30), serves to complete the significance of the tower by capping it with a "boundary" element that can be taken to bridge the classes of functional and non-functional relations on X.

Having been assured of the existence of a largest element $\Phi \in \text{Multi}(X)$, we now proceed to construct it topologically. Let $(\chi_i := f^i(A))_{i \in \mathbb{N}}$ for a subset $A \subseteq X$ that we may take to be the domain of f, correspond to the ordered sequence (10.2.30). Using the notation of (10.2.29), let the sequences $\chi(\mathbb{R}_i) = \cup_{j \geq i} f^j(A)$ for each $i \in \mathbb{N}$ generate the decreasingly nested filter-base

$$\begin{aligned} {}_{\mathrm{F}}\mathcal{B}_\chi &= \left\{ \bigcup_{j \geq i} f^j(A) \right\}_{i \in \mathbb{N}} \\ &= \left\{ \bigcup_{j \geq i} f^j(x) \right\}_{i \in \mathbb{N}}, \qquad \forall x \in A, \end{aligned} \tag{10.2.35}$$

corresponding to the sequence of functional iterates $(f^j)_{j \geq i \in \mathbb{N}}$. The existence of a maximal chain with a maximal element guaranteed by the Hausdorff Maximal Principle and Zorn's Lemma respectively implies a nonempty core of ${}_{\mathrm{F}}\mathcal{B}_\chi$. We now identify this filterbase with the neighbourhood base at Φ and thereby define

$$\begin{aligned} \Phi(A) &\stackrel{\text{def}}{=} \text{adh}({}_{\mathrm{F}}\mathcal{B}_\chi) \\ &= \bigcap_{i \geq 0} \text{Cl}(A_i), \qquad A_i = \{f^i(A), f^{i+1}(A), \cdots\} \end{aligned} \tag{10.2.36}$$

as the attractor of A, where the closure is with respect to the topology of pointwise bi-convergence induced by the neighbourhood filter base ${}_{\mathrm{F}}\mathcal{B}_\chi$. Clearly the attractor as defined here is the graphical limit of the sequence of functions $(f^i)_{i\in\mathbb{N}}$ with respect to the directed sets of Table 10.2. This attractor represents, in the product space $X \times X$, the converged limit of the bi-directional evolutionary dynamics occurring in the kitchen $X \times \mathfrak{X}$ that induces the image $\Phi(A)$ in X. The exclusion space $\mathfrak{X}$ is not directly observable, being composed of complementary *negelements* $\mathfrak{x}$ that correspond in an unique, one-to-one fashion to the corresponding defining observables $x \in X$, just as the negative reals — which are not physically directly observable either — are attached in a one-to-one fashion with their corresponding defining positive counterparts by

$$r + (-r) = 0, \qquad r \in \mathbb{R}_+. \tag{10.2.37}$$

The exclusion space $(\mathfrak{X}, \mathfrak{U})$ introduced next is necessary for the understanding of bi-directional evolutionary process responsible for a synthesis of opposites of two sub-systems competitively collaborating with each other. The basic example of an exclusion space is the negative reals with a *forward* arrow of the *decreasing* negatives resulting from an *exclusion topology* $\mathcal{U}_-$ generated by the topology $\mathcal{U}_+$ of the observable positive reals R_+. This generalization of the additive inverse of the real number system to sets follows.

The Negative Exclusion Space of a Topological Space

Postulate NEG-1. The Negative $\mathfrak{X}$ of a set X.[11] Let X be a set and suppose that for every $x \in X$ there exists a *negative element* $\mathfrak{x} \in \mathfrak{X}$ with the property that

$$\mathfrak{X} \stackrel{\text{def}}{=} \{\mathfrak{x} \colon \{x\} \bigcup \{\mathfrak{x}\} = \emptyset\} \tag{10.2.38a}$$

defines the *negative, or exclusion, set* of X. This means that for every subset A of X there is a complementary neg(ative)set $\mathfrak{A} \subseteq \mathfrak{X}$ *associated with* (*generated by*) it such that

$$A \bigcup \mathfrak{G} \stackrel{\text{def}}{=} A - G, \qquad G \longleftrightarrow \mathfrak{G}, \tag{10.2.38b}$$

implies $A \cup \mathfrak{A} = \emptyset$. Hence a neg-set and its generating set act as relative *discipliners* of each other in restoring a measure of order in the evolving confusion, disquiet and tension, with the intuition of the set-negset pair "undoing", "controlling", or "stabilizing" each other. The complementing neg-element is an unitive inverse of its generating element, with $\emptyset$ the corresponding *identity* and G the *physical manifestation of* $\mathfrak{G}$. Thus for $r > s \in \mathbb{R}_+$, the physical manifestation of any $-s \in \mathfrak{R}_+ (\equiv \mathbb{R}_-)$ is the smaller element $(r - s) \in \mathbb{R}_+$.

As compared with the directed set $(\mathcal{P}(X), \subseteq)$ that induces the natural direction of *decreasing subsets* of Table 10.2, the direction of *increasing supersets* induced by $(\mathcal{P}(X), \supseteq)$ — which understandably finds no ready application in convergence theory — proves useful in generating a co-topology $\mathcal{U}_-$ on $(X, \mathcal{U}_+)$ as follows. Let $(x_0, x_1, x_2, \cdots)$ be a sequence in X converging to $x_* \in X$ with

[11] These quantities will be denoted by fraktur letters.

reference to any of the reverse-inclusion directions of decreasing neighbourhoods of Table 10.2[12], and consider the backward arrow induced at x_* by the directed set $(\mathcal{P}(X), \supseteq)$ of increasing supersets at x_*. As the reverse sequence $(x_*, \cdots, x_{i+1}, x_i, x_{i-1}, \cdots)$ does not converge to x_0 unless it is eventually in every neighbourhood of this initial point, we employ the closed-open subsets

$$N_i - N_j = \begin{cases} (N_i - N_j) \bigcap N_i, & \text{(open)} \\ (N_i - N_j) \bigcap (X - N_j) & \text{(closed)} \end{cases} \tag{10.2.39}$$

$(j > i)$ in the *inclusion* topology $\mathcal{U}_+$ of X with $x_i \in N_i - N_{i+1}$, $N_i \in \mathcal{N}_{x_*}$, to define an additional *exclusion topology* $\mathcal{U}_-$ on $(X, \mathcal{U}_+)$ as follows. First recall that whereas the *x-inclusion topology* $\mathcal{U}_+$ of X comprises, together with $\emptyset$, all subsets of X that *include* x with the neighbourhood system $\mathcal{N}_x$ being just these non-empty subsets of X, the *x-exclusion topology* is, along with X, all the subsets $\mathcal{P}(X - \{x\})$ that *exclude* x. The $A \subseteq X$ exclusion topology $\{\mathcal{P}(X - A), X\}$ therefore consists of all subsets of X that do not intersect A and the $(X - A)$-exclusion topology $\{\mathcal{P}(A), X\}$ comprises, with X, only the subsets of A. Since $\mathcal{N}_x = \{X\}$ and $\mathcal{N}_{y \neq x} = \{\{y\}\}$ are the neighbourhood systems at x and any $y \neq x$ in the x-exclusion topology, it follows that while every net must converge to the defining point of its own topology, only the eventually constant net $\{y, y, y, \cdots\}$ converges to any $y \neq x$[13]. The exclusion topology of x therefore has the remarkable property of compelling every other element of X to either submit to the dictum of x by being in its sphere of influence, or else to effectively isolate any other member of X from establishing its own sphere of influence. All directions with respect to x are consequently rendered equivalent; hence the directions of $\{1/n\}_{n=1}^{\infty}$ and $\{n\}_{n=1}^{\infty}$ are equivalent in $\mathbb{R}_+$ as they converge to 0 in its exclusion topology, and this basic property of the exclusion topology induces an opposing direction in X.

It is now possible to postulate with respect to the directed set ${}_{\mathbb{D}}N_i = \{(N_i, i) : (N_i \in \mathcal{N}_{x_*})(x_i \in N_i)\}$ of Table 10.2 and a sequence $(x_i)_{i \geq 0}$ in $(X, \mathcal{U}_+)$ converging to $x_* = \cap_{i \geq 0} \mathrm{Cl}(N_i) \in X$, that

Postulate NEG-2. The $\mathbf{x_0}$-exclusion topology $\mathcal{U}_-$ of $(\mathbf{X}, \mathcal{U}_+)$. There exists an increasing sequence of negelements $(\mathfrak{x}_i)_{i \geq 0}$ of $\mathfrak{X}$ that converges to $\mathfrak{x}_*$ in the $\mathfrak{x}_*$-inclusion topology $\mathfrak{U}$ of $\mathfrak{X}$ generated by the $\mathfrak{X}$-images of the neighbourhood system $\mathcal{N}_{x_*}$ of $(X, \mathcal{U}_+)$. Since the only manifestation of neg-sets in the observable world is their regulating property, the $\mathfrak{X}$-increasing sequence $(\mathfrak{x}_i)_{i \geq 0}$ converges to $\mathfrak{x}_*$ in $(\mathfrak{X}, \mathfrak{U})$ if and only if the sequence $(x_0, x_1, x_2, \cdots)$ converges to x_* in $(X, \mathcal{U}_+)$. Affinely translated to X, this means that the $\mathfrak{x}_*$-inclusion

[12] We henceforth adopt the convention that the arrow induced by the inclusion topology of the real world is the forward arrow of the system, and the exclusion neg-matter manifests in this real world as its backward arrow. The forward arrow therefore corresponds to the increasing direction of an appropriate pre-ordering of the real physical world.

[13] I thank Joseph T. H. Lo for his clarifications on the subtleties of the exclusion topology, Private Communication, May 2004.

arrow in $(\mathfrak{X}, \mathfrak{U})$ transforms to an x_0-exclusion arrow in $(X, \mathcal{U}_+)$ generating an additional topology $\mathcal{U}_-$ in X that opposes the arrow converging to x_*. This *direction of increasing supersets of* $\{x_*\}$ *excluding* x_0 associated with $\mathcal{U}_-$ of Table 10.3, is to be compared with the *natural direction of decreasing subsets containing* x_* in $(X, \mathcal{U}_+)$, Table 10.2. We take the reference natural direction in $X \cup \mathfrak{X}$ to be that of X pulling the inclusion sequence $(x_0, x_1, x_2, \cdots)$ to x_*; hence the decreasing subset direction in $\mathfrak{X}$ of the inclusion sequence $(\mathfrak{x}_0, \mathfrak{x}_1, \mathfrak{x}_2, \cdots, \mathfrak{x}_*)$ appears in X as an exclusion sequence converging to x_0 because any sequence in an exclusion space must necessarily converge to the defining element in its own topology. In this perspective, the left side of Eq. (10.2.38*b*), read in the more familiar form $a + (-b) = a - b$ with $a, b \in \mathbb{R}_+$ and $-b := \mathfrak{b} \in \mathfrak{R}_+$, represents "+" evolution in the base kitchen of Nature, which is then served in its bi-directional physical-world manifestation on the dining-table of the right side supporting retraction along the "−" direction. At the risk of an apparent "abuse of language", $(\mathfrak{X}, \mathfrak{U})$ will be termed the *exclusion space* of $(X, \mathcal{U}_+)$.

Directed set $\mathbb{D}$	Direction $\preceq$ induced by $\mathbb{D}$
$_{\mathbb{D}}\mathfrak{N} = \{\mathfrak{N} \colon \mathfrak{N} \in \mathcal{N}_{\mathfrak{x}}\}$	$\mathfrak{M} \preceq \mathfrak{N} \Leftrightarrow \mathfrak{M} \subseteq \mathfrak{N}$
$_{\mathbb{D}}\mathfrak{N}_{\mathfrak{t}} = \{(\mathfrak{N}, \mathfrak{t}) \colon (\mathfrak{N} \in \mathcal{N}_{\mathfrak{x}})(\mathfrak{t} \in \mathfrak{N})\}$	$(\mathfrak{M}, \mathfrak{s}) \preceq (\mathfrak{N}, \mathfrak{t}) \Leftrightarrow \mathfrak{M} \subseteq \mathfrak{N}$
$_{\mathbb{D}}\mathfrak{N}_{\beta} = \{(\mathfrak{N}, \beta) \colon (\mathfrak{N} \in \mathcal{N}_{\mathfrak{x}})(\mathfrak{x}_{\beta} \in \mathfrak{N})\}$	$(\mathfrak{M}, \alpha) \leq (\mathfrak{N}, \beta) \Leftrightarrow (\alpha \preceq \beta) \wedge (\mathfrak{M} \subseteq \mathfrak{N})$

Table 10.3. Natural directions of increasing supersets in $(\mathfrak{X}, \mathfrak{U})$ is to be compared with Table 10.2 of the natural reverse directions in $(X, \mathcal{U})$. The direction of coevents in $\mathfrak{X}$ is opposite to that of X in the sense that the temporal sequence of images of events in X opposes that in $\mathfrak{X}$ and the order of occurrence of events induced by the coworld appear to be reversed to the physical observer stationed in X.

Although the backward sequence $(x_j)_{j=\cdots,i+1,i,i-1,\cdots}$ in $(X, \mathcal{U}_+)$ does not converge, the effect of $(\mathfrak{x}_i)_{i\geq 0}$ of $\mathfrak{X}$ on X is to regulate the evolution of the forward arrow $(x_i)_{i\geq 0}$ to an effective state of stasis of dynamical equilibrium, that becomes self-evident on considering for X and $\mathfrak{X}$ the sets of positive and negative reals, and for x_*, $\mathfrak{x}_*$ a positive number r and its negative inverse image $-r$. The existence of a negelement $x \leftrightarrow \mathfrak{x}$ in $\mathfrak{X}$ for every $x \in X$ requires all forward arrows in X to have a matching forward arrow in $\mathfrak{X}$ that actually *appears backward when viewed from* X. It is this opposing complimentary effect of the apparently backward-$\mathfrak{X}$ sequences on X — responsible by (10.2.38*b*) for moderating the normal uni-directional evolution in X — that is useful in establishing a stasis of dynamical balance between the opposing forces generated in the composite of a compound system with its environment. Obviously, the evolutionary process ceases when the opposing influences in X due to it-

	$(X,\mathcal{U}_+)$	$(X,\mathcal{U}_-)$
T_0	$(\forall x \neq y \in X)\,(\exists N \in \mathcal{N}_x : N \cap \{y\} = \emptyset) \vee (\exists M \in \mathcal{N}_y : M \cap \{x\} = \emptyset)$	$(\forall x \neq y \in X)\,(\exists N \in \mathcal{N}_x : N \cap \{y\} \neq \emptyset) \vee (\exists M \in \mathcal{N}_y : M \cap \{x\} \neq \emptyset)$
T_1	$(\forall x \neq y \in X)\,(\exists N \in \mathcal{N}_x : N \cap \{y\} = \emptyset) \wedge (\exists M \in \mathcal{N}_y : M \cap \{x\} = \emptyset)$	$(\forall x \neq y \in X)\,(\exists N \in \mathcal{N}_x : N \cap \{y\} \neq \emptyset) \wedge (\exists M \in \mathcal{N}_y : M \cap \{x\} \neq \emptyset)$
T_2	$(\forall x \neq y \in X)\,(\exists N \in \mathcal{N}_x \wedge M \in \mathcal{N}_y) : (M \cap N = \emptyset)$	$(\forall x \neq y \in X)\,(\exists N \in \mathcal{N}_x \wedge M \in \mathcal{N}_y) : (M \cap N \neq \emptyset)$

Table 10.4. Comparison of the separation properties of $(X,\mathcal{U}_+)$ and its inhibitor $(X,\mathcal{U}_-)$.

self and that of its moderator $\mathfrak{X}$ balance out marking a state of dynamic equilibrium.

It should be noted that the moderating image $\mathfrak{X}$ of X needs to be endowed with inverse inhibiting properties if Eq. (10.2.38*b*) is to be meaningful which leads to the separation properties of the conjugate spaces $(X,\mathcal{U}_+)$ and $(\mathfrak{X},\mathfrak{U})$ as shown in Table 10.4. Significantly, the exclusion space is topologically distinguished in having its sequences converge with respect to the only neighbourhood X of the limit point, a property that leads as already pointed out earlier to the existence of a multiplicity of equivalent limits in large neighbourhoods of x_0 to which the backward sequences in X converges, even when $(X,\mathcal{U})$ is Hausdorff. In the context of iterational evolution of functions that concerns us here, that the function-multifunction asymmetry of (10.1.1) introduced by the non-injectivity of the iterates is directly responsible for the difference in the separation properties of $\mathcal{U}_+$ and $\mathcal{U}_-$, which in turn prohibits the system from annihilating B mentioned earlier and forces it to adopt the forward-backward stasis of opposites. Recalling that non-injectivity of one-dimensional maps translate to pairs of injective branches with *positive* and *negative* slopes, we argue with reference to Fig. 10.3 that whereas branches with positive slope represent matter, those with negative slope correspond to reg(ulating)-matter by Eq. (10.2.38*b*) and the disjoint union of these components represents the compound system of forward-backward opposites. Taking $T_A > T_B$, $p_A > p_B$ and $\mu_A > \mu_B$, the dynamical evolution represented by the shaded boxes would, in the absence of the backward arrow induced by the exclusion space, eventually spread uniformly over the full domain, and equilibrium would be characterized completely by T_A, p_A, μ_A, at the exclusion of B. Denoting matter by 1 and (the effect of) negmatter by 0, the progressively refined partition of $\mathfrak{D}(t)$ induced by the evolving map is indicated in (ii), (iii) and (iv).

As an example, we return to Eqs. (10.2.18) and (10.2.19) for the entropy change due to exchange of resources and its non-linear, irreversible, internal

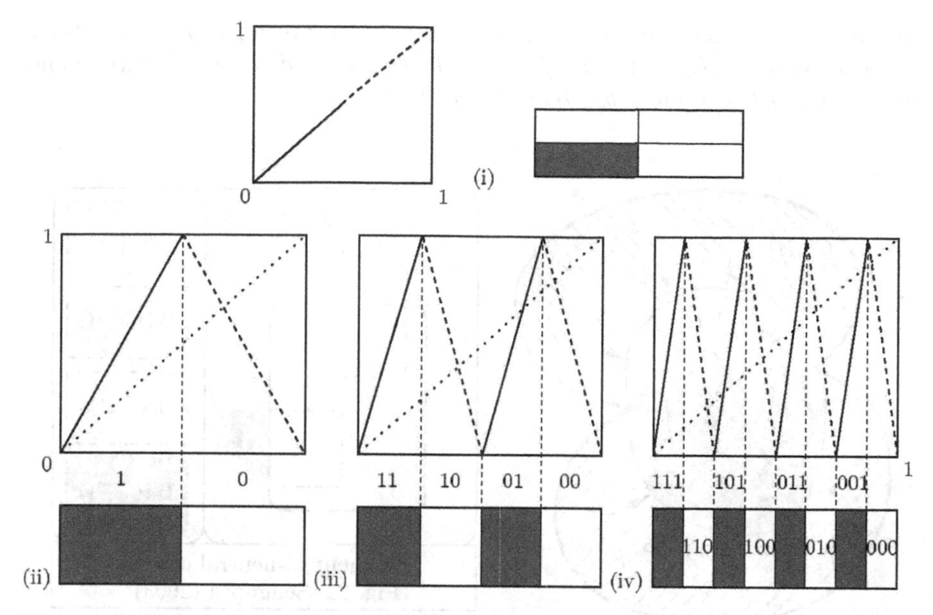

Fig. 10.3. Matter-negmatter synthesis of an evolving system $C = A \cup B$ under the tent interaction. A and B are represented by the solid and dashed lines as injective branches with positive and negative slopes respectively.

generation respectively. The external exchange with the environment leads to a change in the internal state of the system which is then utilized in performing irreversible useful work relative to the environment, conveniently displayed in terms of the *neutral-neutral* convergence mode of a net of Fig. 10.4 adapted from Fig. 22 of [30], which illustrates the irreversible internal generation of entropy in a universe $C = A \cup B$, where A and B are two disjoint components of a system prepared at different initial conditions shown in the figure, with B the physical manifestation of a compatible space $\mathfrak{B}$ endowed with an exclusion topology and a direction opposing that of A. In the real interval $[0, 1]$, notable examples of A and B are $f(x)$ and $f(1-x)$ with B the physical manifestation of $\mathfrak{A}$. This allows us to make the

Definition 10.1 (Interaction Between Two Spaces). *A space* $(A, \mathcal{U})$ *will be said to* interact *with a* disjoint *space* $(B, \mathcal{V})$ *if there exists a function* f *on the compound* disjoint sum $(C = A \cup B, \mathcal{W})$ *where*

$$\begin{aligned}\mathcal{W} &= \{W := U \bigcup V : (U \in \mathcal{U}) \bigwedge (V \in \mathcal{V})\}\\ &= \{W \subseteq C : (W \bigcap A \,\text{is open in}\, A) \bigwedge (W \bigcap B \,\text{is open in}\, B)\},\end{aligned}$$

which evolves graphically to a well defined limit in the topology of pointwise biconvergence on $(C, \mathcal{W})$. *The function* f *will be said to be a* bidirectional interaction *between the subsystems* A *and* B *of* C.[14]

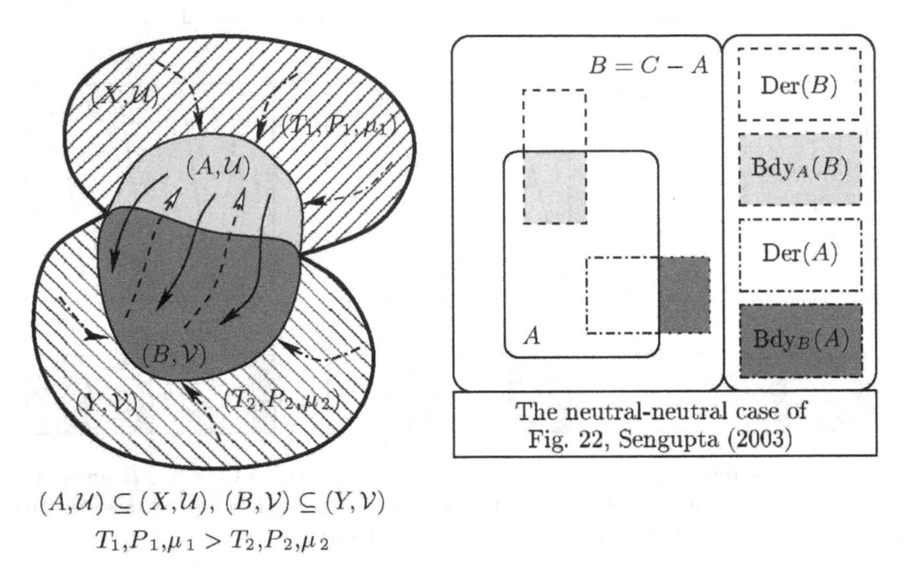

Fig. 10.4. Schematic representation of irreversible entropy generation in $C = A \cup B$ with respect to the universe $X \cup Y$. We will identify the solid arrows in C from the hot body to the cold with inverse limit, neg-entropic self-organization, and the dashed arrows from the cold to hot as direct limit, second law entropic emergence, see Fig. 10.5*b*.

While A and B by themselves need not display any notable features (see Fig. 10.10*c*), the evolution of A in the disjoint compound $(C, \mathcal{W})$, motivated by the inducement of an ininal topology on C, is effectively opposed by the influence of the exclusion topology of B, with the equivalence classes generated in C being responsible for the multi-inverses of the evolving f characterizing the nonlinear state of C following the internal preparation of the system. This irreversible process, indicated in Fig. 10.4 by the nets of full arrows from $(A, \mathcal{U})$ to $(B, \mathcal{V})$ representing transfer of energy, volume, or mass driven

[14] If A and B are not disjoint, then this construction of the compound sum may not work because A and B will generally induce distinct topologies on C; in this case $\mathcal{W}$ is obtained as follows. Endow the disjoint copies $A_1 := A \times \{1\}$ and $B_2 := B \times \{2\}$ of A and B with topologies $\mathcal{U}_1 = \{U \times \{1\} : U \in \mathcal{U}\}$ and $\mathcal{V}_2 = \{V \times \{2\} : V \in \mathcal{V}\}$, which are homeomorphic with their originals with $a \mapsto (a, 1)$ and $b \mapsto (b, 2)$ being the respective homeomorphisms. Then $C = A_1 \cup B_2$ is the *disjoint union (sum)* of A_1 and B_2 with the topology $\mathcal{W} = \{W \subseteq C : W = (U \times \{1\}) \cup (V \times \{2\}) : (U \in \mathcal{U}) \wedge (V \in \mathcal{V})\}$ that induces the subspaces $(A_1, \mathcal{U}_1)$ and $(B_2, \mathcal{V}_2)$.

by an appropriate evolutionary directed set of a thermodynamic force (for instance due to a temperature gradient $T_A > T_B$ inducing the energy transfer), provides the forward impetus for directional transport motivated by ininality. The dashed open arrows show the reverse evolution in C due to its inhibitor $\mathfrak{C}$. The dash-dot arrows stand for the uni-directional transfer of energy from a reservoir that continues till the respective parts of C acquire the characteristics of their reservoirs.

Since physical evolution powered by changes in the internal intensive parameters is represented by convergence of appropriate sequences and nets, it is postulated in keeping with the role of ininality, that equilibrium in uni-directional temporal evolutions like $X \to A \subseteq X$ or $Y \to B \subseteq Y$ sets up A and B as subspaces of X and Y respectively. For bi-directional processes like $A \leftrightarrow B$, the open headed dashed arrows of Fig. 10.4 from B to A represent the backward influence of $(B, \mathcal{V})$ on $(C, \mathcal{W})$. The assumptions

▶ Both the subsets A and B of the compound C are perfect in the sense that $A = \mathrm{Der}(A)$ and $B = \mathrm{Der}(B)$ so that there are no isolated points in A and B with all points of each of the sets accessible by sequences eventually in them, and

▶ $\mathrm{Bdy}_B(A) = B$ and $\mathrm{Bdy}_A(B) = A$ which enables all points of A and B to be directly accessed as limits by sequences in B and A,

imply that any exchange of resource from the environment $E = X \cup Y$ to system C will be evenly dispersed throughout by the irreversible, internal evolution of the system, once C attains equilibrium with E and is allowed to evolve unperturbed thereafter. This *global homogenizing principle of detailed balance*, applicable to evolutionary processes at the micro-level provides a rationale for equilibration in nature that requires every forward arrow to be balanced by a backward, leading to the global equilibrium of thermodynamics. If these influences exactly balance each other resulting in a complete restoration of all the intermediate stages, then the resulting *reversible process* is actually quasi-static with no effective changes; hence nontrivial dynamical equilibrium cannot be generated by reversible processes.

For unimodal maps like the logistic $f_\lambda = \lambda x(1 - x)$ that are defined with respect to the forward-backward, positive-negative slope characteristic, which for a particular λ can be taken to represent the subspace $C \subseteq E$ at equilibrium with its environment E, evolutionary changes in the effective available resources λ induce changes in the internal intensive thermodynamic parameters that follow uni-directional exchanges of C with E. This perturbs the equilibrium between components A and B resulting in further evolutionary iterational interaction between them. The iterational evolution of f_λ is relatively moderated by the reverse effect of the evolution of f_λ^- which suppresses the continual increase of noninjectivity of f_λ that would otherwise lead to a state of maximum noninjective ill-posedness for this λ: note the negative branches of f appear positive to its inverse, and conversely. Measurable global dynamic equilibrium represents a balance between the opposing induced local forces that are determined by, and which in turn determine, the degree of resource exchange λ. The eventual ininality at $\lambda = 4$ represents continual resource utilization from E that is dissipated for the globalizing uniformity of Figs. 10.3 and 10.10a(iii). In the range $3 < \lambda \leq \lambda_* = 3.5699456$, the input is gainfully employed to generate the complex structures that are needed to sustain the process at that level of λ.

Recalling footnote 8, we now summarize the principal features of the nonlinear evolutionary dynamics following interaction of a compound system C with its surroundings.

► If the state of dynamic equilibrium of a composite system $C = A \cup B$ with its surroundings, as represented by the logistic map is disturbed by an interaction between them, forces are set up between the components A and B so as to absorb the effect of this disturbance.

► Consumption of the effects of the exchange is motivated by a simultaneous, non-reductionist drive towards increasing surjectivity and decreasing injectivity of $(f_\lambda)_\lambda$ and its evolved iterated images, that eventually leads to a state of maximal non-injective degeneracy on the domain of f. Owing to the function-multifunction asymmetry of f, such a condition would signify static equilibrium and an end to all further evolutionary processes, a state of dissipative annihilation, burn-out and ininality.

► Since such eventual self-destruction cannot be the stated objective of Nature, this unrelenting thrust toward collapse is opposed by the negworld exclusion effects we have described earlier, generating a reversed sequential direction effectively inhibiting the drive towards self-destruction induced by the simultaneous increase of λ and the increased noninjectivity under iterations. The resulting state of dynamic equilibrium is the observed equilibrium of Nature. Like all others, nature's *kitchen* $C \times \mathfrak{C}$ where the actual dynamical evolution occurs is beyond direct observation; only its disciplining effect in $C \times C$ is perceived by the observer stationed in $\mathfrak{D}(f) = C$.

As an example, consider an isolated system of two parts each locally in equilibrium with its environment as in (10.2.19) that can now be re-written as

$$\begin{aligned}\mathcal{S}_C(t) = \mathcal{S}_C(0) + &\left[N_A c_A \ln\left(\frac{T}{T_A(t)}\right) + N_B c_B \ln\left(\frac{T}{T_B(t)}\right)\right] \\ &- R\left[N_A \ln\left(\frac{P_A}{p_A(t)}\right) + N_B \ln\left(\frac{P_B}{p_B(t)}\right)\right], \qquad (10.2.40)\end{aligned}$$

where we note with reference to Fig. 10.4 that $T_A = T_1$, $T_B = T_2$ are the temperatures of subsystems A and B, $V_A + V_B = V$ is the total volume of C, p_A, p_B are the pressures of A and B, $P_{A,B} := N_{A,B}RT_{A,B}/V$ are their partial pressures with $P = P_A + P_B$ the total pressure exerted by the gases in V, and T is the equilibrium temperature of (10.2.21).

Then

(i) If the parts containing nonidentical ideal gases at different temperatures are brought in contact with each other, the equilibrium state of stasis resulting from the flows of *heat* and *cold* (= negheat) between the bodies lead to the equality of temperature, $T_A = T = T_B$, and the vanishing of the first part of (10.2.40).

(ii) If the gas in the first half expands into the second then equilibrium is reached when the *gas outflow* is exactly balanced by the *vacuum inflow* into it if the second is evacuated, or if it is filled with a nonidentical gas then equalization of pressure of the chambers by outflow of the gases from their respective halves into the other, results in the vanishing of the second term of (10.2.40). In either case, competitive collaboration of the two opposites with unequal resources, rather than annihilation of the weaker by the more resourceful, leads to the state of mutual equilibrium.

In all these instances, the two disjoint opposing parts act in competitive non-reductionist collaboration to generate a moderated and inhibited stasis of the union: this is its only manifestation of the complementary neg-world on its observable physical partner. Thus *cold, vacuum* and *a nonidentical substance* are the negations of *heat* and *matter* — just as $-r \in \mathfrak{R}_+$ is the negation of $r \in \mathbb{R}_+$. These negations as elements of the negworld are no more observable than -5, for example, is to us in our physical world: we cannot collect -5 objects around us or measure the distance between two places to be -100 kilometers; more generally, the set of complex numbers can be considered to constitute the coreals, without which there would have been no zero, no starting initial point in any ordered set, and no "equilibrium" either. Nature, propelled by its unidirectional increasing entropic disorder, without the containing Schrodinger and de Broglie $\lambda = h/p$ waves, would have probably crashed out of existence long ago!

In summary, then, for an interaction $f : C \to C$ and the bijective map $\mathfrak{f}\colon C \to \mathfrak{C}$ corresponding to (10.2.38b), the hierarchal order

$$\begin{aligned}\text{Dynamics of } \mathfrak{f}f\colon C \to \mathfrak{C} \text{ in nature's } \textit{kitchen}\,(C,\mathcal{W}) \times (\mathfrak{C},\mathfrak{W})\\ \longrightarrow \text{Evolution of } f \text{ on } (C,\mathcal{W})^2\\ \longrightarrow \text{Experimental observation in } \mathfrak{D}(f) = C\end{aligned}$$

accompanied by

▶ Increasing iterates of f, driven by ininality of topology generated on C, constitutes the activating sense of the dynamics, that as we see subsequently, corresponds to the backward, entropy increasing, destabilizing direction of the evolutionary process. The function-multifunction asymmetry between f and f^- generates and sustains this unidirectional ininality, and

▶ Decreasing iterates of f corresponds to the forward, entropy decreasing, stabilizing direction in the evolving, competitive collaboration of interactions generated by f and f^-,

defines the state of equilibrated stasis schematized in Fig. 10.4. From the discussion in connection with Fig. 10.1 that ininality is an effective expression of a *non-bijective homeomorphism* when the sequence of evolutions (f^n) become progressively more bijective on the saturated open sets of equivalence classes and their respective images, Eq. (10.1.7), it can be argued that *the incentive towards the resulting effective simplicity of invertibility on the definite classes of sets associated with* (f^n) *is responsible for evolutionary dynamics on* C.

This account of "providing a mechanical (i.e., dynamical) explanation of why classical systems behave thermodynamically" [5] is to be compared with [10], see also [31]. The distinctive feature of the present approach is in its use of difference equations rather than the *microscopic* Hamilton differential equations that yield the Liouville equation of *macroscopic* mechanical systems. As so forcefully inquired by Baranger [2], can the emerging evolutionary properties of strongly nonlinear, emergent, self-organizing systems be described by linear (Hamiltonian) differential equations? By employing functional interactions as solutions to difference equations by the technique of graphical convergence of their iterates, we explicitly invoke the immediate past in determining its future and are thereby able to circumvent the issues of time reversal invariance and Poincare recurrence that are inherently associated with the microscopic dynamics of Hamilton's differential equations. This also enables us to avoid direct reference to statistical and probabilistic arguments except in so far as are inherently implied by the Axiom of Choice.

While our preference for unimodal, single-humped, logistic-like difference equations is based on the understanding that only an appropriate juxtaposition of the opposing directional effects — like that of $x-a$ and $b-x$ in the interval $a \leq x \leq b$ — can lead to meaningful emergence and self-organization, it is also well known [17, 20] that time evolution of a discrete model and its continuous counterpart can be so different as to have no apparent correlation with each other. Thus the logistic differential equation

$$\dot{x} = g(x) := (\lambda - 1)x\left(1 - \frac{\lambda}{\lambda - 1}x\right) \qquad (10.2.41a)$$

having the same equilibrium fixed points $x = 0$, $x_* = (\lambda - 1)/\lambda$ as the discrete version, has the harmless "trivial" solution

$$\begin{aligned} x(t) &= \frac{x_0\, x_* e^{(\lambda-1)t}}{x_* + x_0(e^{(\lambda-1)t} - 1)} \\ &\overset{t\to\infty}{\longrightarrow} x_*. \end{aligned} \qquad (10.2.41b)$$

Compared with the structurally rich multifunctional graphical convergence leading to chaos and entropic drive of the discrete form, the tranquil differential variety can only produce a simple *monotonic* convergence to the basic fixed point x_* which is responsible for the complex dynamics of the former; in fact, linear systems can only admit stable or exponentially growing oscillatory or non-oscillatory solutions. This apparently surprising, though not unexpected, result arises from the fundamental difference in the bifurcation characteristics of these equations: *the availability of additional spatial dimensions allows the dynamical system a greater latitude in its evolution so that the complex hierarchal structure generated by iteration of one-dimensional maps are absent in flows under Hopf bifurcations.* In fact, for Eq. (10.2.41*a*), 0 is unstable and x_* stable for all values of $\lambda > 1$ because $g'(x) = (\lambda - 1) - 2\lambda x$ is positive at $x = 0$ and negative for $x = x_*$. In contrast with the bifurcation dominated rich and varied dynamics of maps, bifurcation-less evolution of vector fields on the real line — capable only of monotonically converging to fixed points or diverging to infinity without any oscillations or other dynamically interesting features — precludes any qualitative change in the evolution of solutions like (10.2.41*b*).

The alert reader would not have failed to notice that our use of the qualifiers "discipliner", "inhibitor", "stasis" signifying a condition of balance among various forces of the forward-backward opposites, can only provisional as the existence of a set of negatives for every positive as postulated in (10.2.38*b*) does not necessarily imply that their natural directions interact to generate a smaller positive. This crucial dynamical manifestation of matter-negmatter is provided by the second law of thermodynamics which is formalized through the concept of inverse and direct limits that incorporates directional arrows in their definitions.

Direct Limit, Inverse Limit, Irreversibility

> *Otherwise put, every "it" — every particle, every field of force, even space-time continuum itself — derives its function, its meaning, its very existence, from answers to yes-no questions, binary choices, bits. "It from bit" symbolizes the idea that every item of the physical world has at bottom — at a very deep bottom, in most instances — an immaterial source and explanation, that which we call reality arises in the last analysis from the posing of yes-no questions; in short, that all things physical are information-theoretic in origin and this is a participatory universe.*
>
> J. A. Wheeler (1990)

These limits also known as colimit and limit respectively, with the confusing terminology arising possibly from the fact that the "natural" direction in convergence theory is a reverse direction where the counting index increases with decreasing size of the defining open sets, is summarized in Fig 10.5*a*.

Direct limit. The direct (or inductive) limit is a general method of taking limits of a "directed families of objects". Let $(\mathbb{D}, \preceq)$ be a directed partially ordered set, $\{X_\kappa\}_{\kappa \in \mathbb{D}}$ a family of spaces, and $\eta_{\alpha\beta} : X_\alpha \to X_\beta$ a family of continuous connecting maps *oriented along* $(\mathbb{D}, \preceq)$ satisfying the properties

$$\eta_{\alpha\alpha}(x) = x, \qquad \text{for all } x \in X_\alpha \tag{10.2.42a}$$

$$\eta_{\alpha\gamma} = \eta_{\beta\gamma} \circ \eta_{\alpha\beta}, \quad \text{for all } \alpha \preceq \beta \preceq \gamma. \tag{10.2.42b}$$

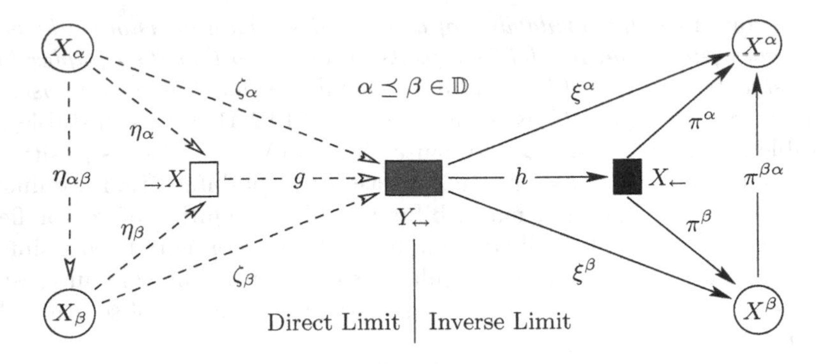

Fig. 10.5*a*. Direct and inverse limits of commutative diagrams. η_α, π^α are projections and $\eta_{\alpha\beta}$, $\pi^{\beta\alpha}$ are connecting maps.

Then the pair $(X_\alpha, \eta_{\alpha\beta})$ is called a *direct* (or *inductive*) *system over* $\mathbb{D}$. The image of a $x_\alpha \in X_\alpha$ under any connecting map is called the *successor of* x_α, and a direct system $(X_\alpha, \eta_{\alpha\beta})$ yields a direct limit space $_\rightarrow X$ as follows. Let $X = \uplus_\kappa X_\kappa$ be the disjoint union of $\{X_\kappa\}$ and let $x_\alpha \in X_\alpha$. The class of elements

$$[x_\alpha] = \{x_\beta \in X_\beta \colon \exists\, \gamma \succeq \alpha,\, \beta \text{ such that } \eta_{\alpha\gamma}(x_\alpha) = \eta_{\beta\gamma}(x_\beta)\} \qquad (10.2.43)$$

with a common successor in the union constitutes an equivalence class of x_α: while reflexivity and symmetry are obvious enough, transitivity of $\sim_{\mathrm{D}}$ follows from

$$\begin{aligned}[x_\alpha \sim_{\mathrm{D}} x_\beta] \bigwedge [x_\beta \sim_{\mathrm{D}} x_\gamma] &\Longrightarrow \exists\, \delta, \epsilon \succeq \alpha,\, \beta,\, \gamma \text{ s.t. } [\eta_{\alpha\delta}(x_\alpha) = \eta_{\beta\delta}(x_\beta)] \\ &\qquad\qquad \bigwedge [\eta_{\beta\epsilon}(x_\beta) = \eta_{\gamma\epsilon}(x_\gamma)] \\ &\Longrightarrow \exists\, \zeta \succeq \delta,\, \epsilon \text{ s.t. } \eta_{\alpha\zeta}(x_\alpha) = \eta_{\gamma\zeta}(x_\gamma) = \eta_{\beta\zeta}(x_\beta),\end{aligned}$$

with two elements in the disjoint union being equivalent iff they are "eventually equal" in the direct system. Then the quotient space

$$_\rightarrow X \overset{\text{def}}{=} \biguplus_\kappa X_\kappa / \sim_{\mathrm{D}} \qquad (10.2.44a)$$

of the disjoint union of $\{X_\kappa\}$ modulo $\sim_{\mathrm{D}}$ is known as the *direct,* or *inductive, limit* of the system $(X_\alpha, \eta_{\alpha\beta})$. The pair $(_\rightarrow X, \eta_\alpha)$ must be universal in the sense that *if there exists* any other such pair $(Y_\leftrightarrow, \zeta_\alpha)$ there is a unique morphism $g \colon {}_\rightarrow X \to Y_\leftrightarrow$ with the respective sub-diagrams commuting for all $\alpha \preceq \beta \in \mathbb{D}$. If

$$p \colon \biguplus_\kappa X_\kappa \to {}_\rightarrow X$$

is the projection, then its restriction

$$\eta_\kappa : X_\kappa \to {}_{\to}X$$

maps each element to its equivalence class, see Fig. 10.5a; hence

$$_{\to}X = \bigcup_\kappa \eta_\kappa(X_\kappa) \tag{10.2.44b}$$

implies that ${}_{\to}X$ is not empty whenever at least one X_α is not empty and the algebraic operations on ${}_{\to}X$ are defined via these maps in an obvious manner. Clearly, $\eta_\alpha = \eta_\beta \eta_{\alpha\beta}$.

If the directed family is a family of disjoint sets $(X_\alpha)_\alpha$, with each X_α the domain of an injective branch of f that partitions $\mathcal{D}(f)$, then the direct limit ${}_{\to}X$ of (X_α) is isomorphic to the basic set X_{B} of Fig. 10.1, where $\eta_{\alpha\beta}(x_\alpha)$ is the element of $[x_\alpha]_f$ in X_β.

Inverse Limit. The inverse (or projective) limit is a construction that allows the "glueing together" of several related objects, the precise nature of the glueing being specified by morphisms between the objects. Let $(\mathbb{D}, \preceq)$ be a directed partially ordered set, $\{X^\alpha\}_{\alpha\in\mathbb{D}}$ a family of spaces, and $\pi^{\beta\alpha} : X^\beta \to X^\alpha$ a family of continuous connecting maps *oriented against* $(\mathbb{D}, \preceq)$ satisfying the properties

$$\pi^{\alpha\alpha}(x) = x, \qquad \text{for all } x \in X^\alpha \tag{10.2.45a}$$

$$\pi^{\gamma\alpha} = \pi^{\beta\alpha} \circ \pi^{\gamma\beta}, \qquad \text{for all } \alpha \preceq \beta \preceq \gamma. \tag{10.2.45b}$$

Then the pair $(X^\alpha, \pi^{\beta\alpha})$ is called an *inverse,* or *projective, system* over $\mathbb{D}$. The image of a $x^\beta \in X^\beta$ under any connecting map is the *predecessor of* x^β and the *inverse,* or *projective, limit*

$$X_{\leftarrow} \overset{\text{def}}{=} \{x \in \prod_\kappa X^\kappa : p^\alpha(x) = \pi^{\beta\alpha} \circ p^\beta(x) \text{ for all } \alpha \preceq \beta \in \mathbb{D}\}, \tag{10.2.45c}$$

of $(X^\alpha, \pi^{\beta\alpha})$, where

$$p^\alpha : \prod_\kappa X^\kappa \to X^\alpha$$

is the projection of the product onto its components, is a subspace of $\prod X^\kappa$ with the property that a point $x = (x^\kappa) \in \prod X^\kappa$ is in $X_{\leftarrow}$ iff its coordinates satisfy $x_\alpha = \pi^{\beta\alpha}(x_\beta)$ for all $\alpha \preceq \beta \in \mathbb{D}$. Every element of $X_{\leftarrow}$ has a unique representation in each X_κ, but an element of X_κ may correspond to many points of the limit. As for direct limits, the pair $(X_{\leftarrow}, \pi^\alpha)$ must be universal such that the existence of any other such pair $(Y_{\leftrightarrow}, \xi^\alpha)$ implies the existence of a unique morphism $h : Y_{\leftrightarrow} \to X_{\leftarrow}$ with the respective sub-diagrams commuting for all $\alpha \preceq \beta \in \mathbb{D}$. The sets $(\pi^\alpha)^{-1}(U)$, $U \subseteq X^\alpha$ open, is a topological basis of $X_{\leftarrow}$, and all pairs of points of $X_{\leftarrow}$ obeying $x_\alpha = \pi^{\beta\alpha}(x_\beta)$ for $\alpha \preceq \beta$ is identical iff their images coincide for every α. The restrictions

$$\pi^\alpha : X_{\leftarrow} \to X^\alpha$$

of p^α is the continuous *canonical morphism* of $X_{\leftarrow}$ into X^α with two points of $X_{\leftarrow}$ being identical iff their images coincide for every α.

Straightforward examples of these limits are

(a) Let $\{X_k\}_{k\in\mathbb{Z}_+}$ be an increasing family of subsets of a set X, and let $\eta_{mn}: X_m \to X_n$ be the inclusion map for $m \leq n$. The direct limit of this system is

$$_{\rightarrow}X = \bigcup_{k=1}^{\infty} X_k \qquad (10.2.46)$$

with the inclusion functions mapping from each X_k into this union. Generally, if $\mathbb{D}$ is any directed partially ordered set with a greatest element ω, then the direct limit of any corresponding direct system is isomorphic to X_ω and the canonical morphism $\eta_\omega: X_\omega \to {}_{\rightarrow}X$ is an isomorphism.

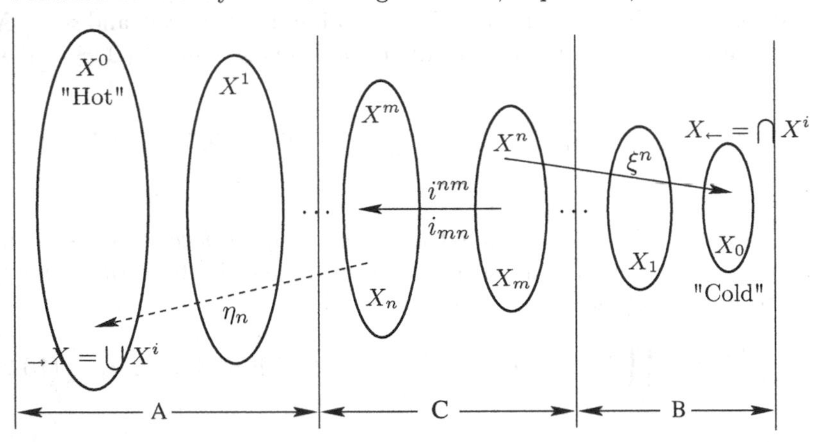

Fig. 10.5*b*. Direct-inverse limits for a family of nested subsets of a set X, with the direction of "order" $\mathbb{O}$ and "disorder" $\mathbb{D}$ — to be understood as implying smaller and larger multiplicities of the state — shown opposite so that all maps of the systems are now in the same direction. In the absence of a direct component, the inverse on its own would cause bottom-up, *self-organized* "cold death" to $X_{\leftarrow}$; if the inverse system were absent, *emergent*, "heat death" from the lone effect of the direct system would follow in $_{\rightarrow}X$, with each acting essentially as a gradient dissipator of the other. The nested decreasing subsets denote stability inspired *expansion* and self-organization as the system's response of utilizing "every possible avenue in sucking orderliness from its environment" to counter attempts to move it away from thermodynamic equilibrium, while the increasing supersets signify instability driven contraction and emergence. Compare Fig. 10.10*a*.

(b) Let $\{X^k\}_{k\in\mathbb{Z}_+}$ be a decreasing family of subsets of X, and let $\pi^{nm}: X^n \to X^m$ be the inclusion map for $m \leq n$. Since the inverse limit consists of only those points of the cartesian product whose "eventual" coordinate can be assigned independently,

$$X_{\leftarrow} \simeq \bigcap_{k=1}^{\infty} X^k \tag{10.2.47}$$

might be empty even though $X^k \neq \emptyset$ for each k, with $\pi^{kk} = \mathbf{1}$ being the identity map. What this result means is that the limit $X_{\leftarrow}$ must have all its components from the intersection only. Thus the inverse limit of $X_1 = [0,1]$, $X_2 = [0,0.6]$ and $X_3 = [0,0.2]$ is of the form $X_{\leftarrow} = \{(x_1,x_2,x_3) : x_1 \in X_3,\ x_2 \in X_3,\ x_3 \in X_3\}$ with the first and second coordinates considered to be elements of X_1 and X_2 respectively, by the inclusion map.

The consequence of these limit constructs in providing a dynamical basis to Postulates NEG-1 and NEG-2 of the exclusion space is contained in the following arguments. For a given resource λ, the inverse and direct limits $X_{\leftarrow}$ and ${}_{\rightarrow}X$, in competitive collaboration with each other, can be taken to represent respectively the *anabolic synthesis* of expansion, order, entropy-decrease and *catabolic analysis* of contraction, disorder, entropy-increase of the corresponding systems[15] leading to the dynamically equilibrated state $X_{\leftrightarrow}$: recall that everything else remaining the same, "hot" objects have higher entropy than "cold" ones, and when two bodies of different resources are brought in contact, entropy of the hot body *decreases* while that of the cold body *increases* such that the entropy decrease in the former is more than compensated by its increases in the later. This spontaneous flow of "heat" is associated with an overall entropy increase that continues till the combined entropy is a maximum. This is the essence of entropy production in the universe at the expense of exergy of the more resourceful constituent that in simple terms represents the opposition of a cold stable system to the urge of a hot unstable component to stabilize at its expense. The second law represents a straightforward stipulation that a part of the useful energy of a closed system must always be wasted as heat with the entropy being a quantitative measure of the amount of thermal energy not available for doing work, of the tendency for all matter and energy in the universe to evolve toward a dead state of inert uniformity. In the absence of the direct limit component, however, the inverse system would proceed to its logical destination of $X_{\leftarrow}$ leading to its

[15] *Metabolism* comprises the chemical processes taking place within a living cell or organism involving consumption and breakdown of complex compounds necessary for the maintenance of life, often accompanied by liberation of energy and waste products. It is the major process of living systems affecting all its chemical processes, consisting of a series of changes in an organism by means of which food is manufactured and utilized, and waste materials are eliminated. Metabolism is broadly subdivided into two opposing parts: *anabolic synthesis* of simple substances into complex materials is its constructive phase, and *catabolic analysis* of complex substances into simpler ones is the destructive.

minimum-entropy frozen "cold death", which translated to practical terms requires the whole system to acquire the unmoderated properties of the infinite colder reservoir. Inverse limits, however, demand the existence of connecting maps *opposing* $\mathbb{O}$; this manifests itself through generation of the reverse direction $\mathbb{D}$ of the direct limit which acting on its own would likewise lead to a maximum-entropy roasted "heat death" condition of $_{\to}X$. In communion with each other, $X_{\leftrightarrow}$ shares properties of both the opposites with the equilibrium representing some intermediate state $X^m = X_n$ of Fig. 10.5*b*. Physically this represents either (i) a hot body A^* interacting with another body B to yield the compound system $A^* (\equiv X_0) + B (\equiv X^0)$ which then evolves with time, or (ii) an infinite reservoir A^* that induces a temperature gradient in B; in this case the heat source remains external to the system. This reading of the dual limits, suggested by the directions of Fig. 10.5*b* representing converging sequences generated by points in the respective $\{X^\alpha\}$ and $\{X_\alpha\}$, can be viewed to be the basis of our postulate of an exclusion space leading to dynamical homeostasis, with the direction of the inverse limit being effectively inhibited by that of the direct limit. A second related interpretation is to consider, by the definition of footnote 7, the family of spaces and the restrictions of the associated projections to generate final and initial topologies on $_{\to}X$ and $X_{\leftarrow}$ respectively. The dynamically equilibrated steady state

$$X_{\leftrightarrow} = X^m = X_n \tag{10.2.48}$$

is therefore in an ininal state because all sub-diagrams of Fig. 10.5*a* must commute and the connecting sequences converge to the respective limits iff these carry the final and initial topologies of the direct and initial systems. Note that the dynamic equilibrium of (10.2.48) is effectively a saddle-node centre manifold, and is in fact the state $_{\text{eq}}$ of Eq. (10.2.2).

A thermodynamic analysis of the preceding heuristic rationale for the existence of a $X_{\leftrightarrow}$ will be given below that reduces the inverse-direct system to a coupled engine-pump dual with the natural inverse-limit engine $E : T_h \to T_c$ generating, under proper condition of irreversibility, a direct-limit pump $P : T_c \to T_h$ such that $X_{\leftrightarrow}$ is characterized by an equilibrium temperature $T \in [T_c, T_h]$.

In applying these considerations to the iterative evolution of maps, we take the domain of the interaction f to be a disjoint union C of a physical space A and an exclusion space B, when f generates bi-directional forward-backward arrows on C that are quite distinct from the catabolic-direct and anabolic-inverse limits. Accordingly two sets of arrows, the forward-inverse and backward-direct, are imposed on an evolving system and the character of the system depends on which of the two plays the role of an activating partner and which the restraining, representing a dynamical balance between the competitive collaboration of forward, self-organization and backward emergence, with new structures appearing only for the first few steps that is subsequently self-organized into a composite whole. This interpretation of the restoring effects implies that with appropriate interactions f, even extreme irreversibilities of non-injective ill-posedness can be effectively reversed with time, fully or partially depending on the nature of f, through internally generated regulating effects. Irreversibilities therefore need not be only wasteful: given adequate interactive support these can actually be utilized to induce higher-level order and discipline in the otherwise naturally occurring emerging entropic disorder, through a regulated process of adaption and self-organization. We employ this basic characteristic of the synthesis of matter and negative matter in formulating the definitions of complexity and "life" below.

The Lorenz Equation

To fully appreciate these observations and arrive at an understanding of the dynamics of difference equations vis-a-vis differential equations, we consider the Lorenz-Rayleigh-Benard model of two-dimensional convection of a horizontal layer of fluid heated from below involving three dynamical variables: x proportional to the circulatory convection velocity of the fluid that produces the flow pattern with positive x indicating clockwise circulation, y proportional to the temperature difference between the ascending warm and descending cold flows at a given height h, and z proportional to the nonlinear deviation of the vertical temperature profile from equilibrium linearity. The Lorenz equations

$$\dot{x} = \sigma(-x + y) \tag{10.2.49a}$$
$$\dot{y} = Rx - y - xz \tag{10.2.49b}$$
$$\dot{z} = xy - bz, \tag{10.2.49c}$$

with σ the Prandtl number (ratio of the kinematic viscosity of the fluid to its thermal diffusivity), $R = r/r_c$ the relative Rayleigh number (where $r := g\alpha d^3 \triangle T/(\kappa\nu)$ is the Rayleigh number — with g acceleration due to gravity, α, κ, ν coefficients of volume expansion, thermal diffusivity, kinematic viscosity, $\triangle T$ temperature difference between the upper and lower surfaces of the fluid separated by a distance d — and $r_c := (a^2 + \pi^2)^3/a^2 = 27\pi^4/4$ is the critical value that defines $a = \pi/\sqrt{2}$ to give the lowest r at which convection starts), and b (ratio of the width to the height of the region in which convection is occurring), represents a state of competing collaboration between the downward stabilizing arrow of gravity and an upward buoyancy-driven instability of viscous friction and conductive heat losses. The equilibrium fixed point $\dot{\mathbf{x}} = 0$ of supercritical pitchfork bifurcation

$$\dot{\mathbf{x}} = 0 \iff x^3 - b(R-1)x = 0$$

has the roots

$$C_0 = (0, 0, 0), \qquad \text{all } R \tag{10.2.50a}$$

$$C_\pm = (\pm\sqrt{b\rho}, \pm\sqrt{b\rho}, \rho), \qquad R > 1, \rho = R - 1. \qquad (10.2.50b)$$

A linear stability analysis about C_0 requires the characteristic polynomial of the combined linearized equation

$$\dot{\mathbf{x}} = \begin{pmatrix} -\sigma & \sigma & 0 \\ R & -1 & 0 \\ 0 & 0 & -b \end{pmatrix} \begin{pmatrix} x \\ y \\ z \end{pmatrix}$$

to satisfy

$$f(\lambda) := (\lambda + b)[\lambda^2 + (1+\sigma)\lambda - \sigma(R-1)] = 0 \qquad (10.2.51)$$

with the real eigenvalues

$$\begin{aligned} \lambda_z &= -b \\ \lambda_\pm &= -\frac{1+\sigma}{2} \pm \frac{1}{2}\sqrt{(1+\sigma)^2 + 4\sigma(R-1)} \end{aligned} \qquad (10.2.52)$$

in which only $\lambda_\pm$ depends on the control parameter R. It can now be verified that for all positive R, σ and b:

(a) $\underline{R < 1}$: All the zeros λ_z, $-(1+\sigma) \le \lambda_- \le -1$ (upper and lower bounds occurring at $R = 1$ and $R = 0$), $-\sigma \le \lambda_+ \le 0$ (bounds occurring at $R = 0$ and $R = 1$), are negative which means that C_0 is a *stable node.*

(b) $\underline{R = 1}$: λ_z, $\lambda_- = -(1+\sigma)$ are negative and $\lambda_+ = 0$ with corresponding eigenvectors $\mathbf{u}_z = (0,0,1)^{\mathrm{T}}$, $\mathbf{u}_- = (-\sigma,1,0)^{\mathrm{T}}$, and $\mathbf{u}_+ = (1,1,0)^{\mathrm{T}}$; hence C_0 is *marginally* (neutrally) *stable*, leading to its *pitchfork bifurcation.* The three real equilibria for $R > 1$ as given in (c) below merge to the single stable node of $R < 1$ at $R = 1$.

(c) $\underline{R > 1}$: λ_z and λ_- are negative, λ_+ is positive; hence C_0 is an unstable fixed point. The flows along the eigenvectors of λ_z and λ_- are stable that become unstable along the of λ_+ direction. Hence C_0 undergoes a *saddle node* in three dimensions in this parameter range.

Linearization about the two other equilibrium points $C_\pm$ according to $x \mapsto x \mp \sqrt{b\rho}$, $y \mapsto y \mp \sqrt{b\rho}$, and $z \mapsto z - \rho$ leads to the eigenvalue equation

$$\begin{aligned} g(\mu) &:= \begin{vmatrix} \sigma+\mu & -\sigma & 0 \\ -1 & 1+\mu & \pm\sqrt{b\rho} \\ \mp\sqrt{b\rho} & \mp\sqrt{b\rho} & b+\mu \end{vmatrix} \\ &= \mu^3 + (1+b+\sigma)\mu^2 + b(\sigma+R)\mu + 2b\sigma(R-1) = 0 \end{aligned} \qquad (10.2.53)$$

Since all its coefficients are positive and $g(0) > 0$ when $R > 1$, there is always a negative real root μ_z of Eq. (10.2.53). At $R = 1$, the three zeros of Eq. (10.2.53) are $\mu_z = -b$, $\mu_- = -(1+\sigma)$ and $\mu_+ = 0$, there are therefore two

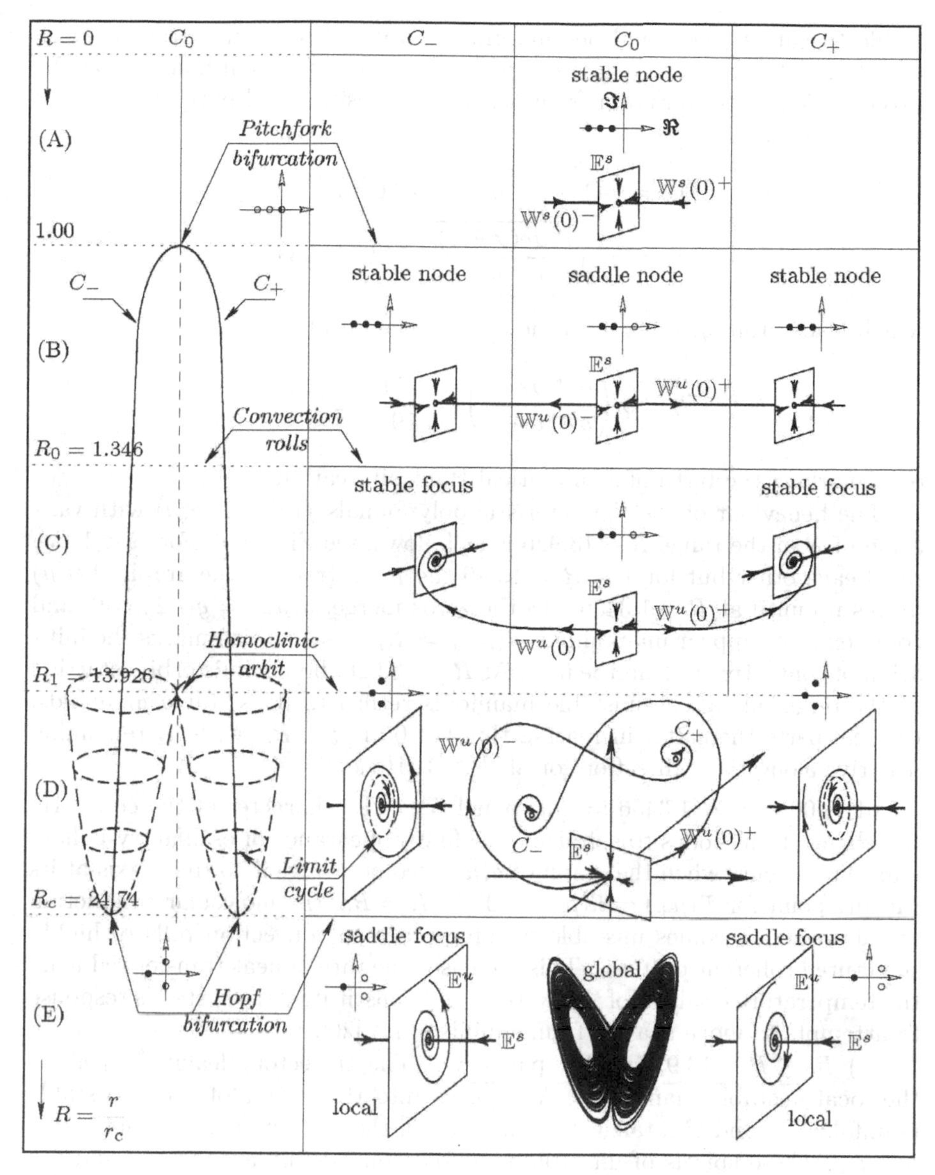

Fig. 10.6. Dynamics of the Lorenz equations. $\mathbb{E}$ and $\mathbb{W}$ are spanned by the respective eigenvectors of $\{\lambda_z, \lambda_-\}$ and λ_+ of Eq. (10.2.52). The local directions of the manifolds in panel (E) are determined by the eigenvectors of $C_\pm = (\pm\sqrt{b\rho}, \pm\sqrt{b\rho}, \rho)$, where $\rho = R - 1$ with R the relative Rayleigh number. Figure adapted from Argyris et al. [1]

stable (negative) roots and one marginally stable 0 root, in agreement with Eq. (10.2.51). From $\mu_z + \Re(\mu_-) + \Re(\mu_+) = -(1+b+\sigma) < 0$, it follows that the two complex roots cross over from negative to positive real parts, for $b = 8/3$ and $\sigma = 10$, when

$$\mu_z = -(1 + b + \sigma) = -13.6667,$$

$$\mu_\pm = \pm i \sqrt{\frac{2\sigma b(\sigma + 1)}{\sigma - b - 1}} = \pm 9.62453\, i,$$

which leads, from $g(\mu_z) = 0$, to the critical magnitude

$$R_c = \sigma \left(\frac{\sigma + b + 3}{\sigma - b - 1} \right) = \frac{470}{19} \simeq 24.7368$$

of R marking the birth of a subcritical Hopf bifurcation.

The behaviour of the characteristic polynomials $f(\lambda)$ and $g(\mu)$ with variation of R in the range $R < 13.926$ is as follows, see Fig. 10.6. For $R < 1$, $\lambda_\pm$ repel each other but for $1 \leq R < 1.346$ the $\mu_\pm$ attract as the graph of $g(\mu)$ moves up until at $R = 1.34561718$ the zeros merge, $g(\mu_-) = g(\mu_+) = 0$, and complex roots appear maintaining $\Re(\mu_-) = \Re(\mu_+) < 0$ which marks the initiation of convective rolls in the flow. At $R = 13.926$, homoclinic orbits starting at the origin along the unstable manifolds return to it as stable manifolds, the real parts thereafter increasing through 0 at $R = R_c$, with μ_z remaining negative along the z-direction for all $R > 1$. Hence

(d) $1.00 \leq R < 1.3456 := R_0$, panel (B). The character of the equilibria $C_\pm$ change from nodes to spirals in the first appearance of oscillatory behaviour. This occurs when the graph of $g(\mu)$ becomes tangent to the μ-axis at its turning point for $\Re(\mu_+) = \Re(\mu_-) < 0$. At $R = R_0$, the molecular conduction of this region becomes unstable yielding place to convection rolls of highly structured coherent patterns. This increases the rate of heat transfer reducing the temperature gradient of the system, and constitutes "the system's response to attempts to move it away from equilibrium", [26].

(e) $R_0 \leq R < 13.926 := R_1$, panel (C). The trajectory leaving C_0 along the local unstable manifold of λ_+ spirals into the nearer of the two stable manifolds C_- and C_+, tangent to the span of the respective eigenfunctions of μ_-, μ_+. These spirals of unstable manifolds on looping around C_- and C_+ increase in size with increasing R, until at

(f) $R = R_1$ they tend toward C_- and C_+ in wide arcs, eventually returning as *homoclinic orbits* to C_0 in the "infinite period limit" $t \to \pm\infty$. While no qualitative changes in the distribution of the zeros of Eq. (10.2.53) occur at this value of R, the emergence of homoclinic orbits can be attributed to the transformation of Eq. (10.2.53) to a monotonically increasing function of μ for all $R > R_1$. This is a significant event in the time evolution of the Lorenz equations that eventually leads to chaos at $R = R_c$. This mechanism to chaotic transition is common in systems modeled by differential equations and is not — unlike for maps — accompanied by any change in the character of fixed points but is due to interaction of the trajectory with various instabilities.

(g) $R_1 \leq R < R_c \simeq 24.7368$, panel (D). As R increases beyond R_1, the monotonically increasing $g(\mu)$ results in the homoclinic orbits transforming to increasing *finite period* unstable orbits that eventually coalesce to disappear in a subcritical Hopf bifurcation at $R = R_c$. These increasingly oscillatory solutions of the *pre-chaotic* range $R_1 < R < 24.06$ travel back and forth between C_- and C_+ many times before finally spiraling into one of them: as R increases in this range, the generated unstable limit cycles repel $\mathbb{W}^u(0)$ so that the branch leaving C_0 in the octant of C_- converges to C_+ and that generated in the octant of C_+ ends up at C_-, with the number of crossings between C_- and C_+ increasing with R before eventually converging to one of them. The unstable limit cycles associated with C_-, C_+ shrink in size as R increases, passing over to a subcritical Hopf bifurcation at $R = R_c$. In the range $24.06 < R < R_c$ although the equilibria $C_\pm$ remain stable, some of the pre-chaotic orbits pass over into true chaos; hence in this region there is a chaotic attractor beside the two spiral attractors. At $R = R_c$, the stable spirals become unstable by absorbing the unstable spirals.

This dynamics of the Lorenz equation summarized in Fig. 10.6 allows us to draw the following correspondences with the logistic interaction $\{f_\lambda\}_{\lambda \in [0,4]}$.

- $0 \leq R < 1.00 \Leftrightarrow 0 \leq \lambda < 1$, panel (A). Heat is transferred from the hot bottom to the cold top by molecular thermal conduction. The tendency of the warm, lighter fluid to rise is inhibited by viscous damping and loss by conduction from the hot fluid to the surrounding cooler medium, and the temperature varies linearly with the height of separation between the plates. Recall that the only logistic fixed point $x_0 = 0$ is stable in this range, like the Lorenz C_0. See Fig. 10.8*a*
- $1.00 \leq R < 1.3456 \Leftrightarrow 1 \leq \lambda < 2$, panel (B). This λ-region of loss of stability of x_0 at $\lambda = 1$ and the simultaneous birth of a new stable fixed point marks the onset of a radial R-interaction between the now unstable C_0 and the new stable pair $C_\pm$, Fig. 10.8*a*.
- $1.3456 \leq R < 13.926 \Leftrightarrow 2 \leq \lambda < 3$, panel (C). Oscillations occur in the stable evolution of the logistic map, Fig. 10.8*a*(iv), corresponding to the appearance of the circular convective rolls in the Lorenz equations along the second angular θ-direction consequent of the appearance of complex roots of $g(\mu)$, Eq. (10.2.53).
- $13.926 \leq R < 24.7368 \Leftrightarrow 3 \leq \lambda < 1 + \sqrt{6} = 3.4495$, panel (D). This region of the initiation of period doubling of the one-dimensional map relates to the homoclinic orbit and the unstable limit cycles representing radial interaction between C_0 and $C_\pm$ that activates the third angular φ-direction at C_0. Note that as in the logistic interaction, this R-region is distinguished by the coexistence of the opposite directions due to the stable fixed points C_- and C_+ corresponding to the stable 2-cycle of the map of Fig. 10.8*b*.

The important point to note here is that *unlike for period doubling of the logistic map, the supercritical pitchfork bifurcation in a multidimensional space enables the unstable C_0 to interact with the stable $C_{\pm}$ by opening up new pathways along the angular coordinate directions.* In the one dimensional logistic case where the luxury of the new directions acting as additional tunable parameters are unavailable, *a tiered hierarchal communication system is established between the unstable and stable points* in order to utilize the additional λ-resource available to carry the evolutionary dynamics forward. In fact, compared to the sufficient conditions

$$f = 0, \quad \frac{\partial f}{\partial x} = 0, \quad (x, \mu) = (0, 0) \tag{10.2.54a}$$

and

$$\begin{aligned} \frac{\partial f}{\partial \mu} &= 0, \quad \frac{\partial^2 f}{\partial x^2} = 0, \\ \frac{\partial^2 f}{\partial x \partial \mu} &\neq 0, \quad \frac{\partial^3 f}{\partial x^3} \neq 0 \end{aligned} \tag{10.2.54b}$$

for non-hyperbolicity and pitchfork bifurcation respectively of a one-parameter, one-dimensional vector field $\dot{x} = f(x, \mu)$, a one-dimensional map $x \mapsto f(x, \mu)$ with non-hyperbolic fixed points

$$f = 0, \quad \frac{\partial f}{\partial x} = \pm 1, \quad (x, \mu) = (0, 0), \tag{10.2.55a}$$

not only undergoes pitchfork bifurcation at $\partial f / \partial x = 1$ for the same conditions as given by Eq. (10.2.54b), but more importantly a period doubling bifurcation appears whenever the non-hyperbolic slope $\partial f / \partial x = -1$ emerges and the second iterate of the map passes through a pitchfork

$$\begin{aligned} \frac{\partial f^2}{\partial x} = 1, \quad \frac{\partial f^2}{\partial \mu} &= 0, \quad \frac{\partial^2 f^2}{\partial x^2} = 0, \\ \frac{\partial^2 f^2}{\partial x \partial \mu} &\neq 0, \quad \frac{\partial^3 f^2}{\partial x^3} \neq 0 \end{aligned} \tag{10.2.55b}$$

at (x, μ). More generally, any increase in λ is gainfully employed by the logistic map through a series of period doublings such that a 2^N cycle is generated to effectively utilize the resource λ in N bifurcations, as can be verified from Figs. 10.8b, c and 10.8d that show how the emerging structure develops in N steps terminating with the period-doubling-pitchfork

$$\frac{\partial f^{2^{N-1}}}{\partial x} = -1, \quad \text{(period-doubling)} \tag{10.2.56a}$$

$$\frac{\partial f^{2^N}}{\partial x} = 1, \quad \text{(pitchfork)} \tag{10.2.56b}$$

combination at 2^{N-1} stable-unstable fixed points marking the complete utilization of λ, with the slopes of f^{2^N} and $f^{2^{N-1}}$ simultaneously moving out of the stable unit interval in *opposite directions* into the unstable region $|x| > 1$, in the classic bidirectional competitive collaboration mode. In the absence of this typical double bound of the stable region for differential equations, the possible structures supported by these dynamical systems are comparatively simpler. Specifically it does not possess the hierarchal towered form that is the characteristic feature of two-component ill-posed maps such as the logistic where Eqs. (10.2.56a,b) actually determine the fixed-point x_* and the corresponding λ-value of the end of period 2^{N-1} and beginning of period 2^N. It is this distinction in the relationship between the stable and unstable points that is responsible for the difference between arbitrary complex systems and dissipative structures made below.

- $\underline{R_c \leq R \Leftrightarrow \lambda_1 < \lambda \leq 4}$, panel (E). This R-ray symbolizing total chaos, is characterized as in the logistic case, by the complete lack of stabilizing effects, as the orbits generated by C_- and C_+ endlessly wander between them. Unlike the one-dimensional map, however, the three dimensional differential system does not display characteristic bifurcations beyond R_c, taking advantage instead of the added dimensional latitude in generating an entangled attractor with non-periodic orbits and sensitivity to initial conditions.

Although it is possible, as has been argued above, to establish an overall correspondence between the dynamics of discrete and continuous systems, a careful consideration reveals some notable fundamentally distinctive characteristics between the two that ultimately reflects on the higher number of space dimensions — (r, θ, φ) in the Lorenz case — available to the differential system[16]. This has the consequence that continuous time evolution governed by differential equations is reductionally well-defined and unique — unlike in the discrete case when ill-posedness and multifunctionality forms its defining character — with the system being severely restrained in its manifestation, not possessing a set of equivalent yet discernible possibilities to choose from. In fact, the dynamics of differential equations cannot generate attractors composed of isolated points like the Cantor set, and *it is our premise that the kitchen of Nature functions in an one-dimensional iterative analogue,*

[16] Thus, for example, as in Eq. (10.2.41a,b), the equivalent *Lorenz difference equation*

$$\begin{aligned} x_{n+1} &= x_n(1-\sigma) + \sigma y_n \\ y_{n+1} &= Rx_n - x_n z_n \\ z_{n+1} &= z_n(1-b) + x_n y_n \end{aligned}$$

would have a distinct and different dynamical evolution that is expected to have little bearing or similarity with the solution of its differential counterpart (10.2.49a–c).

not merely to take advantage of the multiplicities inherent therein, but more importantly to structure its dynamical evolution in a hierarchal canopy, so essential for the evolution of an interactive, non-trivial, complex system. The 3-dimensional serving table of physical space only provides a convenient and palatable presentation of nature's produce in its uni-dimensional kitchen. A closed system can gain overall order while increasing its entropy by some of the system's macroscopic degrees of freedom becoming more organized at the expense of microscopic disorder. In many cases of biological self-assembly, for instance metabolism, the increasing organization of large molecules is more than compensated by the increasing disorder of smaller molecules, especially water. At the level of whole organisms and longer time scales, though, biological systems are open systems feeding on the environment and dumping waste into it.

The special significance of one-dimensional dynamics relative to any other finds an appealing substantiation from the following interpretation of the Sharkovskii Theorem. Recall that the distinguished Sharkovskii ordering

$$3 \succ 5 \succ 7 \succ \cdots \succ 2\cdot 3 \succ 2\cdot 5 \succ 2\cdot 7 \succ \cdots \succ 2^n\cdot 3 \succ 2^n\cdot 5 \succ 2^n\cdot 7 \succ \\ \cdots \succ 2^n \succ \cdots \succ 2^2 \succ 2 \succ 1$$

of positive integers implies the Sharkovskii Theorem which states that if $f:[a,b] \to \mathbb{R}$ is a continuous function having a n-periodic point, and if $n \succ m$, then f also has a m-periodic point: observe the significance of the upper and lower bounds of this ordering. Noting that the periodicity of an f-interaction between two spaces essentially denotes the number of independent degrees of freedom required to completely quantify the dynamics of f, it is inferred that while a fixed point of "dimension" 1 embodies the basic informations of all other periods, a period-3 embodies every other dimension within itself. Hence it can be concluded that dynamics on 1-dimension, by being maximally restrained compared to any other, allows for the greatest emergence of structures as mutifunctional graphical limits, while dimension 3 by being the least restrained is ideally suited for an outward well-defined, and aesthetically appealing, simultaneous expression of the multitude of eventualities that the graphical limits entail.

The convection rotating cells of the Lorenz system that appear spontaneously in the liquid layer when heated from outside is an example of Prigogine's *dissipative structure* [15]. At first when the temperature of the bottom plate T_h is equal to that of the top T_c, the liquid will be in equilibrium with its environment. Then as the temperature of the bottom is increased, the fluid resists the applied temperature gradient $\Delta T = (T_h - T_c) \backsim R$ by setting up a backward arrow of inter-molecular conductive dissipation, and the temperature increases linearly from top to bottom to establish thermal equilibrium in the fluid. If the temperature of the bottom is increased further, there will be a far from equilibrium temperature T_0 corresponding to R_0 of Fig. 10.6 at which the system becomes unstable, the incoherent molecular conduction yields place to coherent convection, and the cells appear increasing the rate of dissipation. The appearance of these ordered convective structures — a "striking example of emergent coherent organization in response to an external energy input" [28] — dissipates more energy than simple conduction, and convection becomes the dominant mode of heat transfer as R increases further. The microscopic random movement of conduction spontaneously becomes macroscopically ordered with a characteristic correlation length generated by convection. The rotation of the cells is stable and alternates between clockwise to counter-clockwise horizontally, and there is spontaneous symmetry breaking.

According to Schneider and Kay [28], the basic role of dissipative structures, like the Lorenz convection cells, is to act as *gradient dissipators* by "continually sucking orderliness from its environment" in hindering motion of the system away from equilibrium due to the increasing temperature gradients. The dissipative structures increase the rate of heat transfer in the fluid thereby utilizing this exergy in performing useful work in generating the structures. With increasing gradient, more work needs to be done to maintain the increased dissipation in the far-from-equilibrium state, more exergy must be destroyed in creating more entropy, the boundary layers become thinner, and the original vertically uniform temperature profile is restored in the bulk of the fluid. The structures developed in the Lorenz system thus organize the disorder of the backward convective cells by dissipation of an increasing amount of exergy in the activating, forward "sucking-orderliness" direction of heating.

Thermodynamics of Bidirectionality: Optimized Adaptation in Engine-Pump Duality

They know enough who know how to learn.

Henry Adams

This subsection is an investigation into the relationship of our steady state $X_{\leftrightarrow}$ to the entropy principle of non-equilibrium thermodynamics. In recent papers Dewar [9] establishes the Maximum Entropy Principle for stationary states of open, non-equilibrium systems by maximizing the path information entropy $S = -\sum_{\Gamma} p_{\Gamma} \ln p_{\Gamma}$ with respect to p_{Γ} subject to the imposed constraints. In this non-equilibrium situation, the maximum entropy principle amounts to finding the most probable history realizable by the largest number of microscopic *paths* rather than microscopic *states* typical of Boltzmann-Gibbs equilibrium statistical mechanics. This approach to non-equilibrium MEP is supported by many investigations: the earth-atmosphere global fluid system, for example, is believed to operate such that it generates maximum potential energy and the steady state of convective fluid systems, like that of the Lorenz model, have been suggested to represent a state of maximum convective heat transfer, [23].

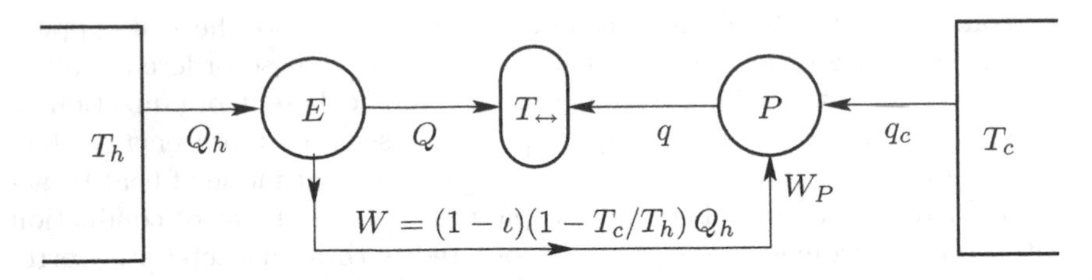

Fig. 10.7. Reduction of the dynamics of opposites of Fig. 10.5*b* to an equivalent engine-pump thermodynamic system. The fraction $W = (1-\iota)W_C$ of the available maximum reversible work $W_C = \eta_C\,Q_h := (1-T_c/T_h)\,Q_h$ of a reversible engine operating between $[T_c, T_h]$ is internally utilized to self-generate a heat pump P to inhibit, by gradient dissipation, the entropy that would otherwise be produced in the system. This permits decoupling natural irreversibility to a reversible engine-pump dual that uses the fraction ι of the available exergy in running the pump. The coefficient of performance $q/W = q/(q-q_c) = T_{\leftrightarrow}/(T_{\leftrightarrow} - T_c)$ of P establishes the reverse arrow of $q := q_c + W$. The two parameters $T_{\leftrightarrow}$ and ι are obtained as described in the text.

An effective reduction of the inverse-direct model of Fig. 10.5*b* as a coupled thermodynamic engine-pump system is illustrated in Fig. 10.7 in which heat transfer between temperatures $T_h > T_c$ is reduced to a engine E-pump P combination operating respectively between temperatures $T < T_h$ and $T_c < T$. We assume that a complex adaptive system is distinguished by the full utilization of the fraction $W := (1-\iota)W_C = (1-\iota)\eta_C Q_h = (1-\iota)(1-T_c/T_h)Q_h$ of the work output of an imaginary reversible engine running between temperatures T_h and T_c, to generate a pump P working in competitive collaboration with a reversible engine E, where the irreversibility index

$$\iota \stackrel{\text{def}}{=} \frac{W_C - W}{W_C} \in [0,1] \tag{10.2.57}$$

accounts for that part ιW_C of available energy (exergy) that cannot be gainfully utilized but must be degraded in increasing the entropy of the universe. The self-induced pump effectively decreases the temperature gradient $T_h - T_c$ operating the engine to a value $T_h - T$, $T_c \le T < T_h$, thereby inducing a degree of dynamic stability to the system.[17] With $q = q_c + (1-\iota)W_C = q_c + W_P$, the coefficient of performance $\zeta_P = q/W_P = T/(T-T_c)$ of P yields

$$q = (1-\iota)Q_h\left(\frac{T}{T_h}\right)\left(\frac{T_h - T_c}{T - T_c}\right).$$

Let the irreversibility ι be computed on the basis of dynamic equilibrium[18]

[17] More generally, W is to be understood to be indicative of the exergy of Eq. (10.2.2).

[18] Note that this is $W_E = W_P = \iota W_C$.

$$Q_h\left(\frac{T_h-T}{T_h}\right):=W_E(T)=W_P(T):=q\left(\frac{T-T_c}{T}\right)$$

of the engine-pump system; hence

$$\begin{aligned}\iota&=\frac{T-T_c}{T_h-T_c}\\&=\frac{(T_h-T_c)-(T_h-T)}{T_h-T_c}\end{aligned}\qquad(10.2.58)$$

where $T_h - T_c$ represents the original reversible work that is split up into the non-entropic $T_h - T$ shaft output internally utilized to generate the pump P, and a $T - T_c$ manifestation of entropic work by P with the equilibrium temperature T defining this recursive dynamics. The irreversibility ι can be taken to have been adapted by the engine-pump system such that the induced instability due to P balances the imposed stabilizing effort of E to the best possible advantage of the system and its surroundings. This the system does by adapting itself to a state that optimizes competitive collaboration for the greatest efficiency consistent with this competitiveness. This distinguishing feature of the non-equilibrium situation with corresponding equilibrium case lies in the mobility of the defining temperature T: for the introverted self-adaptive systems, the dynamics organizes to the prevailing situation by best adjusting itself *internally* for maximum possible global advantage.

Define the equilibrium steady-state representing $X_{\leftrightarrow}$ of optimized E-P adaptability between E and P be given in terms of the adaptability function

$$\alpha_P(T_P):=\eta_E\zeta_P=\left(\frac{T_h-T_P}{T_P-T_c}\right)\left(\frac{T_P}{T_h}\right)$$

that represents an effective adaptive efficiency of the engine-pump system to the environment (T_c, T_h). Hence

$$T_P=\frac{1}{2}\left[(1-\alpha_P)T_h+\sqrt{(1-\alpha_P)^2T_h^2+4\alpha_PT_hT_c}\right].\qquad(10.2.59a)$$

Alternatively if the system induces P to act as a refrigerator rather than a pump then the defining equations, with $\zeta_R := q_c/(q - q_c) = T_c/(T_R - T_c)$, become

$$q_c=(1-\iota)Q_h\left(\frac{T_c}{T_h}\right)\left(\frac{T_h-T_c}{T_R-T_c}\right),$$

with the adaptability criterion

$$\alpha_R(T_R):=\eta_E\zeta_R=\left(\frac{T_h-T_R}{T_R-T_c}\right)\left(\frac{T_c}{T_h}\right)$$

leading to

$$T_R = \frac{(1+\alpha_R)T_hT_c}{T_c+\alpha_R T_h}. \tag{10.2.59b}$$

For the reversible $(\iota = 0) \Rightarrow (T = T_c)$, $\alpha \to \infty$ case, with no entropy production and no generation of P, the resulting inverse-system operates uni-directionally as an ordering agent, while in the absence of E at $(\iota = 1) \Rightarrow (T = T_h)$, $\alpha = 0$, the self-generation of P cannot, infact, occur. An intermediate, non-zero, finite value of α is what the self-emergent system seeks for its optimization that we take to be the maximum at $\alpha := 1 - T_c/T_h$. Hence

$$\alpha = \eta_C \Longrightarrow \begin{cases} T_P = \frac{1}{2}\left[T_c + \sqrt{T_c^2 + 4T_c(T_h - T_c)}\right] \\ T_R = \frac{(2T_h - T_c)T_c}{T_h}, \end{cases} \tag{10.2.60}$$

leads to $\iota_R = T_c/T_h := 1 - \alpha$ for the E-R system. The original temperature gradient $T_h - T_c$ is shared by the $E{-}P$ system in the true spirit of synthetic cohabitation of opposites in the proportion $E : T_h - T$, $P : T - T_c$ thereby optimizing its adaptability to the environment.

	E-P	E-R		E-P	E-R
T	426.5860	412.5000	q_c	19.7791	28.1250
ι	0.7033	0.6250	S_1	0.09375	0.09375
W_C	28.1250	28.1250	$S_{\leftrightarrow}$	0.06593	0.05859
W_E	8.3459	10.5469	η	0.1113	0.1406
Q_c	66.6541	64.4531	$\eta_{\leftrightarrow}$	0.1113	0.1406
Q	55.6541	64.4531	ζ	8.9865	6.1111
q	28.1250	38.6719	$\zeta_{\leftrightarrow}$	3.3699	2.6667

Table 10.5. Comparison of engine-pump and engine-refrigerator bi-directionality. The equations used for E-R are (with corresponding ones for E-P): $\alpha = 0.375$, $W_C = [1 - (T_c/T_h)]Q_h$, $W_E = [1 - (T/T_h)]Q_h$, $Q_c = Q_h - (1-\iota)W_C$, $Q = Q_h - W_E$, $q = (1-\iota)(T/(T-T_c))W_C$, $q_c = (1-\iota)(T_c/(T-T_c))W_C$, $S_1 = W_C/T_c$, $S_{\leftrightarrow} = \iota W_C/T_c = (Q_h/T_h)[(T/T_c) - 1]$, $\eta = (Q_h - Q_c)/Q_h$, $\eta_{\leftrightarrow} = (T_h - T)/T_h$, $\zeta = Q_c/(Q_h - Q_c)$, $\zeta_{\leftrightarrow} = q_c/(q_h - q_c) = T_c/(T_h - T_c)$. The role of the pump as a "gradient dissipator" is to decrease the irreversibility (and chanoxity) index from the metallic conduction value of 1 to $(T - T_c)/(T_h - T_c)$.

As an example, in the conduction of heat along a bar from $T_h = 480°\text{K}$ to $T_c = 300°\text{K}$ for $Q_c = Q_h - W(=0) = 75\,\text{kJ-min}^{-1}$ involving an entropy increase of $S_{\iota=1} = -75/480 + 75/300 = 0.09375\,\text{kJ-(min-K)}^{-1}$. If the bar is replaced by a reversible $\iota = 0$ engine between the same temperatures, then $W_C = 28.125\,\text{kJ-min}^{-1}$, $Q_c = Q_h - W = (W_C) = 46.875\,\text{kJ-min}^{-1}$, and the entropy change of $S_{\iota=0} = -75/480 + 46.875/300 = 0$ precludes any emergence in this reversible case. If, however, bi-directionality of $X_{\leftrightarrow}$ is to be established by an induced pump or refrigerator then the results, summarized in Table 10.5, shows that the actual entropy increases are 70% of the unmoderated value S_1 with an increase of the shaft work to $(1-\iota)W_C$ from 0.

This self-generation of bi-directional stability is to be compared and contrasted with the entropy generation when a hot body is brought in thermal contact with a cold body: As in the bi-directional case, the entropy increase $m_1c_1\ln(T/T_h) + m_2c_2\ln(T/T_c)$ of the universe is maximum at $T = T_h$ and minimum for $T = T_c$. Unlike in self-organizing complexes however, the equilibrium system has a well-defined temperature $T = (m_1c_1T_h + m_2c_2T_c)/(m_1c_1 + m_2c_2)$ that is not amenable to adjustment by the system for its best possible advantage, with the resultant *negative* entropy $m_1c_1\ln(T_c/T_h)$ implying that order must be imported from outside if such a condition is to be physically realizable. Thus for $m_1/m_2 = 30\,\text{kg}/150\,\text{kg}$, $c_1/c_2 = 0.5\,\text{kJ/kg-}^\circ\text{K}/2.5\,\text{kJ/kg-}^\circ\text{K}$, and $T_h/T_c = 480^\circ\text{K}/300^\circ\text{K}$, whereas the equilibrium temperatute of $T = 306.92^\circ\text{K}$ generates $1.8477\,\text{kJ/K}$ of entropy, for a self-organizing system reversibility would impose $T = 305.472^\circ\text{K}$ as the solution of $0 = m_1c_1\ln(T/T_h) + m_2c_2\ln(T/T_c)$, import $7.05\,\text{kJ/K}$ of order from the enlarged environment at $T = T_c$, and export $176.25\,\text{kJ/K}$ of disorder when $T = T_h$.[19]

[19] In a revealing analysis of *What is Life?* [29], the theoretical biologist Robert Rosen contends [25] that it is precisely the duality between "how a given material system changes its own behaviour in response to a force, and how that same system can generate forces that change the behaviour of other systems" that Schrodinger was addressing in the context of Mendelian genes and molecules and "the mode of forcing of phenotypes (the actual physical properties of a molecule) by genotypes (the genetic profile of the molecule)". While the phenotype and genotype are related, they are not necessarily identical with the environment playing an important role in shaping the actual phenotype that results, Rosen proceeds to argue that "We cannot hope for identical relations between inertial and gravitational aspects of a system, such as are found in the very special realms of particle mechanics. Yet, in a sense, this is precisely what Schroedinger essay is about. Delbruck was seeking to literally reify a *forcing* (the Mandelian gene), *something 'gravitational'*, by clothing it in *something with 'inertia'*, by realizing it as a molecule. Schrodinger, on the other hand understood that this was not nearly enough, that we must be able to go the other way and determine the forcings manifested by something characterized inertially: just as we realize a *force* by a *thing*, we must also, perhaps more importantly, be able to realize a *thing* by a *force* (emphasis added). It was in this later connection that Schrodinger put forward the 'principle of order from order' and the 'feeding of negative entropy'. It was here that he was looking for the new physics".

In the Lorenz system, the potential energy of the top-heavy liquid created by the imposed temperature gradient $\Delta T = T_h - T_c$, taking T_c to be fixed, leads to conversion of the input heat energy to mechanical work of convective viscous mixing that acts as a gradient dissipator. Taking $Q_h = 1$, W_{r} corresponds to R and $\iota = (R - R_g)/R$ to that fraction of R that is not utilized in gravitational gradient dissipation through convection. In an arbitrary non-equilibrium steady state, the temperature induced upward potential energy production must be balanced by the dissipations which includes an atmospheric loss component also. In general for the non-equilibrium steady state $X_{\leftrightarrow}$, the increase in internal stability due to viscous dissipation leads to a backward-forward synthesis, when the direct arrow of entropy increasing emergence is moderated by the inverse arrow of order and self-organization. This is when all irreversible motivations guiding the system must cease, and the dead state of a "local non-equilibrium maximum entropy" — of magnitude less than that of the completely irreversible "global" equilibrium conductive state — consistent with the applied constraint of viscous damping, is reached. Refer Fig. 10.5*b*.

The earth-atmosphere system offers another striking example of this non-equilibrium local principle, in which the earth is considered as a two-region body of the hot equator at T_h and the cold poles at T_c, with radiative heat input at the equator and thermal dissipation at the poles. A portion of the corresponding W_{r} is utilized in establishing the P induces pole $\leftrightarrow$ equator atmospheric circulation resulting in internal stabilization, structuring, and inhibitory gradient dissipation. The radiative polar heat loss constitutes the entropy increasing direct arrow that is moderated by the that makes this planet habitable.

As a final illustration, mention can be made to the interesting example of frost heaving [22] as a unique model of a "reverse Lorenz system" where the temperature gradient is *along* the direction of gravity. A regular Lorenz under such conditions would be maximally irreversible, as an effective conductive entity, without any internal generation of P-stabilization. In frost heaving, however, ice and supercooled water are partitioned by a microporous material permeable to the water, the pressure of the ice on the top of the membrane being larger than that exerted by the water below: thus the temperature and pressure of the water below are less than that of the ice above. If the water is sufficiently supercooled however, it flows up against gravity due to P, into the ice layer, freezes and in the process heaves the ice column up.

Thus according to Rosen, Schroedinger supreme contribution in posing his now famous question elevated the object of his inquiry from a passive adjective to an active noun by suggesting the necessity of a "new physics" for investigating how in open, non-equilibrium systems, every forward-indirect arrow of phenotype inertia engine E is necessarily coupled to a backward-direct impulse from some genotype gravity pump P. For Schrodinger while a Mandelian gene was surely a molecule, it was more important to investigate when the molecule becomes a gene.

The non-equilibrium steady-state $X_{\leftrightarrow}$, Equation (10.2.48), is therefore a local maximum-entropy state that the dynamics of the non-linear system seeks as its most gainful eventuality, given the constraint of conflicting and contradictory demands of the universe it inhabits, with the constraints effectively lowering the entropic sum $S = -\sum_j p_j \ln p_j$. Accordingly while the entropy of a partition of unconstrained elementary events in the rolling of a fair die with $\{p_j\}_{j=1}^{6} = 1/6$ is $\ln 6 = 1.7918$, the entropy of a constrained partition satisfying $p_1 + p_3 + p_5 = 0.6$ and $p_2 + p_4 + p_6 = 0.4$ in the appearance of odd and even faces is $0.6 \ln(0.2) + 0.4 \ln(0.1333) = 1.7716$. The applied constraints therefore reduce the number of faces of the die to an unconstrained effective value of $\exp(1.7716) = 5.88$, thereby reducing the disorder of the system, which can be interpreted as a corresponding lowering of the temperature gradient ΔT of the irreversible $\iota = 1$ instance of $W = 0$. In the examples above the respective constraints are the convection rolls, atmospheric convection currents, and anti-gravity frost heaving. Without this component of the energy input, emergent internal structuring in natural systems would be absent. It may therefore be inferred that the two-component decomposition (10.2.1) of entropy corresponds to the break-up we propose here.

10.2.3 An Index of Nonlinearity

At the moment there is no formalization of complexity that enables it to overcome its current rather confused state and to achieve the objective of first becoming a method and then a bonafide scientific theory. The complexity approach that has recently appeared in modern scientific circles is generally still limited to an empirical phase in which the concepts are not abundantly clear and the methods and techniques are noticeable lacking. This can lead to the abuse of the term "complexity" which is sometimes used in various contexts, in senses that are very different from one another, to describe situations in which the system does not even display complex characteristics.

Formalizing complexity would enable a set of empirical observations, which is what complexity is now, to be transformed into a real hypothetical-deductive theory or into an empirical science. Therefore, at least for the moment, there is no unified theory of complexity able to express the structures and the processes that are common to the different phenomena that can be grouped under the general heading of complexity. There are several evident shortcomings in modern mathematics which make the application of a complexity theory of little effect. Basically this can be put down to the fact that mathematics is generally linear.

We are now faced with the following problem. We are not able to describe chaotic phenomenology or even that type of organized chaos that is complexity by means of adequate general laws; consequently we are not able to formulate effective long-term predictions on the evolution of complex systems. The mathematics that is available to us does not enable us to do this in an adequate manner, as the techniques of such mathematics were essentially developed to describe linear phenomena in which there are no mechanisms that unevenly amplify any initial uncertainty or perturbation.

Bertuglia and Vaio [3]

With ininality in the cartesian space $C \times C$ serving as the engine for the increase of evolutionary entropic disorder, we now examine how a specifically nonlinear index can be ascribed to chaos, nonlinearity and complexity to serve as the benchmark for chanoxity. For this, we first recall two non-calculus formulations of entropy that measure the complexity of dynamics of evolution of a map f.

Let $\mathcal{A} = \{A_i\}_{i=1}^{I}$ be a disjoint partition of non-empty subsets of a set X; thus $\bigcup_{i=1}^{I} A_i = X$. The entropy

$$S(\mathcal{A}) = -\sum_{i=1}^{I} \mu(A_i) \ln(\mu(A_i)), \qquad \sum_{i=1}^{I} \mu(A_i) = 1 \tag{10.2.61}$$

of the partition $\mathcal{A}$, where $\mu(A_i)$ is some normalized invariant measure of the elements of the partition, quantifies the uncertainty of the outcome of an experiment on the occurrence of any element A_i of the partition $\mathcal{A}$. A refinement $\mathcal{B} = \{B_j\}_{j=1}^{J \geq I}$ of the partition $\mathcal{A}$ is another partition such that every B_j is a subset of some $A_i \in \mathcal{A}$, and the largest common refinement

$$\mathcal{A} \bullet \mathcal{B} = \{C \colon C = A_i \bigcap B_j \text{ for some } A_i \in \mathcal{A}, \text{ and } B_j \in \mathcal{B}\}$$

of $\mathcal{A}$ and $\mathcal{B}$ is the partition whose elements are intersections of those of $\mathcal{A}$ and $\mathcal{B}$. The entropy of $\mathcal{A} \bullet \mathcal{B}$ is given by

$$\begin{aligned} S(\mathcal{A} \bullet \mathcal{B}) &= S(\mathcal{A}) + S(\mathcal{B} \mid \mathcal{A}) \\ &= S(\mathcal{B}) + S(\mathcal{A} \mid \mathcal{B}), \end{aligned} \tag{10.2.62}$$

where the weighted average

$$S(\mathcal{B} \mid \mathcal{A}) = \sum_{i=1}^{I} P(A_i)\, S(\mathcal{B} \mid A_i) \tag{10.2.63a}$$

of the conditional entropy

$$S(\mathcal{B} \mid A_i) = -\sum_{j=1}^{J} P(B_j \mid A_i) \ln(P(B_j \mid A_i)) \tag{10.2.63b}$$

of $\mathcal{B}$ given $A_i \in \mathcal{A}$, is a measure of the uncertainty of $\mathcal{B}$ if at each trial it is known which among the events A_i has occurred, and

$$P(B_j \mid A_i) = \frac{P(B_j \cap A_i)}{P(A_i)} \tag{10.2.63c}$$

yields the probability measure $P(B_j \cap A_i)$ from the conditional probability $P(B_j \mid A_i)$ of B_j given A_i, with $P(A)$ the probability measure of event A.

The entropy (10.2.61) of the refinement $\mathcal{A}^n$, rather than (10.2.62), that has been used by Kolmogorov in the form

$$h_{\mathrm{KS}}(f;\mu) = \sup_{\mathcal{A}_0}\left(\lim_{n\to\infty}\frac{1}{n}S(\mathcal{A}^n)\right) \tag{10.2.64}$$

to represent the complexity of the map as measuring the time rate of creation of information with evolution, yields $\ln 2$ for the tent transformation. Another measure — the topological entropy $h_{\mathrm{T}}(f) := \sup_{\mathcal{A}_0}\lim_{n\to\infty}(\ln N_n(\mathcal{A}_0)/n)$ with $N_n(\mathcal{A}_0)$ the number of divisions of the partition $\mathcal{A}^n$ derived from $\mathcal{A}_0$, that reduces to

$$h_{\mathrm{T}}(f) = \lim_{n\to\infty}\frac{1}{n}\ln \mathcal{I}(f^n) \tag{10.2.65}$$

in terms of the number of injective branches $\mathcal{I}(f^n)$ of f^n for partitions generated by piecewise monotone functions — also yields $\ln 2$ for the entropy of the tent map. For the logistic map,

$$\mathcal{I}(f^n) = \mathcal{I}(f^{n-1}) + \left\langle\{x\colon x = f^{-(n-1)}(0.5)\}\right\rangle \tag{10.2.66}$$

is the number of injective branches arising from the solutions of

$$\begin{aligned} 0 = \frac{df^n(x)}{dx} &= \frac{df(f^{n-1})}{df^{n-1}}\frac{df^{n-1}(x)}{dx} \\ &= \frac{df(f^{n-1})}{df^{n-1}}\frac{df(f^{n-2})}{df^{n-2}}\cdots\frac{df(f)}{df}\frac{df(x)}{dx} \end{aligned}$$

that yields

$$x = f^{-}(\cdots(f^{-}(f^{-}(0.5)))\cdots)$$

where $\langle\{\cdots\}\rangle$ is the cardinality of set $\{\cdots\}$. Note that in the context of the topological entropy, $\mathcal{I}(f)$ is only a tool for generating a partition on $\mathcal{D}(f)$ by the iterates of f.

Example 10.1. (1) In a fair-die experiment, if $\mathcal{A} = \{\text{even, odd}\}$ and the refinement $\mathcal{B} = \{j\}_{j=1}^{6}$ is the set of the six faces of the die, then for $i = 1, 2$

$$P(B_j \mid A_i) = \begin{cases} \dfrac{1}{3}, & j \in A_i \\ 0, & j \notin A_i, \end{cases}$$

and $S(\mathcal{B} \mid A_1) = \ln 3 = S(\mathcal{B} \mid A_2)$ by (10.2.63*b*). Hence the conditional entropy of $\mathcal{B}$ given $\mathcal{A}$, using $P(A_1) = 0.5 = P(A_2)$ and Eq. (10.2.63*a*), is $S(\mathcal{B} \mid \mathcal{A}) = \ln 3$. Hence

$$\begin{aligned} S(\mathcal{A} \bullet \mathcal{B}) &= S(\mathcal{A}) + S(\mathcal{B} \mid \mathcal{A}) \\ &= \ln 6. \end{aligned}$$

If we have access only to partition $\mathcal{B}$ and not to $\mathcal{A}$, then $S(\mathcal{B}) = \ln 6$ is the amount of information gained about the partition $\mathcal{B}$ when we are told which face showed up in a rolling of the die; if on the other hand the only partition available is $\mathcal{A}$, then $S(\mathcal{A}) = \ln 2$ measures the information gained about $\mathcal{A}$ on the knowledge of the appearance of an even or odd face.

(2) The dynamical evolution of Fig. 10.3 provides an example of conditional probability and conditional entropy. Here the refinements of basic partition $\mathcal{A}_0 = \{\text{matter, negmatter}\} = \{A_{01}, A_{00}\}$ generated by the inverses of the tent map, are denoted as $\mathcal{A}_n = \{t^{-n}(A_{0i})\}_{0,1}$ for $n = 1, 2, \cdots$ to yield the largest common refinements

$$\mathcal{A}^n = \mathcal{A}_0 \bullet \mathcal{A}_1 \bullet \mathcal{A}_2 \bullet \cdots \bullet \mathcal{A}_n, \qquad n \in \mathbb{N}, \tag{10.2.67}$$

where the refinements are denoted as indicated in the figure, and $\mathcal{A}^n = \mathcal{A}_n$. Taking the measure of the elements of a partition to be its euclidean length, gives

$$P(A_{nj} \mid A_{0i}) = \begin{cases} \dfrac{1}{2^{n-1}}, & j \in A_{0i} \\ 0, & j \notin A_{0i}, \end{cases}$$

$S(\mathcal{A}_n \mid A_{0i}) = (n-1)\ln 2$, $i = 0, 1$, (Equation 10.2.63b), $S(\mathcal{A}_n \mid \mathcal{A}_0) = (n-1)\ln 2$, and finally $S(\mathcal{A}_n \bullet \mathcal{A}_0) = n \ln 2$. In case the initial partition $\mathcal{A}_0$ is taken to be the whole of $\mathfrak{D}(t)$, then (10.2.61) gives directly $S(\mathcal{A}_n) = n \ln 2$.

(3) Logistic map $f_\lambda(x) = \lambda x(1-x)$, [21]. For $0 \le \lambda < 3$, Fig. 10.8a, the dynamics can be subdivided into two broad categories. In the first, for $0 \le \lambda \le 2$, $\mathfrak{I}(f_\lambda^n) = 2$ gives $h_{\mathrm{T}}(f_\lambda) = 0$. This is illustrated in Fig. 10.8a (i), (ii), and (iii) which show how the number of subsets generated on X by the increasing iterates of the map tend from 2 to 1 in the first case and to the set $\{\{0\}, (0,1), \{1\}\}$ for the other two. The figure demonstrates that while in (a) the dynamics eventually collapses and dies out, the other two cases are equally uneventful in the sense that the converged multifunctional limits — of $(0, [0, 1/2]) \cup ((0, 1), 1/2) \cup (1, [0, 1/2])\}$ in figure (iii), for example — are as much passive and has no real "life"; this is quantified by the constancy of the lap number and the corresponding topological entropy $h_{\mathrm{T}}(f) = 0$. Although the partition induced on $X = [0, 1]$ by the evolving map in (iv) is refined with time, the *stability of the fixed point* $x_* = 0.6656$ prevents the dynamics from acquiring any meaningful evolutionary significance with its multifunctional graphical limit being of the same type as in (ii) and (iii): as will be evident in what follows, *instability of fixed points is essential for the evolution of meaningful complexity.* $\lambda^{(0)} = 2$ of (iii) — obtained by solving the equation $f_\lambda(0.5) = 0.5$ — is special because its super-stable fixed point $x = 0.5$ is the only point in $\mathfrak{D}(f)$ at which f is injective and therefore well-posed by this criterion.

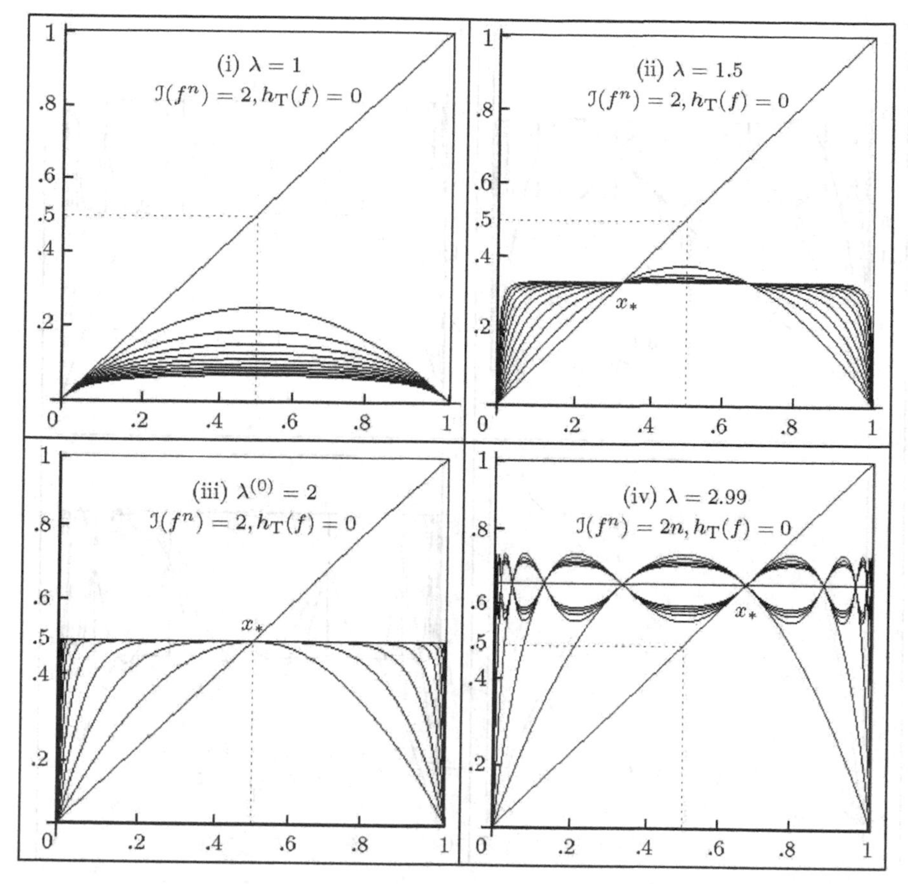

Fig. 10.8*a*. Non-life dynamics of the first 10 iterates of the logistic map $f_\lambda = \lambda x(1-x)$ generated by its only stable fixed point $x_* = (\lambda - 1)/\lambda$.

For $3 \leq \lambda \leq 4$, $h_T(f_\lambda) = 0$ whenever $\mathfrak{I}(f_\lambda) \leq 2n$ which occurs, from Fig. 10.8*b*, for $\lambda \leq \lambda^{(1)} = 1 + \sqrt{5} = 3.23607$; here $\lambda^{(m)}$ is the λ value at which a super-stable 2^m-cycle appears. The super-stable λ for which $x = 0.5$ is fixed for f^n, $n = 2^m$, $m = 0, 1, 2, \cdots$ leads to a simplification of the dynamics of the map, possessing as they do, the property of the stable horizontal parts of the graphically converged multifunction being actually tangential to all the turning points of every iterate of f. The immediate consequence of this is that for a given $3 < \lambda < \lambda_* = 3.5699456$, the dynamics of f attains a state of basic evolutionary stability after only the first $\{2^m\}_{m=0,1,\cdots}$ time steps in the sense that no new spatial structures *emerge* after this period, any further temporal evolution being fully utilized in spatially *self-organizing* this basic structure throughout the system by the generation of equivalence classes of the initial 2^m time steps. As seen in Fig. 10.8*b*, the unstable fixed point x_* is directly linked to its stable partners of f^2 that report back to x_*. Compared

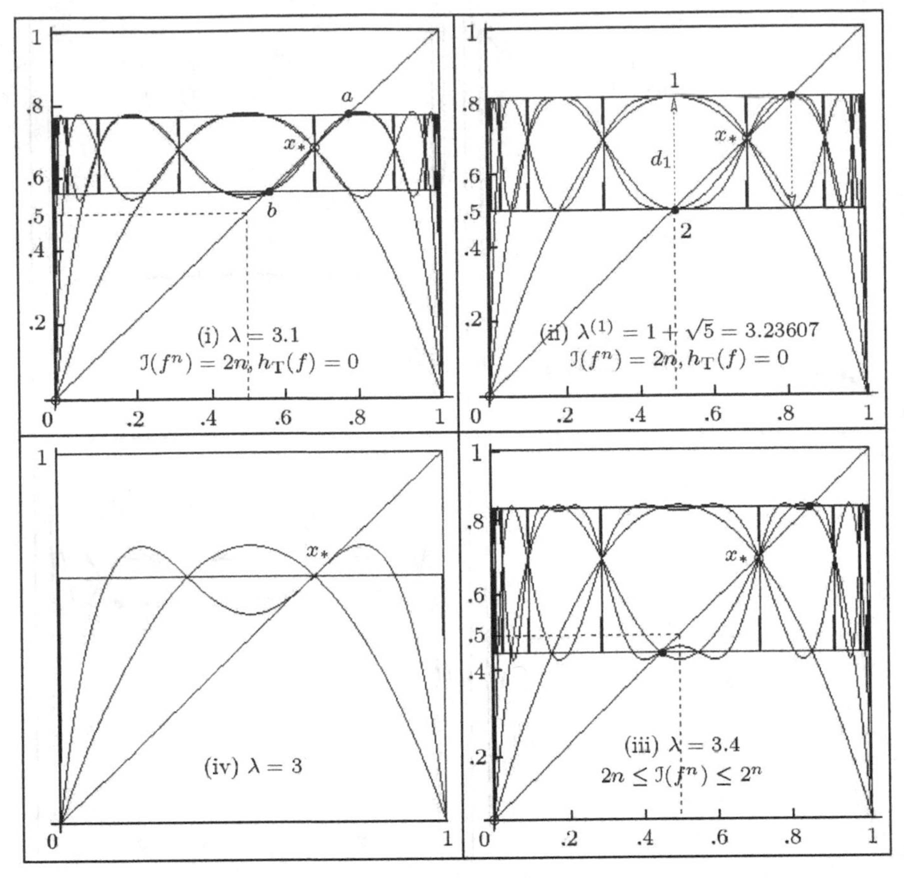

Fig. 10.8*b*. Dynamics of stable 2-cycle of the logistic map, where each panel displays the first four iterates superposed on the graphically converged multifunction represented by iterates 1001 and 1002. Panel (iv) in this and the following two figures, illustrates Eqs. (10.2.56*a*, *b*) in the birth of new period doubling cycles. The $d_i := f_{\lambda_i}^{2^{i-1}}(0.5) - 0.5$ in these figures define the universal Feigenbaum constant $-\alpha := \lim_{i\to\infty} d_i/d_{i+1} = 2.502907\cdots$, while the super-cyclic parameters $(\lambda_i)_i$ generate the second constant $\delta := \lim_{i\to\infty}(\lambda_i - \lambda_{i-1})/(\lambda_{i+1} - \lambda_i) = 4.669201\cdots$ of period doubling.

to (i) however, where the relative simplicity of the instability of x_* allows its stable partners to behave monotonically as in Fig. 10.8*a* (ii), the instability of 10.8*b* (iii) is strong enough to induce the oscillatory mode of convergence of 10.8*a* (iv). Case (ii) of the super-stable cycle for $\lambda^{(1)} = 1 + \sqrt{5}$ — obtained by solving the equation $f_\lambda^2(0.5) = 0.5$ — reflecting well-posedness of f at $x = 0.5$ represents, as in Fig 10.8*a* (iii), a mean of the relative simplicity of (i) and the complex instability of (iii) that grows with increasing λ.

When $\lambda > \lambda^{(1)}$ as in Figs. 10.8*b* (iii) and 10.8*c*, the number of injective branches lie in the range $2n \leq \mathcal{I}(f_\lambda^n) \leq 2^n$ and the difficulty in actually obtaining these numbers for large values of n is apparent from Eq. (10.2.66). The unstable basic fixed point x_* in Fig. 10.8*c* is now linked to its *un*stable partners denoted by open circles arising from f^2, who report back to the overall controller x_* the information they receive from their respective stable subcommittees. Compared to the 2-cycle of Fig. 10.8*b*, the instability of principal x_* is now serious enough to require sharing of the responsibility by two other instability governed partners who are further constrained to delegate authority to the subcommittees mentioned above. Case (ii) of the super-stable cycle for $\lambda^{(2)} = 3.49856$ is obtained by solving $f_\lambda^4(0.5) = 0.5$ denotes as before the mean of the relative simplicity of (i) and the large instability of (iii). For $\lambda = 4$, however $\mathcal{I}(f_4^n) = 2^n$ and the topological entropy reduces to the simple $h(f_4) = \ln 2$; $h_{\mathrm{T}}(f) > 0$ is sufficient condition for f_λ to be chaotic. The tent map behaves similarly and has an identical topological entropy, see Fig. 10.10*a*.

The difficulty in evaluating $\mathcal{I}(f^n)$ for large values of n and the open question of the utility of the number of injective branches of a map in actually measuring the complex dynamics of nonlinear evolution, suggests the significance of the role of *evolution of the graphs of the iterates of* f_λ in defining the dynamics of natural processes. It is also implied that the dynamics can be simulated through *the partitions induced on* $\mathcal{D}(f)$ *by the evolving map* as described by graphical convergence of the functions in accordance with our philosophy that the dynamics on C derives from the evolution of f in C^2 as observed in $\mathcal{D}(f)$. The following subsection carries out this line of reasoning, to define a new index of chaos, nonlinearity and complexity, that is of *chanoxity*.

ChaNoXity

The really interesting comparison (of Windows) *is with Linux, a product of comparable complexity developed by an independent, dispersed community of programmers who communicate mainly over the internet. How can they outperform a stupendously rich company that can afford to employ very smart people and give them all the resources they need? Here is a possible answer: Complexity.*

Microsoft's problem with Windows may be an indicator that operating systems are getting beyond the capacity of any single organization to handle them. Therein may lie the real significance of Open Source. Open Source is not a software or a unique group of hackers. It is a way of building complex things. Microsoft's struggles with Vista suggests it may be the only way to do operating systems in future.

John Naughton, Guardian Newspapers Limited, May 2006.

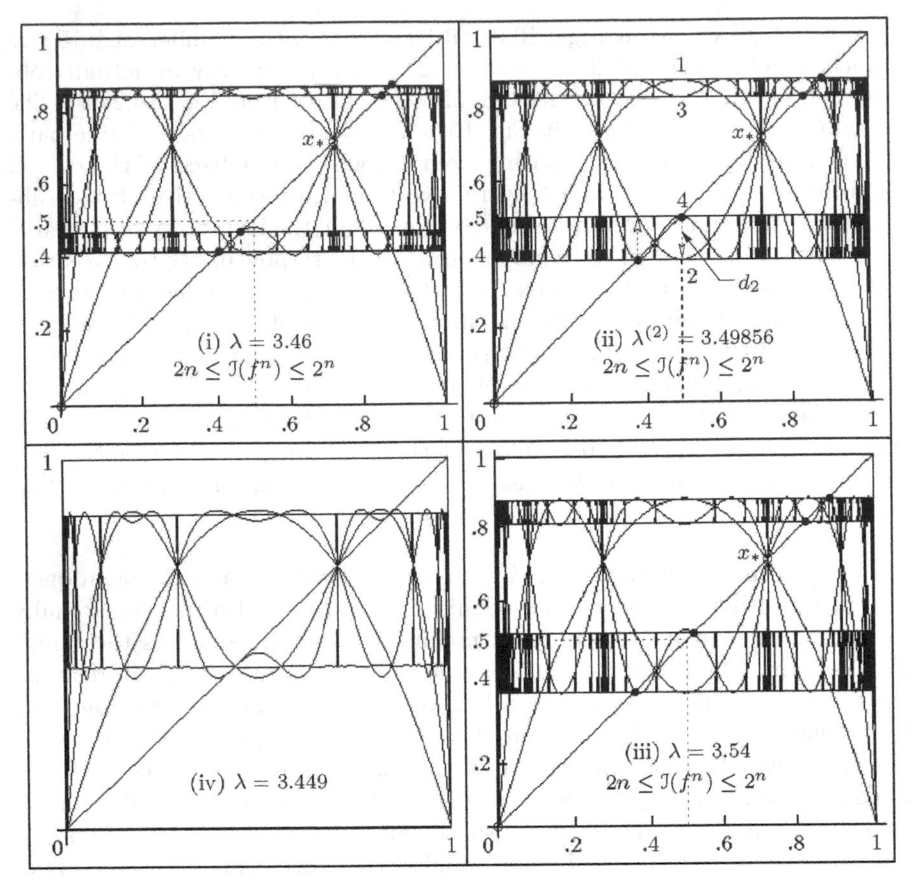

Fig. 10.8*c*. Dynamics of stable 4-cycle of the logistic map, where each panel displays the first four iterates superposed on the graphically "converged" multifunction represented by iterates 1001-1004.

The magnified view of the *stable* 8-cycle, Fig. 10.8*d*, graphically illustrates evolutionary dynamics of the logistic interaction. The 2^3 *unstable* fixed points marked by open circles interact among themselves as indicated in the figure to generate the stable periodic cycle, providing thereby a vivid illustration of competitive collaboration between matter-negmatter effects. The increasing iterations of irreversible urge toward bijective simplicity of ininality constitutes the activating backward-direct direction of increasing entropic disorder that is effectively balanced by restraining forward-inverse exergy destruction of expansion, increasing order, and self-organization that eventually leads to the stable periodic orbit. The activating effect of the direct limit appears in the figure as the negative slope associated with each unstable fixed points except the first at $x = 0$ which must now be paired with its equivalent image at $x = 1$. Display (iii) of the partially superimposed limit graphs 1001-1008 on the first 8 iterates — that remain invariant with further temporal evolution — illustrate that while nothing new emerges after this initial period, *further increasing temporal evolution propagates the associated changes throughout the system as self-generated equivalence classes guiding the system to a state of local* (that is spatial, for the given λ) *periodic stasis.* As compared to Fig. 10.3 for the tent interaction, this manifestation of coeffects in the logistic for $\lambda < \lambda_* = 3.5699456$ has a feature that deserves special mention: while in the former the negative branch belongs to distinct fixed points of equivalence classes, in the later matter-negmatter competitive-collaboration is associated with each of the 2^N generating branches possessing bi-directional characteristics with the activating effect of negmatter actually initiating the generation of the equivalence class. In the observable physical world of $\mathcal{D}(f)$, this has the interesting consequence that whereas the tent interaction generates matter-negmatter intermingling of disjoint components to produce the homogenization of Fig. 10.3, for the logistic interaction the resulting behaviour is a consequence of a deeper interplay of the opposing forces leading to a higher level of complexity than can be achieved by the tent interaction.

This distinction reflects in the interaction pair $(f, \mathfrak{f})$ that can be represented as

$$x \longmapsto 2x \longmapsto \begin{cases} 2x, & \text{if } 0 \le x < 0.5 \\ 2(1-x), & \text{if } 0.5 \le x \le 1 \end{cases}, \qquad x \longmapsto 2x \longmapsto 4x(1-x), \tag{10.2.68}$$

which leads — despite that "researchers from many disciplines now grapple with the term *complexity*, yet their views are often restricted to their own specialties, their focus non-unifying; few can agree on either a qualitative or quantitative use of the term" [6] — to the

Definition 10.2 (Complex System, Complexity). *The couple* $((X, \mathcal{U}), f)$ *of a compound topological space* $(X, \mathcal{U})$ *and an interaction* f *on it is a* complex system $\mathcal{C}$ *if* (see Fig. 10.9 and Eq. (10.2.72))

(CS1) *The* algebraic structure *of* $\mathcal{D}(f)$ *is defined by a* finite *family* $\{A_j\}_{j=0}^{n}$, $n = 1, 2, \cdots, N$, *of progressively refined hierarchal partitions of non-empty subsets induced by the iterates of* f, *with increasing evolution building on this foundation the overall configuration of the system.*

This family interacts with each other through

(CS2) *The* topology of $(\mathcal{D}, \mathcal{U})$ *such that the subbasis of* $\mathcal{U}$ *at any level of refinement is the union of the open sets of its immediate coarser partition and that generated by the partition under consideration where all open sets are saturated sets of equivalence classes generated by the evolving iterates of the interaction.*

The complexity of a system is a measure of the interaction between the different levels of partitions that are generated on $\mathcal{D}(f)$ under the induced topology on X. Thus as a result of the constraint imposed by (CS1), under

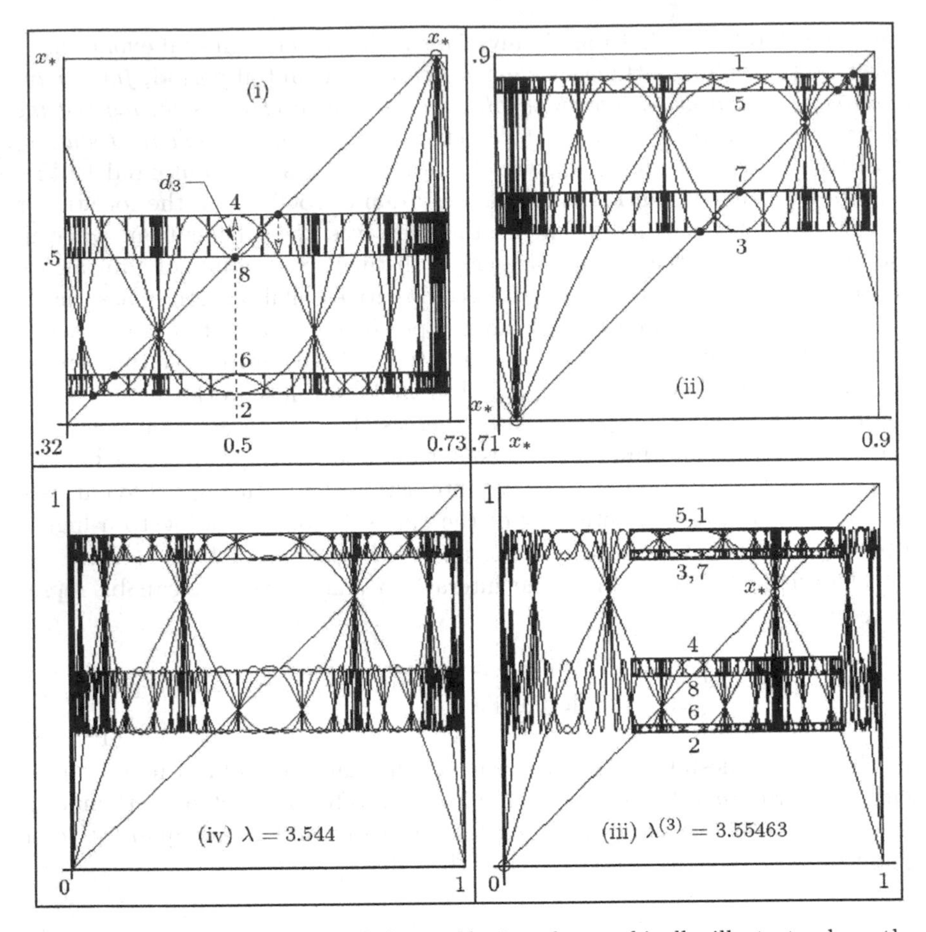

Fig. 10.8*d*. Magnified view of the *stable* 8-cycle graphically illustrates how the evolutionary dynamics of the logistic interaction, under the synthetic influence of its stable-unstable components, spontaneously produces for any given resource $3 \leq \lambda < \lambda_*$, a set of 2^n uniquely stable configurations between which it periodically oscillates. Thus in this case the "unpredictability" of nonlinear interactions manifests as a "surprise" in the autonomous generation of a set of well-defined stable states, which as we shall see defines the "complexity" of the system.

the logistic interaction complex structures can emerge only for $3 \leq \lambda < \lambda_*$ which in the case of the stable 2-cycle of Fig. 10.8*b*(ii), reduces to just the first 2 time steps that is subsequently propagated throughout the system by the increasing ill-posedness, thereby establishing the global structure as seen in Fig. 10.9. With increasing λ the complexity of the dynamics increases as revealed in the succeeding plots of 4- and 8-cycles: compared to the single refinement for the 2-cycle, there are respectively 2 and 3 stages of refinements in the 4- and 8-cycles and in general there will be N refining partitions of

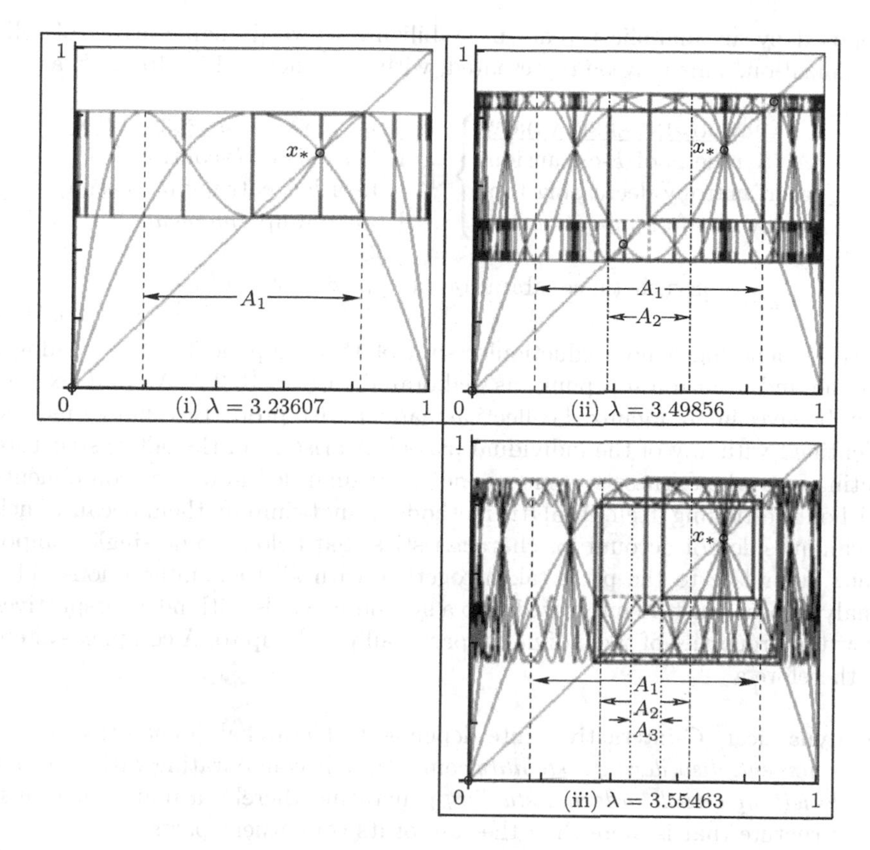

Fig. 10.9. The role of unstable fixed points in generating the partitions $\{A_j\}_{j=0}^n$, $n = 1, 2, \cdots, N$ required in the definition of complexity, where the $\{A_j\}$ are appropriately defined as the inverse images of $f_{i,j} := \left|f^i(0.5) - f^j(0.5)\right|$, refer Eq. (10.2.72), and $A_0 = \mathcal{D}(f)$. The open circles in (i) and (ii) represent the unstable fixed points that have been omitted from (iii) for the sake of clarity. The converged multifunctional graphical limits are also shown for the 2- and 4-cycles.

$\mathcal{D}(f)$ for the 2^N-stable cycle. The equilibrated $X_{\leftrightarrow}$, by Fig. 10.5*b* and the subsequent discussion, corresponds to the $\{\mathcal{D}, f(\mathcal{D}), \cdots, f^{2^N}(\mathcal{D})\}$ on $\mathcal{D}(f)$. Below $\lambda = 3$, absence of instabilities allows no emergence of new features, while above $\lambda = \lambda_*$ the absence of stabilizing effects prevent self-organization from moderating the dynamics of the system.The *motivating* saturated open sets of X on $\mathcal{D}(f)$ and $\mathcal{R}(f)$ are the projections of the boxes of the converged multi-limits in Figs. 10.8*b*, *c*, *d* onto the x- and y−axes, with their boundary being represented by the members of the equivalence class $[x_*]$ of the unstable fixed point x_*.

Complexity therefore, represents a state of dynamical balance between a catabolic emergent, destabilizing, backward, bottom-up pump direction, opposed by an anabolic top-down, stabilizing forward engine arrow of self-organization. This may be represented, with reference to Fig. 10.5a, b, as

$$\left.\begin{array}{r}\underline{\text{FORWARD-INVERSE ARROW}}\\ \text{Synthesis of } E\text{-expansion,}\\ \text{order, entropy decreasing top-}\\ \text{down } \textit{self-organization } C_{\leftarrow}\end{array}\right\} \oplus \left\{\begin{array}{l}\underline{\text{BACKWARD-DIRECT ARROW}}\\ \text{Analysis of } P\text{-contraction,}\\ \text{disorder, entropy increasing}\\ \text{bottom-up } \textit{emergence } {}_{\rightarrow}C\end{array}\right.$$
$$\Longleftrightarrow$$
$$\text{Synthetic cohabitation of opposites } \mathfrak{C} = C_{\leftrightarrow}, \qquad (10.2.69)$$

with $\oplus$ denoting a non-reductionist sum of the components of a top-down engine and a bottom-up pump as elaborated in Sec. 10.2.2. A complex system behaves in an organized collective manner with properties that cannot be identified with any of the individual parts but arise from the entire structure acting as a whole: these systems cannot be dismantled into their components without destroying itself. Analytic methods cannot simplify them because such techniques do not account for characteristics that belong to no single component but relate to the parts taken together with all their interactions. This analytic base must be integrated into a synthetic whole with new perspectives that the properties of the individual parts fail to add up to. A complex system is therefore a

▶ dynamical, C-interactive, interdependent, hierarchal homeostasy of *P-emergent, disordering instability* competitively collaborating with adaptive *E-self-organized, ordering stability* generating thereby a non-reductionist structure that is more than the sum of its constituent parts.

Emergence implies instability inspired (and therefore "destructive", anti-stabilizing) generation of overall characteristics that do not reduce to a linear composition of the interacting parts: complexity is a result of the "failure of the Newtonian paradigm to be a general schema through which to understand the world", [3], and in fact "if there were only Newton's laws, there could never have been any motion in the earth" [22].[20] As noted earlier, complexity can be distinguished into two subclasses depending on which of the two limits of Eq. (10.2.69) serve as *activating* and which *restraining* and our classification of "life" will be based on this distinction.

A complexity supporting interaction will be distinguished as C-interaction. Examples of C- and non-C-interactions that will be particularly illuminating

[20] Darwinian theory of *natural selection* is different from complexity generated emergence and self-organization. Selection represents a competition between different systems for the limited resources at their disposal: it signifies an *externally* directed selection between competing states of equilibria that serves to maximize the "fitness" of the system with respect to its environment. Complexity, on the other hand, typifies an *internally* generated process of "continuous tension between competition and cooperation".

in our work are respectively the λ-logistic map and its "bifurcated" $(\lambda/2)$-tent counterpart

$$\lambda x(1-x) \longmapsto \begin{cases} \frac{\lambda}{2}x, & 0 \le x \le 0.5 \\ \frac{\lambda}{2}(1-x), & 0.5 \le x \le 1. \end{cases} \tag{10.2.70}$$

It will be convenient to denote a complex system $\mathfrak{C}$ simply as $(A, \mathfrak{B})$, with the interaction understood from the context. The distinguishing point of difference between the dissipative structures $\mathcal{D}$ of multi-dimensional differential system and evolutionary complex dynamics of a C-interaction is that the former need not possess any of the hierarchal configuration of the later. This tiered structure of a complex system is an immediate consequence of the partitioning refinements imposed by the interaction on the dynamics of the system with emergence and self-organization being the natural outcome when these refinements, working independently within the global framework of the interaction, are assembled together in a unifying whole. Hence it is possible to make the distinction

- ► a *dissipative structure* $\mathcal{D}$ is a special system of spatially multidimensional, non-tiered, forward-backward synthesis of opposites that attains dynamic equilibrium largely through self-organization without significant instability inspired emergence

from a general complex system.

A Measure of ChaNoXity

The above considerations allow us to define, with reference to Fig. 10.5*b*, the *chanoxity index* of the interaction to be the *constant* $0 \le \chi \le 1$ that satisfies

$$f(x) = x^{1-\chi}, \qquad x \in \mathcal{D}(f). \tag{10.2.71a}$$

Thus if $\langle f(x) \rangle$ and $\langle x \rangle$ are measures that permit (10.2.71*a*), then in

$$\chi = 1 - \frac{\ln \langle f(x) \rangle}{\ln \langle x \rangle} \tag{10.2.71b}$$

we take

(a) $\langle x \rangle$ to be the number of *basic unstable fixed points* of f responsible for *emergence.* Thus for $1 < \lambda \le 3$ there is no *basic* unstable fixed point at $x = 0$, followed by the familiar sequence of $\langle x \rangle = 2^N$ points until at $\lambda = \lambda_*$ it is infinite.

(b) for $f(x)$ the estimate

$$\langle f(x) \rangle = 2f_1 + \sum_{j=1}^{N} \sum_{i=1}^{2^{j-1}} f_{i,i+2^{j-1}}, \qquad N = 1, 2, \cdots, \tag{10.2.72}$$

λ	N	$\langle f(0.5)\rangle$	χ_N	λ	N	$\langle f(0.5)\rangle$	χ_N
(1, 3]	–	1.000000	0.000000	3.5699442	9	3.047727	0.821363
3.2360680	1	1.927051	0.053605	3.5699454	10	3.053571	0.838950
3.4985617	2	2.404128	0.367243	3.5699456	11	3.056931	0.853447
3.5546439	3	2.680955	0.525751	3.5699457	12	3.058842	0.865585
3.5666676	4	2.842128	0.623257	3.5699457	13	3.059855	0.875887
3.5692435	5	2.935294	0.689299	3.5699457	14	3.060524	0.884730
3.5697953	6	2.988959	0.736726	↓	↓	↓	↓?
3.5699135	7	3.019815	0.772220	λ_*	∞	3.??????	1.000000
3.5699388	8	3.037543	0.799637				

Table 10.6*a*. In the passage to full chaoticity, the system becomes increasingly complex and nonlinear (remember: chaos is maximal nonlinearity) such that at the critical value $\lambda = \lambda_* = 3.5699456$, the system is fully chaotic and complex with $\chi = 1$. For $1 < \lambda \leq 3$ with no generated instability of which $\lambda = 2$ is representative, $\chi = 1 - \ln(1/2 + 1/2)/0 = 0$. The expression for $\langle f(x)\rangle$ reduces to $2f_1 + f_{12}$, $2f_1 + f_{12} + (f_{13} + f_{24})$, $2f_1 + f_{12} + (f_{13} + f_{24}) + (f_{15} + f_{26} + f_{37} + f_{48})$ for $N = 1, 2, 3$ respectively.

with $f_i = f^i(0.5)$ and $f_{i,j} = |f^i(0.5) - f^j(0.5)|$, to get the measure of chanoxity as

$$\chi_N = 1 - \frac{1}{N \ln 2} \ln \left[2f_1 + \sum_{j=1}^{N} \sum_{i=1}^{2^{j-1}} f_{i,i+2^{j-1}} \right], \tag{10.2.73}$$

that we call the *dimensional chanoxity of* f_λ[21]; notice how Eq. (10.2.72) effectively divides the range of f into partitions that progressively refine with increasing N. In the calculations reported here, λ is taken to correspond to the respective superstable periodic cycle, where we note from Figs. 10.8*b*, *c* and *d*, that the corresponding super-stable dynamics faithfully reproduces the features of emergence during the first N iterates, followed by self-organization of the emerging structure for all times larger than N.

The numerical results of Table 10.6*a* suggest that

$$\lim_{N\to\infty} \chi_N = 1$$

[21] Recall that the fractal dimension of an object is formally defined very similarly:

$$D = \frac{\ln(\#\text{ self-similar pieces into which the object can be decomposed})}{\ln(\text{magnification factor that restores each piece to the original})}.$$

at the critical $\lambda = \lambda_* = 3.5699456$. Since $\chi = 0$ gives the simplest linear relation for f, a value of $\chi = 1$ indicates the largest non-linearly emergent complexity so that the logistic interaction is maximally complex at the transition to the fully chaotic region. It is only in this region $3 \leq \lambda < \lambda_*$ of resources that a global synthesis of stability inspired self-organization and instability driven emergence lead to the appearance of a complex structure.

λ		N						
		12	14	16	18	20	$\rightarrow$	∞
3.5700	$\langle f(0.5) \rangle$	5.057857	10.69732	31.38651	119.4162	468.8398		
	χ_N	0.805123	0.755773	0.689245	0.616675	0.556352	$\stackrel{?}{\rightarrow}$	0.0000
3.6000	$\langle f(0.5) \rangle$	275.7782	1125.908	4480.310	17996.46	72205.91		
	χ_N	0.324386	0.275938	0.241914	0.214699	0.193009	$\stackrel{?}{\rightarrow}$	0.0000
3.7000	$\langle f(0.5) \rangle$	885.4386	3683.121	14863.74	59511.41	236942.7		
	χ_N	0.184146	0.153806	0.133781	0.118840	0.107291	$\stackrel{?}{\rightarrow}$	0.0000
3.8000	$\langle f(0.5) \rangle$	1167.886	4597.633	18266.08	73197.48	293016.6		
	χ_N	0.150860	0.130952	0.115195	0.102249	0.091969	$\stackrel{?}{\rightarrow}$	0.0000
3.9000	$\langle f(0.5) \rangle$	1381.043	5595.363	22404.29	89472.39	358001.9		
	χ_N	0.130705	0.110713	0.096782	0.086158	0.077520	$\stackrel{?}{\rightarrow}$	0.0000
3.9999	$\langle f(0.5) \rangle$	1691.944	6625.197	26525.88	106254.9	424020.1		
	χ_N	0.106294	0.093304	0.081555	0.072379	0.065311	$\stackrel{?}{\rightarrow}$	0.0000
4.0000	$\langle f(0.5) \rangle$	14.00000	16.00000	18.00000	20.00000	22.00000	$\rightarrow$	$N+2$
	χ_N	0.682720	0.714286	0.739380	0.759893	0.777028	$\rightarrow$	1

Table 10.6*b*. Illustrates how the fully chaotic region of $\lambda_* < \lambda < 4$ is effectively "linear" with no self-organization, and only emergence. The jump discontinuity in χ at λ_* reflects a qualitative change in the dynamics, with the energy input for $\lambda \leq \lambda_*$ being fully utilized in the generation of complex internal structures of the system of emerging patterns and no self-organization .

What happens for $\lambda > \lambda_*$ in the fully chaotic region where emergence persists for all times $N \to \infty$ with no self-organization, is shown in Table 10.6*b* which indicates that on crossing the chaotic edge, the system abruptly transforms to a state of *effective linear simplicity* that can be interpreted to result from the drive toward ininality and effective bijectivity on saturated sets and on the component image space of f. This jump discontinuity in χ demarcates order from chaos, linearity from (extreme) nonlinearity, and simplicity from complexity. This non-organizing region $\lambda > \lambda_*$ of deceptive simplicity characterized by dissipation and irreversible "frictional losses", is to be compared with the nonlinearly complex domain $3 \leq \lambda < \lambda_*$ where irreversibility generates self-organizing useful changes in the internal structure of the system in order to attain the levels of complexity needed in the evolution. While the state of eventual evolutionary homeostasy appears only in $3 \leq \lambda < \lambda_*$, the relative linear simplicity of $\lambda > \lambda_*$ arising from the dissipative losses characteristic of this region conceals the resulting self-organizing thrust of the higher periodic windows of this region, with the smallest period 3 appearing at $\lambda = 1 + \sqrt{8} = 3.828427$. By the Sarkovskii ordering of natural numbers, there is embedded in this fully chaotic region a backward arrow that induces a chaotic tunnelling to lower periodic stability eventually terminating with the period doubling sequence in $3 \leq \lambda < \lambda_*$. This decrease in λ in the face of the prevalent increasing disorder in the over-heated scorching $\lambda > \lambda_*$ region reflecting the negmatter effect of "letting off steam", is schematically indicated in Fig. 10.10a and is expressible as

$$x \longrightarrow f_\lambda(x) \left\{ \begin{array}{r} \text{self-organizing complex system} \\ 3 \leq \lambda < \lambda_*,\ 0 < \chi \leq 1, \\ \\ \overset{\text{inality}}{\longrightarrow}\ \lambda_* \leq \lambda \leq 4,\ \chi = 0, \\ \text{chaotic complex system} \end{array} \right\} \begin{array}{c} \overset{\text{regulating}}{\longleftarrow} \\ \text{Sarkovskii} \uparrow \\ \underset{\text{effects}}{\longrightarrow} \end{array} \qquad (10.2.74)$$

Under normal circumstances dynamical equilibrium is attained, as argued above, within the temporal, iterational, self-organizing component of the loop above. If, however, the system is spatially driven by an increasing λ into the fully chaotic region, the global negworld effects of its periodic stable windows acts as a deterrent and, prompted by the Sarkovskii ordering induces the system back to its self-organizing region of equilibration. This condition of dynamical homeostasy is thus marked by a balance of *both the spatial and temporal effects,* with each interacting synergetically with the other to generate an optimum dynamical state of stability, with Figs. 10.8b, c, d clearly illustrating how new, distinguished and non-trivial features of the evolutionary dynamics occur only at the 2^N unstable fixed points of f_λ, leading to emerging patterns that characterize the net resources λ available to the interaction.

Panels (i), (ii), and (iii) of Fig. 10.10a magnifies these features of the defining fixed points and their classes for $3 \leq \lambda < \lambda_*$ that generates the stable-unstable signature in the graphically convergent limit of $t \to \infty$, essentially reflecting the synthetic cohabitation of the matter-negmatter components associated with these points. This in turn introduces a sense of symmetry with respect to the input-output axes of the interaction that, as shown in panel (iii), is broken when $\lambda > \lambda_*$ with the boundary at the critical $\lambda = \lambda_*$ signaling this physical disruption with a discontinuity in the value of the chanoxity index χ. Fig. 10.10b which summarizes these observations, identifying the self-organizing emergent region $3 \leq \lambda < \lambda_*$ as the "life" supporting complex domain of the logistic interaction f_λ. Below $\lambda = 3$, the resources of f_λ are insufficient in generating complexity, while above $\lambda = \lambda_*$ too much "heat" is

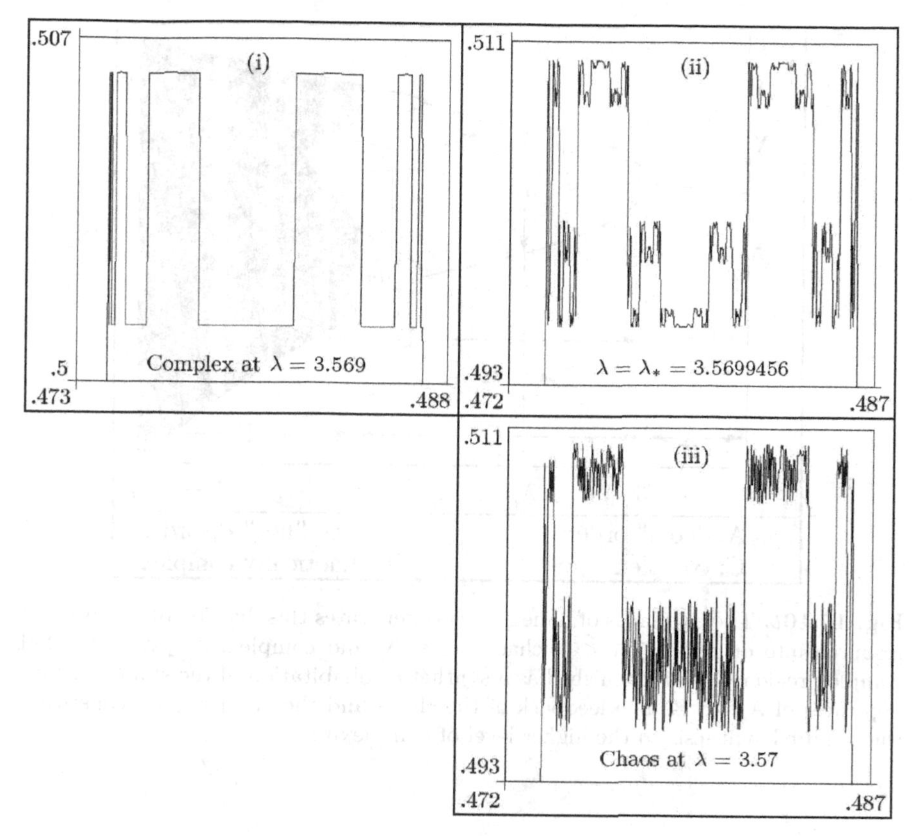

Fig. 10.10*a*. In contrast with the relatively tame (i) and (ii), panel (iii) illustrates the property of fully chaotic maximal ill-posedness and instability.

produced for support of constructive competition between the opposing directions, with the drive toward uniformity of ininality effectively nullifying the reverse competition. χ is in fact the irreversibility index ι in the complexity range $3 \leq \lambda < \lambda_*$. Both these parameters lie in the identical unit interval [0,1], with absence of disorder-inducing P at $(\iota = 0)(T = T_c)$ corresponding to the order-freezing $\lambda = 3$ and absence of order-generating E at $(\iota = 1)(T = T_h)$ consistent with the disorder-disintegrating $\lambda = \lambda_*$. The later case is effectively indistinguishable from the former because when the engine is not present no pump can be generated that shows up as an identical $\chi = 0$ for $\lambda > \lambda_*$. Significantly, however, while the former represents stability with reference to $\mathcal{D}(f)$ the later is stability with respect to $\mathcal{R}(f)$, and in the absence of an engine direction at $\iota = 1$ with increasing irreversibility and chanoxity, control effectively passes from the forward stabilizing direction to the backward destabilizing sense, thereby bringing the complementary neg-world effects into greater prominence through the appearance of singularities with respect to $\mathcal{D}(f)$. Finally, Fig. 10.10*c* which is a plot of the individual increasing and de-

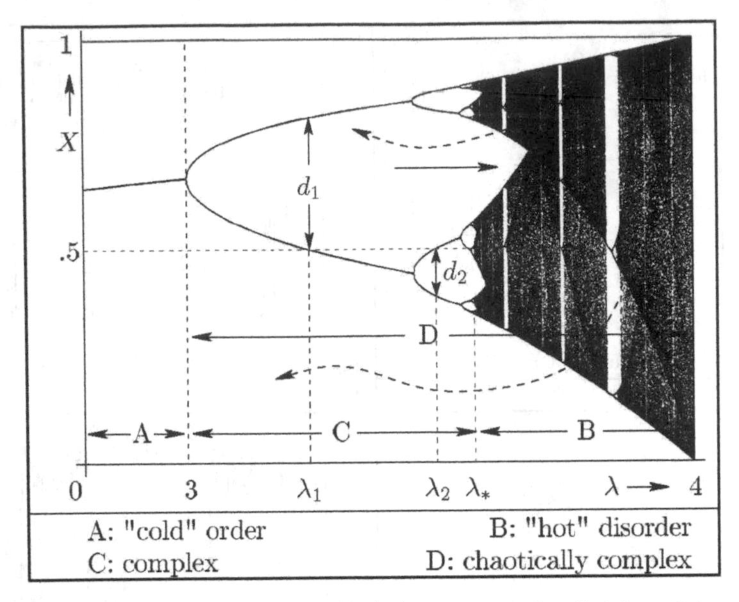

Fig. 10.10*b*. The dynamics of panels (i)-(iii) generates this division of the available resource into order, $0 \leq \lambda < 3$; chaos, $\lambda_* \leq \lambda$; and complex, $3 \leq \lambda < \lambda_*$. This complex region C is distinguished as a synthetic cohabitation of the stable-unstable opposites of A and B. The feedback of the chaos and the order regions constitutes the required synthesis to the higher level of complexity.

creasing parts of the logistic map confirms the observation that independent reductionist evolution of the component parts of a system cannot generate chaos or complexity. This figure, illustrating the unique role of non-injective ill-posedness in defining chaos, complexity and "life", clearly shows how the individual parts acting on their own in the reductionist framework and not in competitive collaboration, leads to an entirely different simple, non-complex, dynamics.

The figures of the dynamics in regions $\lambda < 3$ and $\lambda_* < \lambda$ of *actual* and *deceptive* simplicity can be interpreted in terms of symmetry arguments as follows [3]. In the former stable case of symmetry in the position of the individual parts of the system, the larger the group of transformations with respect to which the system is invariant the smaller is the size of the part that can be used to reconstruct the whole, and *symmetry is due to stability in the positions.* By comparison, the unstable chaotic region displays *statistical symmetry in the sense of equal probability of each component part that, without any fixed position, finds itself anywhere in the whole,* and symmetry is in the spatial or spatio-temporal averages.

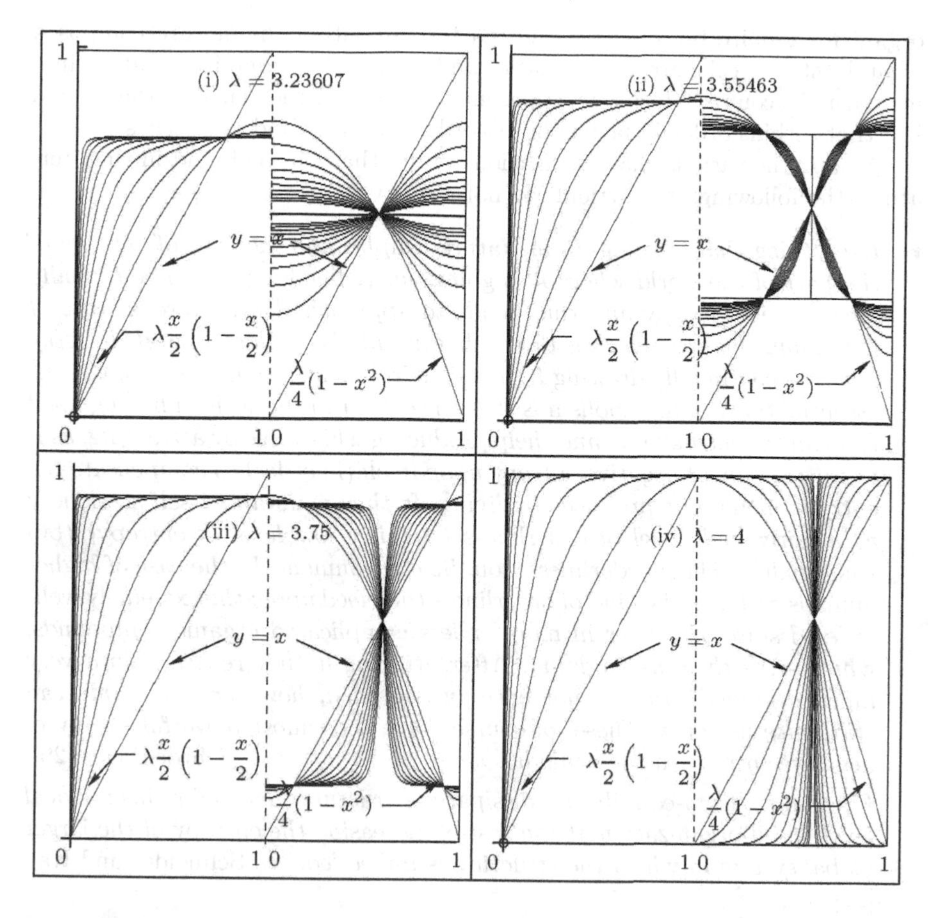

Fig. 10.10c. Reductionism cannot generate chaos or complexity or "life". This figure clearly illustrates the unique role of non-injective ill-posedness in defining chaos, complexity and "life", how the individual parts acting independently on their own in the reductionist framework not in competitive collaboration, leads to an entirely different simple, non-complex, dynamics.

10.3 What Is Life?

This 1944 question of Erwin Schroedinger [29, "one of the great science classics of the twentieth century"] credited with "inspiring a generation of physicists and biologists to seek the fundamental character of living systems" [13], suggests that "the essential thing in metabolism is that the organism succeeds in freeing itself from all the entropy it cannot help producing while alive", thereby maintaining order by consuming the available free energy in generating high entropy waste. In biology, "life" might mean the ongoing process of which living things are a part, or the period between birth and death of an organism, or the state of something that has been born and is yet to die. Living organisms require both energy and matter to continue living, are composed of at least one cell, are homeostatic, and evolve; life organizes matter into increasingly complex forms in apparent violation of the tenet of the second law that forbids order in favour of discord, instability and lawlessness.

Among the various characterizations of life that can be found in the literature, the following are particularly noteworthy.

- *Everything that is going on in Nature (implies) an increase of entropy of the part of the world where it is going on. A living organism continually increases its entropy and thus tends to approach the dangerous state of maximum entropy, which is death. It can only keep aloof from it, i.e. stay alive, by continually drawing from its environment "negative entropy". The essential thing in metabolism is that the organism succeeds in freeing itself from all the entropy it cannot help producing while alive by attracting, as it were, a stream of negative entropy upon itself (in order) to compensate the entropy increase it produces by living. It thus maintains itself stationery at a fairly high level of orderliness (= fairly low level of entropy) (by) continually sucking orderliness from its environment. In the case of higher animals we know the kind of orderliness they feed upon: the extremely well-ordered state of matter in more or less complicated organic compounds, which serve them as foodstuff. After utilizing it they return it in a very much degraded form — not entirely degraded, however, for plants can still make use of it. These, of course, have their most powerful supply of negative entropy in the sunlight.* Schroedinger [29].
- *Life is a far-from-equilibrium dissipative structure that maintains its local level of self-organization at the cost of increasing the entropy of the larger global system in which the structure is imbedded.* Schneider and Kay [27].
- *A living individual is defined within the cybernetic paradigm as a system of inferior negative feedbacks subordinated to (being at the service of) a superior positive feedback.* Korzeniewski [16].
- *Living things are systems that tend to respond to changes in their environment, and inside themselves, in such a way as to promote their own continuation; this may be interpreted to mean that a living system continuously computes the solution to the problem of its own continued existence through a process of internal adjustments to external causation.* Morales [19].

The message of bidirectional homeostasy implicit in the above passages forms the basis of Cinquin and Demongeot's *Positive and Negative Feedback: Striking a Balance Between Necessary Antagonists* [7] in a wide class of biological systems that possess multiple steady states. To deal with such classes of nonequilibrium systems, Schneider and Kay's [27] reformulation of Kestin's Unified Principle of Thermodynamics [14] implies that thermodynamic gradients drive self-organization, and chemical gradients lead to autocatalytic self-organizing dissipative reactions with positive feedback, with the activity of the reaction augmenting itself in self-reinforcing reactions, stimulating the global activity of the whole. Seen in this perspective, "life is a balance between the imperatives of survival and energy degradation" identifiable respectively with the backward and forward directions of Eq. (10.2.69). In the present context, it is more convenient and informative to view these arrows not by affine translation as was done in Sec. 10.2.2, but by considering the two worlds in their own reference frames with their forward arrows opposing each other and establishing a one-to-one correspondence between them; the activating and its regulating spaces are then equivalent[22]. This equivalence of the forward with its corresponding backward will serve to differentiate "life" from the normal complex system as suggested below.

All multicellular organisms are descendants of one original cell, the fertilized egg (or zygote) with the potential to form an entire organism through a process of bifurcation called mitosis. The function of mitosis is to first destabilize the zygote by constructing an exact copy of each chromosome and then to distribute, through division of the original (mother) cell, an identical set of chromosomes to each of the two progeny (daughter) cells. The two opposites involved in this process are the male — modeled by the increasing positive slope half of the logistic map — sperm cell (represented by the fixed point $x_{\mathrm{M}} = 0$) and the female — modeled by the decreasing, negative slope of the map — egg (represented by the fixed point $x_{\mathrm{F}} = (\lambda - 1)/\lambda$). The first cell division of the fertilized egg for $\lambda = 3$, initiates a chain of some 50 bifurcations to generate the approximately 10^{14} cells in an adult human, with each division occurring at equal intervals of approximately twenty hours. All of the approximately 200 distinct types of cells are derived from the single fertilized egg x_{F} through a process known as differentiation and specialization by which an unspecialized cell specializes into one of the many cooperating types, such as the heart, liver and muscle, each with its own individually distinctive role collaborating with the others to make up the whole living system. During this intricately regulated stage of self-organization, certain genes are turned on, or become activated, while other genes are switched off, or deactivated, so that a differentiated cell develops specific characteristics and performs specific functions. Differentiation involves changes in numerous aspects of cell physiology: size, shape, polarity, metabolic activity, responsiveness to signals, and gene expression profiles can all change during differentiation. Compare this with the emerging patterns of partitioning induced by the logistic map for number of iterates $\leq N$ in the 2^N stable cycle that resulted in the definition (10.2.73) of the chanoxity index in Sec. 10.2.3, followed by the self-organizing iterates for times larger than N. This sequence of destabilizing-stabilizing cell divisions

[22] Thus in $\mathbb{R}$, $\mid a \mid = \mid -a \mid$ defines an equivalence, and if $a < b$ then $-b < -a$ when viewed from $\mathbb{R}_+$, but $\mathfrak{a} < \mathfrak{b}$ in the context of $\mathbb{R}_-$. The basic fact used here is that two sets are "of the same size", or *equipotent*, iff there is a one-to-one correspondence between them.

represent emerging self-organization in the bidirectional synthetical organization (10.2.69) of a complex system: *through cell cooperation, the organism becomes more than merely the sum of its component parts.*

Abnormal growth of cells leading to cancer occur because of malfunctioning of the mechanism that controls cell growth and differentiation, and the level of cellular differentiation is sometimes used as a measure of cancer progression. A cell is constantly faced with problems of proliferation, differentiation, and death. The bidirectional control mechanism responsible for this decision is a stasis between cell regeneration and growth on the one hand and restraining inhibition on the other. Mutations are considered to be the driving force of evolution, where less favorable mutations are removed by natural selection, while more favorable ones tend to accumulate. Under healthy and normal conditions, cells grow and divide to form new cells only when the body needs them. When cells grow old and die, new cells take their place. Mutations can sometimes disrupt this orderly process, however. New cells form when the body does not need them, and old cells do not die when they should. Each mutation alters the behavior of the cell somewhat. This cancerous bifurcation, which is ultimately a disease of genes, is represented by the chaotic region $\lambda \geq \lambda_*$ where no stabilizing effects exist. Typically, a series of several mutations is required before a cell becomes a cancer cell, the process involving both oncogenes that promote cancer when "switched on" by a mutation, and tumor suppressor genes that prevent cancer unless "switched off" by a mutation.

Life is a specialized complex system of homeostasis between these opposites, distinguishing itself by being "alive" in its response to an ensemble of stratified hierarchal units exchanging information among themselves so as to maintain its entropy lower than the maximal possible for times larger than the "natural" time for decay of the information-bearing substrates. Like normal complex systems, living matter respond to changes in their environment to promote their own continued existence by resisting "the gradients responsible for the nonequilibrium condition". A little reflection however suggests that unlike normal complex systems, the activating direction in living systems corresponds not to the forward-inverse arrow of the physical world but to the backward-direct component with its increase of entropic disorder generating collaborative support from the restraining self-organizing effect of the forward component in an equilibrium of opposites. Thus it is the receptor "yin" egg x_{F} that defines the activating direction of evolution in collaboration with the donor "yang" sperm x_{M}, quite unlike the dynamics of the Lorenz equation, for example, that is determined by the activating temperature gradient acting along the forward arrow of the physical world.

In the present context, let us identify the backward-direct, catabolic, yin component $\mathfrak{M}$ of life $\mathfrak{L} := \{B, \mathfrak{M}\}$ as its *mind* collaborating competitively with the forward-inverse, anabolic, yang *body* B, and define

Definition 10.3 (Life). *Life is a special complex system of activating mind and restraining body.*

In this terminology, a non-life complex system (respectively, a dissipative structure) is a hierarchal (respectively, non-hierarchal) compound system with activating body and restraining mind. To identify these directions, the following illustrative examples should be helpful.

Example 10.2. (a) In the Lorenz model the forward-inverse arrow in the direction of the positive z-axis is, according to Fig. 10.5*b*, the activating direction of increasing order and self-organization. The opposing gravitational direction, by setting up the convection cells that reduces the temperature gradient by increasing the disorder of the cold liquid, marks the direction of entropy increase. Since the forward-inverse body direction is the activating direction, the Lorenz system denotes a non-life complex system. Apart from these organizing rolls representing "the system's response to move it away from equilibrium", availability of the angular variables prevents the Lorenz system from generating any additional emerging structures in the body of the fluid.

The familiar prototypical example of uni-directional entropy increase required by equilibrium Second law of Thermodynamics of the gravity dominated egg crashing off the table never to reassemble again is explained, in terms of Fig. 10.5*b*, as an "infinitely hot reservoir" dictating terms leading to eventual "heat death": unlike in the Lorenz case, the gravitational effect is not moderated here for example by the floor rising up to meet the level of the table, with the degree of disorder of the crashed egg depending on the height of the table.

(b) For the logistic map in the complexity region of λ, the activating backward-direct arrow $\{\mathfrak{D}, \{\mathfrak{D}, f(\mathfrak{D})\}, \{\mathfrak{D}, f(\mathfrak{D}), f^2(\mathfrak{D})\} \cdots\}$ is of increasing iterations, disorder, and entropy, while the restraining, expanding direction of self organization corresponds to decreasing non-injectivity of the increasing inverse iterates. Because the activating direction is that of the mind, the logistic dynamics is life-like.

The dominance of the physical realization M of $\mathfrak{M}$ as the *brain* in determining the dynamics of $\mathfrak{L}$ is reflected by the significance of sleep in all living matter. While there is much debate and little understanding of the evolutionary origins and purposes of sleep, there appears, nevertheless, to be a consensus that one of the major functions of sleep is consolidation and optimization of memories. However, this does not explain why sleep appears to be so essential or why mental functions are so grossly impaired by sleep deprivation. One idea is that sleep is an anabolic state marked by physiological processes of growth and rejuvenation of the organism's immune and nervous systems. Studies suggest sleep restores neurons and increases production of brain proteins and certain hormones. In this view, the state of wakefulness is a temporary hyperactive catabolic state during which the organism acquires nourishment and procreates: "sleep is the essential state of life itself". Anything that an organism does while awake is superfluous to the understanding of life's metabolic processes, of the balancing states of sleep and wakefulness. In support of this idea, one can argue that adequate rest and a properly functioning immune system are closely related, and that sleep deprivation compromises the immune system by altering the blood levels of the immune cells, resulting in a greater than normal chance of infections. However, this view is not without its critics who point out that the human body appears perfectly able to rejuvenate itself while awake and that the changes in physiology and the immune system during sleep appear to be minor. Nevertheless the fact that the brain seems to be equally — and at times more — active during sleep than when it is awake, suggests that the sleeping phase is not just designed for relaxation and rest. Experiments of prolonged sleep deprivation in rats led to their unregulated body temperature and subsequent death, is believed to be due to a lack of REM sleep of the dreaming phase. Although it is not clear to what extent these results generalize to humans, it is universally recognized that sleep deprivation has serious and diverse biological consequences, not excluding death. In the context of our two-component activating-regulating formulation of homeostasy and evolution, it is speculated that sleep, particularly its dreaming REM period, constitutes a change of guard that hands over charge of $\mathfrak{L}$ to its catabolic $\mathfrak{G}$ component from the anabolic M that rules the wakeful period. It is to be realized that all living matter are constantly in touch with their past through the mind; thus anything non-trivial that we successfully perform now depends on our ability to relate the present to the past involving that subject. In fact an index of the quality life depends on its ability to map the past onto the present and project it to the future, and the fact that a living body is born, grow and flourish without perishing (which an uni-directional second law would have), thanks to anabolic synthesis due to its immune system, is a living testimony to the bi-directionality of the direct-inverse arrow manifesting within the framework of the backward-forward completeness of the living world.

10.4 Conclusions: The Mechanics of Thermodynamics

In this paper we have presented a new approach to the nonlinear dynamics of evolutionary processes based on the mathematical framework and structure of multifunctional graphical convergence introduced in [30]. The basic point we make here is that the *macroscopic* dynamics of evolutionary systems is in general governed by strongly nonlinear, non-differential laws rather than by the Newtonian Hamilton's linear differential equations of motion

$$\frac{d\mathbf{x}_i}{dt} = \frac{\partial H(\mathbf{x})}{\partial \mathbf{p}_i}, \quad \frac{d\mathbf{p}_i}{dt} = -\frac{\partial H(\mathbf{x})}{\partial \mathbf{x}_i}, \qquad -\infty < t < \infty \tag{10.4.1}$$

of an N particle isolated (classical) system in its phase space of microstates $\mathbf{x}(t) = (\mathbf{x}_i(t), \mathbf{p}_i(t))_{i=1}^N$. As is well known, Hamiltonian dynamics leads directly to the microscopic-macroscopic paradoxes of Loschmidt's time-reversal invariance of Eq. (10.4.1), according to which all forward processes of mechanical system evolving according to this law must necessarily allow a time-reversal that would require, for example, that the Boltzmann H-function decreases with time just as it increases, and Zarmelo's Poincare recurrence paradox which postulates that almost all initial states of isolated bounded mechanical system must recur in future, as closely as desired. One approach — [10], [24] — to the resolution of these paradoxes require

(1) A "fantastically enlarged " phase space volume as the causative entropy increasing drive. Thus, for example, a gas in one half of a box equilibrates on removal of the partition to reach a state in which the phase space volume is almost as large as the total phase space available to the system under the imposed constraints, when the number of particles in the two halves becomes essentially the same. In this situation, *for a dilute gas* of N particles in a container of volume V under weak two-body repulsive forces satisfying the linearity condition $V/N \gg b^3$ with b the range of the force, Boltzmann identifies the thermodynamic Clausius entropy with $S_B = k \ln |\Gamma(M)|$, where $\Gamma(M)$ is the region in $6N$-dimensional Lioville phase space of the microstates belonging to the equilibrium macrostate M in question; the second law of thermodynamics then simply implies that an observed macrostate is the most probable in the sense that it is realizable in more ways than any other state. When the system is not in equilibrium, however, the phase space arguments imply that the relative volume of the set of microstates corresponding to a given macrostate for which evolution leads to a macroscopic decrease in the Boltzmann entropy *typically* goes exponentially to zero as the number of atoms in the system increases. Hence for a macroscopic system "the fraction of microstates for which the evolution leads to macrostates with larger Boltzmann entropy is so close to one that such behaviour is exactly what should be seen to always happen", [18]. A more recent interpretation[9] is to consider not the number of microstates of a macrostate M, but the most probable macroscopic history as that which can be realized by the greatest number of microscopic paths compatible with the imposed constraints. Paths, rather than states, are more significant in non-equilibrium systems because of the non-zero macroscopic fluxes whose statistical description requires consideration of the temporal causative microscopic behaviour.

(2) The statistical techniques implicit in the foregoing interpretation of macroscopic irreversibility in the context of microscopic reversibility of Newtonian mechanics rely fundamentally on the conservation of Lioville measures of sets in phase space under evolution. This means that if a state $M(t)$ evolves as $M(t_1) \stackrel{t_1<t_2}{\longrightarrow} M(t_2)$ such that the evolved phase space $\Gamma_{t_2}(M(t_1))$ of $M(t_1)$ is necessarily contained in $\Gamma(M(t_2))$ by the arguments in (1), then the preservation of measures requires that $\Gamma_{t_2}(M(t_1)) \subseteq \Gamma(M(t_2))$ by the law of increasing

S_B. Conversely, even as $M(t_2) \overset{t_1<t_2}{\longrightarrow} M(t_1)$ is not prohibited by the microscopic laws of motion, the exact identification of the subset $\Gamma_{t_2}(M(t_1)) \subseteq \Gamma(M(t_2))$ cannot be ensured *a priori* to enable the system to eventually end up in $\Gamma(M(t_1))$; although the macroscopic reverse process is permissible, it is improbable enough never to have actually occurred. Identifying the macrostate of a system with our image $f(x)$ of a microstate x in "phase space" $\mathfrak{D}(f)$ that generates the equivalence class $[x]$ of microstates, invariance of phase space volume can be interpreted to be a direct consequence of the *linearity assumption of the Boltzmann interaction for dilute gases* that is also inherent in his *stosszahlansatz* assumption of molecular chaos which neglects all correlations between the particles.

(3) Various other arguments like cosmological big bang and the relevance of initial conditions preferring the forward arrow to the reverse are invoked to argue a justification for macroscopic irreversibility, that in the ultimate analysis is a "consequence of the great disparity between microscopic and macroscopic scales, together with the fact (or very reasonable assumption) that what we observe in nature is typical behaviour, corresponding to typical initial conditions", [10].

In comparison the multifunctional graphical convergence techniques, founded on difference rather than differential equations, adapted here avoids much of the paradoxical problems of calculus-based Hamiltonian mechanics, and suggests an alternate specifically nonlinear dynamical framework for the dissipative dynamical evolution of Nature supporting self-organization, adaption, and emergence in complex systems in a natural manner. The significant contribution of the difference equations is that evolution at any time depends explicitly on its immediate predecessor — and thereby on all its predecessors — leading to non-reductionism, self-emergence, and complexity.

To conclude, we recall the following passages from Jordan [11] as a graphic testimony to chanoxity:

Approximately one hundred participants met for three days at a conference entitled "Uncertainty and Surprise: Questions on Working with the Unexpected and unknowable". The diversity of the conference was vital (as) bringing together people with very different views strengthened the probability of extraordinary explanatory behaviour and the hope of producing entirely new structures, capabilities, and ideas. Out of our interconnections might emerge the kind of representation of the world that none of the participants, individually, possess or could possess. One purpose of the conference was to develop the capacity to respond to our changing science and to new ideas about the nature of the world as they relate to the unexpected and unknowable.

Participants recognized early on their difficulties in communicating with one another across the diversity of their backgrounds. One of the issues the group tried to resolve was differences in levels of understanding and experience related to the theme of uncertainty and surprise. The desire for a common language was a reoccurring theme among conference participants as they tried to work out questions and ambiguities regarding even the fundamental themes of the conference, including the definitions of complexity, emergence, and uncertainty. Can we name or label what complexity is? Emergence was an idea that wove itself throughout much of the informal conversation, yet emergence as a term created confusion among the participants. There was acknowledgment of a need to state more clearly our assumptions with regard to fixed structure versus emergence. If you use "emergence" to mean in the complexity sense, it implies some sort of scale shift having to do with a fundamentally different structure of the organization of interactions, or a shift in the nature of the network, or of knowing, or awareness. Some conference participants cautioned the group not to equate emergence with miraculous magic.

(It was) recognized that there are tendencies toward stability and tendencies toward variance. Our assumption about the value of stability may lead us to to our assumption of the value of permanence. There is evidence that the value of permanence may be a socially constructed Western trap that is not shared by Eastern philosophies. Complexity science leads us to understand that the degree of variability in the distribution of fluctuations in system dynamics is more important than any average quantity, which is counter to the traditional paradigms of medicine, management, and scientific research. We used to believe that equilibrium was the optimal for systems. Complexity science leads us to believe that stability is death and survivality is in variability. The tension between stability and variability is similar to the tension in the social sciences between exploitation and exploration. We often think of exploitation as a strategy for maintaining stability and exploration as a strategy for exploiting variability. We may need a balance between exploration and exploitation, stability and variability, convergence and divergence within a state.

An issue that resurfaced several times throughout the conference was the relationship between individual elements and collective elements. Traditionally Western thought has tended toward the individual over the collective; the opposite view is often taken by Eastern thought. It is not a question of either the individual or the collective, but the interaction of the two that is needed; ⋯ the individual and the group are the singular and plural of the same process. In order to honor the tension between the individual and the collective, a good model might be "If you win I win; if I lose, you lose". One participant felt that you can design an organization in such a way that people profited or lost together based upon how well they all did. One of our best levers for facing uncertainty and surprise might be to encourage quasi-autonomy (individuality) but at the same time willingness to cooperate across disciplines because this kind of collaboration gives us more capabilities and skills.

References

[1] Argyris, J., Faust, G. and Haase, M. (1994) *An Exploration of Chaos* North-Holland, Amsterdam.
[2] Baranger, M. (2000) *Chaos, Complexity, and Entropy: A physics talk for non-physiscts,* http://necsi.org/projects/baranger/cce.pdf.
[3] Bertuglia, C. S. and Vaio, F. (2005) *Nonlinearity, Chaos and Complexity. The Dynamics of Natural and Social Systems* Oxford University Press Inc., New York.
[4] Callen, H. B. (1985) *Thermodynamics and Introduction to Thermostatics* John Wiley and Sons.
[5] Callender, C. (1999) Reducing Thermodynamics to Statistical Mechanics: The Case of Entropy. *Jour. Philosophy* **96**, 348–373.
[6] Chaisson, E. J. (2005) Non-Equilibrium Thermodynamics in an Energy-Rich Universe In *Non-Equilibrium Thermodynamics and the Production of Entropy: Life, Earth, and Beyond*, ed. A. Kleidon and R. D. Lorenz, Springer-Verlag, Berlin.
[7] Cinquin, O. and Demongeot, J. (2002) Positive and Negative Feedback: Striking a Balance Between Necessary Antagonists. *J. Theor. Biol.* **216**, 229–241.
[8] Coveney, P. V. (2003) Self-Organization and Complexity: A New Age for Theory, Computation and Experiment. In Proceedings of the Nobel Symposium on Self-Organization held at Karolinska Institutet, Stockholm, August 25-27, 2002. *Phil. Trans. R. Soc. Lond. A.* **361**, 1057–1079.
[9] Dewar, R. (2003) Information Theory explanation of the Fluctuation Theorem, Maximum Entropy Production and Self-Organized Criticality in Non-Equilibrium Stationary States. *J. Phys. A: Math. Gen.* **36**, 631–541. **Also:** Maximum Entropy Production and the Fluctuation Theorem, *J. Phys. A: Math. Gen.* **38**, L371–L381(2005).
[10] Goldstein, S. and Lebowitz, J. L. (2004) On the (Boltzmann) Entropy of Non-Equilibrium Systems. *Physica D* **193**, 53–66.
[11] Jordan, M. E. (2005) Uncertainty and Surprise: Ideas from the Open Discussion In *Uncertainty and Surprise in Complex Systems*, ed. R. R. McDaniel and D. J. Driebe, Springer-Verlag, Berlin.
[12] Katchalsky, A. and Curran, P. F. (1965) *Nonequilibrium Thermodynamics in Biophysics* Harvard University Press, Massachusetts.
[13] Kauffman, S. (2000) *Investigations* Oxford University Press, New York.
[14] Kestin, J. (1968) *A Course in Thermodynamics* Hemisphere Press, New York.
[15] Kondepudi, D. K. and Prigogine, I. (1998) *Modern Thermodynamics* John Wiley and Sons, Chichester.
[16] Korzeniewski, B. (2001) Cybernetic Formulation of the Definition of Life. *J. Theor. Biol* **209**, 275–286.
[17] Kot, M. (2001) *Elements of Mathematical Ecology* Cambridge University Press, Cambridge.

[18] Lebowitz, J. L. (1999) Microscopic Origins of Irreversible Macroscopic Behaviour. *Physica A* **263**, 516–527.
[19] Morales, J. (n.d.) *The Definitions of Life* http://baharna.com/philos/life.htm.
[20] Murray, J. D. (2002) *Mathematical Biology I. An Introduction* Springer-Verlag, New York.
[21] Nagashima, H. and Baba, Y. (1999) *Introduction to Chaos* Institute of Physics Publishing, Bristol.
[22] Ozawa, H. (1997) Thermodynamics of Frost Heaving: A thermodynamic proposition for dynamic phenomena. *Phys. Rev. E* **56**, 2811–2816.
[23] Ozawa, H., Ohmura, A., Lorenz, R. D. and Pujol, T. (2003) The Second Law of Thermodynamics and the Global Climate System: A Review of the Maximum Entropy Production Principle. *Reviews of Geophysics* **41**, (4)1018.
[24] Price, H. (2004) On the Origins of the Arrow of Time: Why There is Still a Puzzle about the Low Entropy Past In *Contemporary Debates in Philosophy of Science*, ed. C. Hitchcock, Blackwell.
[25] Rosen, R. (2000) *Essays on Life Itself* Columbia University Press, New York.
[26] Schneider, E. D. and Kay, J. J. (1995) Order from Disorder, the Thermodynamics of Complexity in Biology In *What is Life, the Next Fifty Years*, ed. M. P. Murphy and L. A. J. O'Neill, Cambridge University Press.
[27] Schneider, E. D. and Kay, J. J. (1994*a*) Complexity and Thermodynamics: Towards a New Ecology. *Futures* **24**, 626–247.
[28] Schneider, E. D. and Kay, J. J. (1994*b*) Life as a Manifestation of the Second Law of Thermodynamics. *Mathematical and Computer Modelling* **19**, 25–48.
[29] Schroedinger, E. (1992) *What is Life?* Cambridge University Press, Canto Edition.
[30] Sengupta, A. (2003) Toward a Theory of Chaos. *International Journal of Bifurcation and Chaos* **13**, 3147–3233.
[31] Sklar, L. (1993) *Physics and Chance: Philosophical Issues in the Foundations of Statistical Mechanics* Cambridge University Press, New York.

Index

F. Balibrea: *Chaos, Periodicity and Complexity on Dynamical Systems*, StudFuzz **206**, 353–358 (2006)
www.springerlink.com